Phytoplankton Ecology
STRUCTURE, FUNCTION AND FLUCTUATION

Phytoplankton Ecology

STRUCTURE, FUNCTION AND FLUCTUATION

Graham P. Harris

Divisions of Fisheries Research and Oceanography
CSIRO Marine Laboratories
Hobart
Tasmania

LONDON   NEW YORK

CHAPMAN AND HALL

First published in 1986 by
Chapman and Hall Ltd
11 New Fetter Lane, London EC4P 4EE
Published in the USA by
Chapman and Hall
29 West 35th Street, New York NY 10001
First issued as a paperback 1987

Printed in Great Britain at the University Press, Cambridge

ISBN 0 412 30690 5

British Library Cataloguing in Publication Data

Harris, Graham P.
  Phytoplankton ecology: structure, function and
  fluctuation.
  1. Phytoplankton—Ecology
  I. Title
  589.4   QK933
  ISBN 0-412-30690-5

Library of Congress Cataloging in Publication Data

Harris, Graham P.
  Phytoplankton ecology.
  Bibliography: p.
  Includes index.
  1. Phytoplankton—Ecology. I. Title.
  QK933.H37   1986   589.4'045'263   85-24254
  ISBN 0-412-30690-5

To
Richard Vollenweider

who contributed so much to my education over a
number of years. Many of the ideas in this book were
first spawned at our regular Friday lunches. It always
took me some months to appreciate the value of the
ideas that Richard threw away in a moment.

Contents

Acknowledgements

The author wishes to acknowledge the assistance of all those who, knowingly or otherwise, contributed to this book. To Chris, Paul and Jonathan; thanks for all your help in so many ways. To all those who helped me to understand the issues and the concepts; my thanks also. It is difficult to single out a few for special thanks but I must acknowledge the help of Colin Reynolds, Ivan Heaney, Jack Talling, John Lund, Joel Goldman, Dick Eppley and David Rollo. Thanks to Britta Hansen for drawing most of the figures. Thanks to Beryl Piccinin and my students in Canada who acted as sounding boards for the earlier versions of what is here and who did much of the field work in Hamilton Harbour and Lake Ontario.

CHAPTER 1

Preamble

Many text books of limnology and oceanography begin by reminding the reader that two thirds of the surface of the Earth is covered by water. Perhaps the most striking image that drives this point home is the Apollo mission view of the Earth taken from directly over the Pacific Ocean, which shows little land and one complete third of the surface of the Earth covered with water. Phytoplankton are the 'grass' of the surface waters of lakes and the oceans, so this book is concerned with the ecology of a group of organisms which are responsible for the process of primary production over much of the surface of the Earth. This is a book about phytoplankton but it is also, I hope, a book about some general ecological principles. I believe that phytoplankton have much to teach us about the way this world works and the lessons we may learn should be as widely applied as possible. I take a certain pride in using phytoplankton as model organisms in an ecology text because phytoplankton have long been regarded as paradoxical. Most of the standard theory of ecology has not included phytoplankton. Phytoplankton do not appear to fit most of the standard explanations and examples from phytoplankton data are missing from most of the standard literature.

So how can phytoplankton, a group of microscopic photosynthetic organisms, be useful as models of general principles? In my opinion, the problems with ecological theory in the past have lain in an incorrect appreciation of scale and an unrealistic reliance on equilibrium theory. Scale is a measure of the way organisms perceive their environment. We, as human beings of a characteristic size and life span, tend to see the world in an anthropocentric way (Allen and Starr, 1982). We have trouble coming to grips with the long time scale components of what we see today, with such concepts as succession and evolution. At the same time we can easily comprehend time scales of minutes, days and seasons as these fit easily into the working span of the average ecologist and into the time span of most research grants. In scales of size the converse is true. With maps and expeditions to the far corners of the globe we may comprehend the larger scale, global distributions of organisms and ecosystems. It is at very small spatial scales that we have had problems. We have tended to see the world in terms of cubic metres and kilometres, convenient scales for us, but entire universes for organisms such as bacteria and phytoplankton. This has been particularly true in the study of marine phytoplankton where the size of the oceans and the ships in use has led oceanographers to concentrate on scales of many kilometres and to neglect scales more relevant to the organisms (Harris, 1980a). We have regarded phytoplankton as paradoxical because we have looked at the environment of the organisms at too large a scale.

The climate of intellectual opinion keeps intruding on the world view and this sometimes encourages us to accept explanations which may be at variance with reality. There has been too much reliance on equilibrium theory in ecology with the result that while the mathematics of the theory has been simplified, the application of the theory to the real world has become more difficult. Clearly there is a great deal of spatial and temporal variability in the real world. Equilibrium theory, by its very nature, considers the results of competition and other environmental interactions at steady state. But what if this steady state is perturbed by external events? What are the consequences of such perturbations? If we consider such questions then equilibrium theory is reduced to a special limiting case of a broader theoretical framework which seems to be slowly emerging. The extremes are equilibrium and chaos: where do real world events lie?

There is one way to deal with the complexity of the real world. In a recent book, Allen and Starr (1982) interpreted ecological events in terms of hierarchy theory. They classified systems into three categories: small, middle and large number systems. Small number systems may be thought of as being akin to billiard ball physics; representable as differential equations. Large number systems are the biological equivalents of physical gas laws. Allen and Starr (1982) asserted that the real world may be thought of as complex middle number systems: systems in which the number of significant elements are too many to be treated by reductionist approaches but too few to be treated by statistical approaches. Koestler (1967) stated that biological systems may 'be regarded as a multi-level hierarchy of semi-autonomous sub-wholes, branching into sub-wholes of lower order and so on. Sub-wholes on any level of the hierarchy are referred to as holons' (Fig. 1.1). Koestler (1967) was discussing the organization of organisms but the analogy to supra-organismic (ecological) organizations had already been made by von Bertalanffy (1952) who discussed the hierarchy of parts and of processes in biological systems. Thus it is important to remember that holons may be both discrete structural units (organelles, organs, organisms) and discrete units of process which may cut across structural boundaries. The evident complexity in the behaviour of biological systems may be analysed by decomposing them into a fully, or partially, nested hierarchy of holons. This approach to biology is characteristic of von Bertalanffy's 'General Systems Theory'.

The definition of the holon contains within it a dichotomy, as the holon

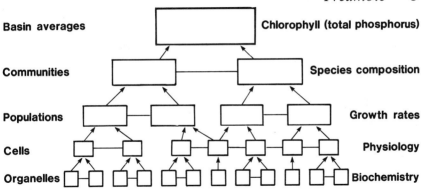

Fig. 1.1 A diagrammatic representation of the hierarchy of process and organization in phytoplankton ecology.

displays the dependent properties of a part as well as the autonomous properties of a whole. Koestler (1967) called this dichotomy the Janus effect. Allen and Starr (1982) define the holon as

> The representation of an entity as a two-way window through which the environment influences the parts, through which the parts communicate as a unit to the rest of the universe. Holons have characteristic rates for their behaviour, and this places particular holons at certain levels in a hierarchy of holons.

The Janus effect is a reference to the passage of information through the two-way window – the holon boundary. Allen and Starr assumed that the world may be thought of as an hierarchy of holons with small, fast holons at the base and large, slow holons at the apex. They did not assume that the world really was such a system, it may merely be conveniently thought of in this way. This is no place for a deep discussion of Marxist dialectics. The hierarchy of holons is nearly decomposable if the holons exhibit apparent disjunctions when viewed in the appropriate way. Allen and Starr assumed that the hierarchy of spatial and temporal processes in nature was continuous and that the relative intensities of the interactions between levels could be used to decompose the continuum into more or less discrete holons. The nested hierarchy of holons is a particular type of hierarchy in which the upper level holons actually contain the lower levels. This need not always be the case.

The hierarchical approach will be used throughout this book as a means of classifying ecological, physical and chemical processes. From the smallest scales of nutrient uptake and cellular physiology to the largest scales of interannual variability there are a number of important interactions between physics and biology. At each level I will attempt to describe the physical and chemical processes in operation and their effects on biological processes.

The view of Allen and Starr is a radical change from the usual ecological approach, where reductionism is rampant and questions about emergent properties have become the purview of metaphysics. The analytical and summative approach of much biological research was criticised by von Bertalanffy (1952) who showed that such methods could not account for the properties of organized wholes. His approach stressed the organized properties of living systems and stressed the role of information exchange in maintaining the organization. He allowed that supra-organismic levels of organization are much less well coordinated than those at the organismic level but the same principles apply. Different properties are expected to emerge as the level in the hierarchy alters; whether the emergent properties are a simple, additive function of the lower holons will depend on the completeness of the data sets, the point of view of the observer and the presence of properties which are not, in themselves, derivable from the behaviour of the components. This world view also obviates all discussion about the difference between equilibrium and non-equilibrium ecology, about the effects of endogenous and exogenous factors and about the discreteness, or otherwise, of ecological communities.

We are thus faced with the problem of classifying temporal and spatial scales of variability in terms of the organisms of interest. What is small and fast for one organism may be large and slow for another. We must also be consistent in our treatment of variability; it is not correct to combine processes which operate at different scales in one discussion. This is as true for biological processes as it is for environmental variables such as fluctuations in light and nutrient availability. Thus we have to contend with a hierarchy of spatial and temporal variance in a number of relevant ecological parameters. What is really interesting about a correct classification and treatment of the hierarchy of variance in ecosystems is the fact that 'noise' at one level may contribute to the predictable behaviour of ensemble averages at higher levels. Thus there are statistical properties of the cascade of variance in ecosystems.

I believe that this cascade of variance in ecologically relevant parameters from large and slow processes to small and fast processes is a vitally important factor in determining what we see. The diversity of life on Earth depends on it. The variability we observe is not just something that we can average out and equilibrium solutions to ecological problems only apply for some organisms in some cases. I shall review the basic tenets of equilibrium theory in order to show that these cannot apply in many instances.

If we appreciate the true scale of interaction in phytoplankton populations they become extremely useful model organisms. There are a number of reasons for this. Phytoplankton are small and they grow very rapidly so that many generations may pass during a year. We may therefore observe the seasonal succession of species which is, in many respects, analogous to that in forest successions. Instead of taking hundreds of years the seasonal succession of phytoplankton may be complete in one hundred days and is thus more amenable to study. In the terminology of holons, we may study more levels and more holons than in most other ecological disciplines as we may study holons which span the range from nutrient uptake kinetics, through physiology, population dynamics and communities to biomass and we may average over minutes and centimetres or whole basins and years. I shall go to some length to show that many phytoplankton populations are, in fact, nowhere near equilibrium, but that there are statistical properties of assemblages of species that allow high level, averaged properties to be discerned. Such high level, statistical properties of ecological systems are akin to physical gas laws as they are characteristic of large number systems. I believe that the study of phytoplankton can reveal such statistical properties more easily than most other systems as the small size and rapid growth of the organisms allows properties to be averaged at a very high level. Allen and Starr point out that changing the scale of observation often reveals important aspects of the functioning of ecological systems. I shall show that this is, indeed, the case.

Another feature of the ecology of phytoplankton which makes them useful model organisms lies in the enormous range of different sized waters which they inhabit. Phytoplankton may be found in water bodies ranging in size from rain water puddles to the oceans. This provides the ecologist with a wide ranging set of environments and it is possible to study the response of the organisms to physical and chemical processes operating at a wide range of spatial and temporal scales. Many textbooks deal with the marine and freshwater environments separately but I recognize no such distinction here. The differences between freshwater and marine environments are essentially only those of scale as Margalef (1978b) has demonstrated. The oceans, because of their size, are dominated by large scale horizontal motions (ocean currents) but in many respects the physics of the surface mixed layer of lakes and the oceans is very similar. One respect in which lakes and the oceans differ is in the fact that lakes are bounded systems which may be treated as wholes for statistical purposes. This makes the study of populations within lakes much easier (Weatherley 1972) and means that the properties of different lakes may be compared by calculating average properties for each lake.

This will not be a compendium of information on phytoplankton: no-one could hope to better the work of Hutchinson (1967). This book is not so much descriptive as process oriented. I wish to make some specific points about the ecology of the organisms and to show how the study of the organisms relates to ecology in general. There is a reason for attempting to do this at this time as in recent years the study of phytoplankton has undergone something of a 'revolution' ('sensu Kuhn') as ecologists and physiologists realized that the standard theory and methodology was not always applicable to the real world. We failed to appreciate that ecologically significant events were occurring at very small temporal and spatial scales; scales which were not fully resolved by traditional methods. In short it has become apparent of late that our knowledge of the ecology of planktonic ecosystems requires reinterpretation. A more dynamic approach has recently emerged (Legendre and Demers, 1984). There will be some general themes in this book that will discuss the relationship between theory and practice in aquatic ecology and the problems of the interpretation of field and experimental data.

Phytoplankton have, in the past, been regarded as paradoxical by virtue of the fact that small samples of water contain many coexisting species. Phytoplankton are very small organisms and they are not to be thought of merely as small flowering plants: in many respects they behave as micro-organisms. For example, one of the major debates at present concerns the growth rates of phytoplankton in the central ocean areas where nutrients are apparently lacking. For years it has been assumed that no nutrients meant no growth, as a simple curvilinear relationship between nutrient concentration and growth rate could be demonstrated in laboratory cultures. Now we have begun to realize that an apparent lack of nutrients may not lead to a suppression of growth in the field. There may be rapid uptake of nutrients and rapid growth by the phytoplankton if the uptake and growth rates are balanced by equally rapid grazing and regeneration of the nutrients. Thus it is not the concentration of nutrient in the water which is the important parameter but the flux rates between the various compartments in the system. For this explanation to be valid we must invoke rapid nutrient uptake and storage by the phytoplankton from small patches of regenerated nutrient in the water. The interpretation of data from laboratory cultures and experiments requires a knowledge of the limitations of the methods used, and the interpretation of data from similar methods in the field requires a knowledge of the temporal and spatial scales of the processes in operation. The scales of observation and natural process must be understood and correctly matched (Harris, 1980b).

There have been a number of recent papers and books about the relationships between physiology, methodology, and the interpretation of productivity and growth rate measurements, many of which serve to illustrate the need for the revision of some basic ideas in the field. Little has been written about the effects of a revised paradigm on our understanding of population dynamics and community structure. Inevitably, any discussion of population dynamics must include a discussion of growth rates and in order to measure growth rates we must make kinetic measurements. Thus I will need to discuss the relationship between theory, kinetic measurements, observed growth rates and population studies

In lakes, the seasonal succession of communities in surface waters may be managed by virtue of the fact that the physical and chemical environment may be manipulated. Thus we may consider both the intellectual implications of our results as well as the practical implications for human intervention and mangement. Phytoplankton are relatively easy to grow in culture so the field data may be supported by experimental data. This is of some considerable importance when it is remembered that the great variability in the real world makes it very difficult to perform controlled experiments with natural populations and ecosystems. The resources required by phytoplankton may also be studied in culture and, while it is not always easy to extrapolate from the constant conditions of culture to the real world, many useful insights have been obtained. Competition between phytoplankton has been studied in culture and such studies have contributed significantly to the development of equilibrium theory. As we shall see, there is good reason to question the role which competition plays in the formation of natural communities in a fluctuating environment. If competition is less important in the real world than in culture this will have considerable implications for our ability to predict the biological responses to changed environmental conditions.

1.1 A brief introduction to the organisms

The term plankton refers to the group of organisms which float in the surface waters of rivers, lakes and the oceans. The term plankton has its roots in the Ancient Greek adjective meaning wanderer. The modern adjective normally used is planktonic but there is some debate as to the correct form. Planktic may be etymologically correct (Rodhe, 1974), whereas planktonic may be preferred for reasons of euphony and common usage (Hutchinson, 1974). As the term implies, planktonic organisms float freely in the water and live at the mercy of water movements. Many phytoplankton are curious and beautiful organisms and microscopic examination of water samples reveals a great diversity of forms (Fig. 1.2). While many planktonic organisms are themselves immobile others have a limited capacity to swim through the water and hence have the ability to change their position in the water column. There is a range of swimming ability depending largely on the size of the organism.

The term phytoplankton is used for the large group of planktonic plants that live in surface waters. There is always some debate as to exactly which organisms to include in this group as, among the single-celled and simple multicellular forms, there is uncertainty about the best form of classification and it is sometimes difficult to distinguish between animals and plants. The vast majority of phytoplankton are algae and belong to a diverse group of lower, non-flowering plants. Some phytoplankton may strictly be described as bacteria as they are prokaryotes while others, by virtue of their mobility

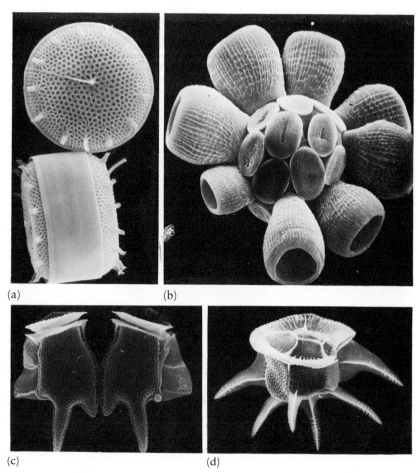

(a)   (b)

(c)   (d)

Fig. 1.2 Scanning electron micrographs of some marine phytoplankton. I am indebted to Gustaff Hallegraeff for the original of this figure.
(a) *Thalassiosira allenii*, a chain forming diatom.
(b) *Scyphosphaera apsteinii*, a coccolithophorid.
(c) *Dinophysis tripos*, a dinoflagellate, dividing pair.
(d) *Ceratocorys horrida*, a tropical dinoflagellate.

and their ability to live on complex organic substrates in the dark, have some distinctly animal-like characteristics. The organisms exist as single cells or simple multicellular forms and are, by the standards of human experience, small. Within the phytoplankton there is a large range of cell size and growth rates. The species range in size from small prokaryotic and eukaryotic cells equivalent in size to bacteria to the largest dinoflagellates which are visible to the naked eye. Thus there is a size (volume) range of at least five orders of magnitude (Malone, 1980a). Their range in growth rates is somewhat less, ranging from a few doublings per day for the fastest growing species, to one doubling every week or ten days for the slowest species (Eppley, 1972). What the range in size and growth rates really means is that we are dealing with a group of organisms that are very small and which (by normal ecological standards) grow very rapidly. In many respects the ecology of phytoplankton is similar to the ecology of bacteria, only the bacteria show comparable sizes, growth rates and metabolic flexibility. We cannot regard phytoplankton as small 'higher' plants (Allen, 1977), and the time scale of important processes is much more rapid than that exhibited by higher plants (Harris, 1980a).

The strict definition of phytoplankton is further confused by the presence of species which normally live on the sediment surface and which become suspended in the water by turbulence (Hutchinson, 1967). Thus some of the species present in the water column are not truly part of the planktonic community. In some species the life cycle includes an encysted or resting phase which settles on the bottom and remains there for a period of months or years, so only a part of the life cycle is planktonic. Such species are obviously more common in shallow waters: in the deep waters of the oceans such a life cycle is clearly impossible as the resting stage would never be resuspended. Thus it is not easy to produce a strict definition of the organisms to be discussed in this book. For the purposes of these arguments it will be sufficient to restrict the discussion to those algae which commonly occur in surface waters and/or those which complete a significant portion of their life cycles in such waters.

Most of the major algal groups have planktonic representatives (Table 1.1). It is evident that planktonic forms of the diatoms and chrysophytes, green algae, crytophytes and dinoflagellates are common but there are very few planktonic red algae and no planktonic brown algae or charophytes (Bold and Wynne, 1978; Sournia, 1982). There is some debate over the best means of classification of the algae and the classification of Bold and Wynne (1978) is but one of many. This classification treats the blue-green algae as algae, even though they are prokaryotes and thus structurally similar to bacteria. The justification for their inclusion here is that they are a very important group of phytoplankton which play a significant role in water quality problems in lakes. As can be seen from the table the comparative lack of morphological characters (many phytoplankton are referred to as 'little round green things' or LRGTs) has led taxonomists to use a wide range of structural, biochemical and other cellular characters as a means of classifying these organisms. The evolutionary relationships between the major groups can be clearly seen in the structure of the flagellae (if present), in the pigment composition, the structure of the chloroplast and the relationship between the chloroplast and the nuclear envelope (Coombs and Greenwood, 1976).

There are significant differences between the dominant groups of phytoplankton in marine and freshwater systems in that, while dinoflagellates are

Table 1.1 Characteristics of the major groups of algae

Group	Cyanochloronta (blue-green algae)	Rhodophycophyta (red algae)	Chlorophycophyta (green algae)	Euglenophycophyta (euglenoids)
Pigments	chlorophyll a, C-phycocyanin, allophycocyanin, C-phycoerythrin, B-carotene, xanthophylls	chlorophyll a, phycocyanin, allophycocyanin, phycoerythrin, carotenes, several xanthophylls	chlorophyll a, b; a, B, carotenes and several xanthophylls	chlorophyll a, b; B-carotene several xanthophylls
Chloroplast organization	single thylakoids	single thylakoids	2–5 (or more) thylakoids in a stack, grana as in higher plants	2–6 thylakoids per stack, sometimes many
Enclosing membranes	absent	2, chloroplasts in cytoplasm	2, chloroplasts in cytoplasm	3, chloroplast in sub-compartment of cytoplasm
Storage products	cyanophycean granules, polyglucose	floridean starch	starch	paramylon
Cell wall	organic	cellulose, several sulphated polysaccharides, calcified in some	cellulose, calcified in some	absent
Flagella	absent	absent	1–many, equal, apical	1–3(–7) apical, subapical
Ejectile bodies	absent	absent	absent	rare(?), status questionable
Habitat	common in freshwater plankton, rare in marine plankton (although more forms discovered recently), terrestrial	exceptionally rare in marine plankton, benthic macrophytes	common in freshwater plankton, minor importance in marine plankton, terrestrial	common in freshwater plankton, not common though locally abundant in marine plankton

Table 1.1 (continued)

Group	Phaeophycophyta (brown algae)	Chrysophycophyta (golden and yellow-green algae including diatoms)	Pyrrophycophyta (dinoflagellates)	Cryptophycophyta (cryptomonads)
Pigments	chlorophyll a, c; B-carotene, fucoaxanthin, several other xanthophylls	chlorophyll a, c; carotenes, several xanthophylls, fucoxanthin	chlorophyll a, c; B-carotene, several xanthophylls peridinin	chlorophyll a, c; carotenes distinctive xanthophylls, phycobilins
Chloroplast organization	2–6 thylakoids per stack, girdle lamella	3 thylakoids per stack, girdle lamella	3 thylakoids per stack	2 thylakoids per stack
Enclosing membranes	4, chloroplast in subcompartment of cytoplasm	4, chloroplast in subcompartment of cytoplasm	3, chloroplast in subcompartment of cytoplasm	4, chloroplast in subcompartment of cytoplasm
Storage products	Laminarin, mannitol	Chrysolaminarin, oil	starch (oil in some)	starch
Cell wall	cellulose, alginic acid, fucoidan	cellulose, silica	cellulose	absent
Flagella	2, unequal, lateral	1-2 unequal, or equal apical	2, one trailing, one girdling	2, unequal, subapical
Ejectile bodies	absent(?)	rare or questionable; absent in diatoms	present	present
Habitat	benthic macrophytes; not planktonic, very rare in freshwater, abundant in marine habitats	common in freshwater and marine plankton	found in both freshwater and marine plankton more diverse in marine habitats	common in freshwater, minor component of marine plankton

more diverse in the oceans, the diversity of green and blue-green algae is higher in freshwater. Diatoms are common in both habitats. Until recently it was thought that the blue-green algae were almost exclusively a freshwater group but the recent discovery of many small coccoid blue-green algae in the oceans has added to the list of known marine species. Green algae are certainly less diverse in the oceans but, again, recent work on marine microflagellates has revealed a surprising diversity of small forms. It is clear that our knowledge of the smallest phytoplankton, particularly in the oceans, is very scant indeed.

The basic features of phytoplankton cells have recently been reviewed by Taylor (1980). The cellular organization of the phytoplankton is usually very simple and most species occur as single cells or as small colonies of cells. It is rare for the colonial or coenobial forms to have more than 32 or 64 cells in a colony and the species of *Volvox*, with between 1000 and 50 000 cells in a colony are quite unusual. The taxonomy and biology of the algae has been discussed in a number of standard texts such as those of West and Fritsch (1927), Fritsch (1935) and the more modern work of Morris (1967), Prescott (1968), Round (1973, 1981) and Bold and Wynne (1978). The phytoplankton are, in general, photoautotrophic prokaryotes and eukaryotes. Thus even those with a simple bacterial cell structure are photosynthetic and obtain most of their energy from sunlight. Some species have rather complex nutritional requirements and may be partially dependant on organic substrates (heterotrophic) so that obligate autotrophy is difficult to demonstrate (Droop, 1974). Most species do, however, derive most of their energy from photosynthesis so that the major ecological factors are those which influence all photosynthetic organisms: light, temperature and the supply of the major nutrient ions. The physiology and biochemistry of the algae has been well reviewed in a number of standard works, notably those of Lewin (1962), Stewart (1974), Morris (1980a) and Platt (1981). Much important physiological data has béen obtained from the growth of algae in culture and two recent works have reviewed the basic culture methods and ancilliary techniques (Stein 1973, Hellebust and Craigie, 1978).

Protoplasmic connections between the cells are rare; for all practical purposes the ecology of phytoplankton may be considered to be the ecology of a group of single-celled organisms. The single-celled structure of phytoplankton means that any study of productivity, growth and reproduction is a special case of a larger ecological problem. Normally, organisms increase in biomass by an increase in size as well as an increase in abundance. Organisms that do not change in size significantly and reproduce by cell division show the closest links between productivity, changes in abundance and increases in biomass. Reproduction by rapid cell division also eliminates the need to determine the age structure of the population, an important requirement in studies of the ecology of multicellular organisms.

The single-celled organization places some important constraints on the organisms as they have no morphological mechanism for combating the vagaries of the planktonic environment, other than adaptations which alter their hydrodynamic properties and sinking rates (Hutchinson, 1967). It might therefore be expected that the organisms would have rather flexible physiological characteristics as this is the only way in which they can adapt to changing environmental conditions (Harris, 1978). At the level of the individual, any buffering mechanism to protect against unfavourable conditions (such as the storage of essential metabolites) must be an intracellular mechanism as the single-celled organization precludes the evolution of specialized storage organs. Once again the single-celled organization of the phytoplankton provides the ecologist with a simplified system in which to study the ability of the organisms to track environmental fluctuations. By 'tracking' I mean the ability of the organism to sense environmental changes and to adapt to them or at least to respond in a way that enhances its chances of survival. Implicit in this definition is the assumption that organisms possess both ways of sensing environmental changes and the means by which to respond. Even at the unicellular level this seems to be the case. The cellular holon senses and integrates external stimuli. I do not assume that the response is necessarily optimal: no organism can track environmental changes immediately and perfectly as there must be a time-lag between the stimulus and the response. The degree of perfection will depend on the speed of the environmental change and the speed of the biological response.

The utility of the study and understanding of tracking mechanisms will become clear as the following arguments unfold. I will attempt to show that many ecological problems may be thought of as tracking problems and that thinking along such lines clarifies a number of basic ecological concepts. As Allen and Starr (1982) note 'Spectral analysis techniques and time series analysis provide means for identifying time constants, lag relationships, transfer functions (i.e. filters), and cyclic behaviour in system data. They have so far found limited use in ecological problems, but they hold considerable promise. . .'. Such techniques will be widely used in what follows. At the community level there are not only feedbacks and time-lags. A form of feedforward or preadaptation is also possible. Many species of phytoplankton form resting cysts which allow the organisms to persist during periods of adverse conditions. The presence of such cysts pre-adapts the population, enabling the rapid exploitation of favourable conditions when such conditions return, and to an extent predetermines future community structures.

The ecology of phytoplankton has previously been reviewed by a number of authors, notably Raymont (1963, 1980), Bougis (1974), Round (1981), Reynolds (1984b) and the classic work of Hutchinson (1967). The factors which influence the distribution of phytoplankton are, as already noted, the same as those which influence all other photosynthetic organisms. The major difference between the terrestrial and aquatic environments lies in the turbulent nature of the surface mixed layer and the three-dimensional nature of the motions induced. The number of coexisting species normally found in surface waters is, indeed, large (50 to 100 is not uncommon) so the phytoplankton assemblage is unexpectedly diverse. Hutchinson (1967) discussed the nature of phytoplankton associations and noted that 'the most important feature of phytoplankton associations, though it is seldom appreciated and perhaps has never been fully explained, is that in nearly every case, even if only a few cubic centimetres of the environment are taken as a sample, a number of different species are present'. Ever since Hutchinson's classic paper of 1961 on 'the paradox of the plankton' (Hutchinson, 1961) there have been attempts to explain the unexpected diversity of phytoplankton communities. These explanations have been set in the context of the ecological theory of the day and have reflected the bias of the prevalent viewpoint. The paradoxical diversity of phytoplankton communities has been explained by both equilibrium and non-equilibrium models. It will be necessary to critically examine the assumptions of both types of models in order to provide a clear theoretical framework for what follows.

So what is this book all about? Basically it is about the population dynamics of a group of very small organisms. These organisms inhabit all the waters of the Earth and are the primary source of energy for all planktonic food chains. There is a broad range of pure and applied research problems which face those interested in the ecology of phytoplankton, and many of these problems have a direct bearing on a number of pressing environmental concerns. Many of the questions that confront phytoplankton ecologists are the same as those which face ecologists in other disciplines. I will argue in this book that phytoplankton are good model organisms to use to answer both specifically aquatic questions and those of a more general nature. Because they are small and grow quickly they may be studied easily and things happen rapidly. Because they are simple organisms their physiology may be studied easily and its relevance to ecology easily discerned. It is my belief that in recent years we have made considerable progress towards answering a number of basic questions about how populations and communities of organisms are organized and that such answers are relevant to our understanding of why the world is the way it is.

This is an exciting time to be a phytoplankton ecologist as ideas are changing and there is a heated debate amongst the proponents of the various camps. This book is one author's attempt to make some sense of the discoveries of recent years, to find a framework to consolidate what we know and to apply the knowledge so gained. The errors of omission and commission in what follows are mine alone. All books of this type contain a

considerable amount of personal bias: I hope the reader will bear with me
long enough to grasp the essentials of my argument.

CHAPTER 2

Ecological theory

The very large number of factors which operate simultaneously in nature (both biological and environmental) make it very difficult to do well-controlled experiments. Thus the relationship between theory and experiment in ecology is much less well defined than in many other sciences. Much ecological theory is concerned with small number systems that can be modelled by means of differential equations. In the sense of Platt (1964), ecology is but poor science as many of the basic theories and concepts turn out to be untestable (Murdoch, 1966). This point has been made more recently by Rigler (1975, 1982) and Peters (1976). Furthermore there is confusion between fact and concept (Ehrlich and Holm, 1962) and there is a strong tendency to explain proximate events by ultimate causes (Mayr, 1961). The distinction between proximate and ultimate causes lies in the inevitable temporal component which pervades all ecological process. In short, it comes down to whether or not a given community structure is the way it is because of present day (proximate) events like competition, or because of evolutionary (ultimate) causes that occurred in the distant past. To make things even more complicated it might be a bit of both (Williamson, 1981a)! Clearly one of the reasons why ecology has made little progress in the last twenty years in comparison to molecular biology, is the poor status of theory and experiment. This statement should not be interpreted as perjorative: if anything, it should stimulate ecologists to do their best to rectify the situation. Platt's (1964) syndromes of 'the eternal surveyor' and 'the all-encompassing theory which can never be falsified' apply to ecology only too often. If we accept that science progresses by refutation then progress in ecology is inevitably slow. It must be remembered that merely finding evidence to support an hypothesis does not constitute an adequate test of its validity.

2.1 An historical perspective: the concept of plenitude

The modern theories of population dynamics and community structure have their roots in the writings of the great nineteenth century naturalists. Their attempts to explain the diversity of life on Earth and to provide a unifying theory culminated in the publication of the *Origin of Species* in 1859. Darwin's theories on evolution and ecology sprang from the intellectual climate of the day; he adopted some crucial ideas from earlier authors and was influenced by the concepts of a *'Scala Naturae'* and of plenitude (Ruse, 1979; Stanley, 1981).

The *'Scala Naturae'* or 'Great Chain of Being' was an idea which had Greek roots and it was supposed that there was continuity in all forms of life and that the 'Great Chain' linked all forms of life in a graded series. Furthermore, the 'Great Chain', as a perfect structure, was without gaps. Ruse (1979) documents the links between these early Lamarkian ideas, the writings of Lyell, and Lyell's influence on Darwin. The associated concept, that of plenitude, became a central concept in Darwin's philosophy and it has become one of the most important concepts in modern ecological theory. The idea behind the concept of plenitude was that the world was brim full of life and that the organisms stood shoulder to shoulder, or in ecological terms niche against niche. There was, in short, no room for new species without others making way. Such ideas finally led Darwin to propose a gradual model of evolution where change was slow and minimal at each stage. Hence the famous dictum *'Natura non facit saltum'*; i.e. if the world was brim full of life, there was no room to jump into. Darwin had some problems explaining the mechanisms of change in a world so full of life (Stanley, 1981), as change, though vitally important, had to be kept to a minimum. Darwin began by speculating about one step, saltatory, species changes (under the influence of Lyell) and the importance of speciation in isolation (Ruse, 1979), but later abandoned these ideas for a much more gradual model. The concept of plenitude did, however, lead directly to Darwin's assertion that interactions between organisms were the driving forces of selection (Diamond, 1978). Darwin was influenced by the Manchester school of economics and the ideas of Malthus. Thus competitive interactions between organisms were seen as the dominant forces driving evolution (ultimate causes of organic diversity), and it therefore followed that such interactions were the determinants of ecological events (proximate causes) also.

Chapter Three of the *Origin of Species* is still one of the best statements of many basic ecological problems and, even in that early work it is possible to identify the beginnings of a basic dichotomy in the approach to understanding the factors which influence population size and community structure. Darwin emphasized the forces of selection which arise from interactions between organisms. He envisaged the geometric growth of population size and the subsequent stresses due to overpopulation and competition (Mayr, 1977) and he wrote of 'a force like a hundred thousand wedges trying (to) force every kind of adapted structure into the gaps in the oeconomy of nature' (Ruse, 1979). Darwin made his ideas quite clear in a letter to Asa Gray in 1857 when he wrote about the process of selection which operated as a result of the 'mutual action of a different set of inhabitants, which I believe to be far more important to the life of each being than mere climate' (Darwin, 1857). Wallace, on the other hand, writing at the same time as Darwin, stressed the importance of interactions between the organisms and the environment and, while acknowledging the importance of food supply to the survival of organisms, saw the struggle for survival under unfavourable conditions as the major force of evolution. The dichotomy is clearly visible in the papers by Darwin and Wallace which were published by the Linnaean Society in 1858. Are population size and community structure primarily controlled by competitive interactions between organisms or are organisms more heavily influenced by external forces, such as climatic events and catastrophes? The ideas which stress the importance of biotic interactions are essentially Darwinian, equilibrium concepts as they depend on density dependent controls on population growth, on competition, and on the coevolution of communities. The ideas which stress the importance of climatic factors are essentially non-equilibrium concepts as population size is envisaged to be controlled by catastrophic events, and community structure is not determined by competition. This dichotomy, which has been present in the literature from the first, is basic to much ecological and evolutionary thought and, while Darwinian ideas have dominated biological thinking from their inception, it is not difficult to trace the development of the two lines of argument to the present day.

The Darwinian concept of evolution and the theory of ecology are inextricably interwoven. Both theories assume that competition plays an important role in determining the outcome of both proximate and ultimate events (Stanley, 1979). Wiens (1976, 1977) discussed the basic assumptions of equilibrium theory and the link to the gradualist explanation of evolution is clear. The 'Modern Synthesis' (Eldredge and Gould, 1972; Stanley, 1981) has brought population genetics, palaeontology, evolutionary theory and ecological theory together around the concepts of plenitude, equilibrium and optimization. This was, in effect, rampant reductionism. Natural selection is presumed to maximize individual fitness and to produce the optimal adaptive response (Cody, 1974a). 'Selection is viewed as incessantly trimming all but the best adapted or optimal phenotypes from a population, and competition is usually presumed to be the major driving force of this selection' (Wiens, 1976). Equilibrium theory has the following limiting assumptions: the population/environment system is in equilibrium; at least some resources are constantly limiting; the environment is fully occupied or saturated with species so there is no temporary relaxation of selection favouring the optimal response; the optimum is always the same in time; the theoretical optimum is always attainable; founder effects are not important (i.e. the identity and genetic make up of the founding individuals are unimportant); there is no effect from the past history of the population. This is effectively the gradualist explanation of evolution in that the species is assumed to be in equilibrium with the environment and with the other species present. The question, of course, is whether or not these limiting assumptions can be said to hold in the real world (Wiens, 1977).

Connell (1978) reconsidered the equilibrium explanation of ecological diversity in coral reefs and tropical rain forests and showed that a non-equilibrium explanation of such systems was preferable. The non-equilibrium approach to ecology has also received support from a number of quarters: from theoretical studies (Huston, 1979); from observation (Shugart and West, 1981; Fox, 1979, 1981) and from the existing literature (Hutchinson, 1941, 1953, 1961, 1967). Non-equilibrium systems in ecology are assumed to be characterized by the following properties: environmental fluctuations, high (density-independent) mortality, lack of competition, absence of limiting resources, environments not saturated with species, adaptive optima that change with time, the presence of founder effects and a strong historical component (Wiens, 1976, 1977). All the above properties are characteristic of the Eldredge and Gould (1972) model of punctuated evolution. The basic difference between the gradualist and punctuated theories of evolution lies in the gradualist assumption of plenitude and the corollary that competition is a crucial factor which selects against individuals with maladaptive features. The adherents to a punctuated evolutionary mechanism assume a lack of continuous competition brought about by the fluctuations in the environment in the real world which disrupt the equilibrium with sufficient frequency to allow coexistence.

The importance of all this for ecology lies in the attack on the concepts of plenitude, optimization and equilibrium (Fig. 2.1). While it is not necessarily possible to use ultimate explanations to account for proximate events, the explanatory mechanisms of ecology and evolution must be consistent as ecology must be viewed as evolution in action. To this end it is interesting that in both ecology and evolution the same basic assumptions have recently been questioned. Modern theoretical ecology, which assumes equilibrium, is in the habit of invoking coevolution to explain observed phenomena (Janzen, 1980, 1981). This amounts to a blind reliance on ultimate explanations of proximate events and has been criticized by Janzen on these grounds. Of course, in a standard, Darwinian, gradualist approach such an explanation is perfectly valid. At the other extreme, if we assume non-equilibrium conditions, however, then founder effects become very important and coevolution cannot be used as a valid explanation. In a changing world, coevolution cannot necessarily be invoked as a theory that accounts for everything. The apparent widespread occurrence of stasis and punctuated change in the fossil record must indicate a fairly widespread occurrence of non-equilibrium ecology in evolutionary time.

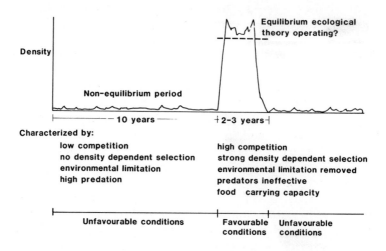

Fig. 2.1 Equilibrium and non-equilibrium ecology are but two ends of a spectrum of possibilities. Conditions may switch from one to the other. Time scales depend on the organisms and may vary from weeks to years. David Rollo drew the first sketch of this figure.

The equilibrium and non-equilibrium viewpoints may simply be viewed as the two ends of an ecological spectrum. In unstable or unfavourable environments growth rates may be reduced and populations are limited by environmental factors. In more stable environments, as a result of higher growth rates, population size may rise until, as envisaged by Malthus, competition becomes important. It is entirely possible that any given species may be limited by environmental factors at some times and by competition at others. The important point to grasp is that natural environments are not constant in time and that the interaction between the environmental fluctuations and the ability of the various species to cope with and adapt to those fluctuations will determine which viewpoint is the more suitable. The two ecological viewpoints may also be linked by viewing the world as a hierarchy of holons. Lower holons (populations) may show much dynamic behaviour whereas the higher holons (communities) may show more deterministic behaviour. The extent to which the hierarchy is influenced by external events, and the level at which the perturbations are felt, will determine the apparent nature of the interactions. Intermittent fires in forest ecosystems are strong perturbations for local populations and may disrupt the local approach to competitive equilibrium. The fires are, however, essential for the survival of some species. In the coral reef, the intermediate disturbance model of Connell (1978) may be interpreted in the same way. The storms which disrupt the approach to competitive equilibrium on the reef are strong perturbations for individual populations but they are essential for the survival of many species and for the maintenance of community diversity.

Tilman (1982) recently pointed out that there is a continuum of models from equilibrium to non-equilibrium, in which disturbances play an increasing role. From equilibrium models of competitive exclusion with no disturbance, the first step is to models in which the disturbance becomes 'a process that influences the relative supply rates of the resources for which competition occurs' (Tilman, 1982). Finally, with increased disturbance the process of competition is disrupted and becomes less and less important (Connell, 1978).

2.2 Ecology and evolution

The gradualist explanation of evolution and the equilibrium theory of ecology are now being challenged. Eldredge and Gould (1972) attacked the gradualist view of evolution and proposed a 'punctuated equilibrium' model instead. The tempo and mode of evolution were reconsidered. Their revised model of evolution received support from a number of quarters: from the existing literature (Goldschmidt, 1940; Mayr, 1958); from reevaluations of the existing fossil record (Stanley, 1979, 1981) and from new research (Williamson, 1981a). There is really no debate between the two evolutionary camps about the mechanisms of evolution. Contrary to popular belief we do not need to invoke new modes of selection to account for punctuated evolution. The real debate is about the persistence of change, not about the rates of change (Schindel, 1982). If we envisage the environment as a fluctuating multidimensional system and the organisms as having varying abilities to cope with such fluctuations then competition may become significant only occasionally when ecological 'crunches' occur (Wiens, 1977). In other words selection pressures may set boundaries within which no single optimum may occur (Fig. 2.2). As Schindel (1982) points out, in evolutionary terms 'stasis may result if directional trends are short-lived and frequently reversed. Long term trends may be processions of morphologically distinct and separate populations rather than gradual modifications within a single standing population'. Thus if the pressure for change persists, persistent changes will result. If the environment fluctuates about a long term mean no changes may be observed.

Environmental fluctuations must be seen as variability or patchiness in space and time. Any population consists of a number of subunits scattered in space and time, each experiencing different environmental conditions. The evolutionary and ecological response of the populations will depend on the heterogeneity of the environment, the way it is perceived by the organisms and the long term temporal trends in both components (Levins, 1968).

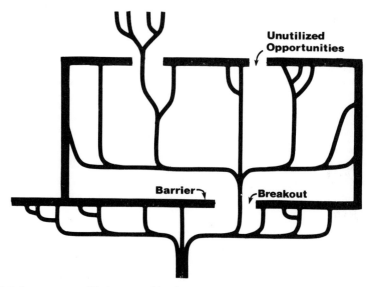

Fig. 2.2 In a non-equilibrium world selection may only set broad limits and may only be important at times of ecological 'crunches'. Ecological breakouts are possible. Unutilized resources may result from a lack of time or a lack of colonizing organisms.

In such a scheme, founder effects become important as speciation may occur rapidly in isolated populations undergoing heavy selection at the geographical or physiological limits of the species. Environmental patchiness is also indicative of a possible absence of plenitude. The organisms must migrate to all the suitable environments before saturation can occur. Thus saturation is only likely in rather invariant environments where the species have a sufficiently long time for extensive migration to occur (Williamson, 1981a). The equilibrium theory of ecology assumes that either there is no environmental patchiness or that all possible patches are saturated with all possible species. Non-equilibrium theory assumes that patches open and close faster than the species can migrate from one to the other. Again the two viewpoints are two sides of the same coin and in reality there is a distribution of patches in space and time.

Patchiness and dispersal are integral components of a non-equilibrium model (Caswell, 1978). A (gradual Darwinian) equilibrium model assumes that all possible opportunities are occupied all the time. A punctuated model, on the other hand, does not assume that all possible opportunities are being exploited and that rapid change may ensue when species break out into new areas and exploit new resources. New areas and resources may become available through migration and/or the interaction of changing environmental heterogeneity with the patchy distribution of the species.

Organisms differ in size, mobility and in their adaptive capacity, so that different organisms will perceive the same patchy environment in different ways. In the terminology of Allen and Starr, (1982) holons at different levels lag and filter the external environment in different ways. We should not expect different organisms to all fall into one world viewpoint or the other.

2.3 Equilibrium theory

As outlined in Chapter One, the growth of phytoplankton populations is a simplified case of a broader ecological problem, so the models which describe the exponential increase of phytoplankton populations are a simplified subset of a larger class of ecological models. As there is no age structure in the populations and the generations in natural populations generally overlap then simple differential equations may be used to model the growth process. It is assumed, following Darwin, that populations do not grow exponentially for any length of time, so an equation which describes population growth in a limited environment is required. The relationship normally used is the basic Verhulst–Pearl logistic equation. This, and the modification of the equation for competition between two, or more, species forms the basis of modern ecological theory (Levins, 1968; MacArthur, 1972; Pianka, 1978). The competition equations were first used by Lotka (1925) and Volterra (1926).

From the earliest work of Lotka and Volterra the theoretical framework of models of population growth and of competition between species has been explicitly an equilibrium theory. This appears to be not only due to the intellectual satisfaction of assuming order and balance in the universe and to the influence of Darwinian ideas of plenitude, but also to the reinforcing effect of the mathematics of the theory applied. Volterra's models led to the experiments of Gause (1964) and the formulation of the 'one species – one niche' axiom. Thus from the earliest times there was a direct connection between the theory of the niche and of competitive exclusion. This axiom of competitive exclusion and the niche has done a lot to popularize the concept of the niche (Vandermeer, 1972) because of its deceptive simplicity. The concept of the niche is one of the most basic concepts in ecology and was formulated in the 1920s (Vandermeer, 1972).

Much has been written about the niche over the years (Whittaker and Levin, 1975) and it remains a basic ecological concept, but despite the efforts to measure niche dimensions directly (Levins, 1968) it remains essentially undefined. Hutchinson's (1957a) use of set theory to distinguish between the fundamental and the realized niche is in accord with the equilibrium model and is a useful way to grasp the concept. Morowitz (1980) has examined the n dimensions of Hutchinson's niche and has concluded that n should not exceed six and may be significantly less in some cases. Thus, while the niche is a very complex concept, the multiple dimensions which are the bane of all ecology may not be far beyond the conceptual grasp of the human mind. (One great problem yet to be faced is that, in a fluctuating environment, the fluctuations themselves become a resource (Levins, 1979) and thus a characterization of the fluctuations is as important as a measurement of the average value of a resource (Harris, 1980a). But more of this anon.)

The theoretical bases of equilibrium models of competition are the Lotka–Volterra models expanded to include the interaction between two, or more, species competing for the same resource. These are models of small number systems, with few species. The derivation of the equations is given in many text books of population biology (MacArthur, 1972; Krebs, 1978; Pianka, 1978; Hutchinson, 1978; Begon and Mortimer, 1981) and, while the equations appear deceptively simple, they have some complex mathematical properties. A simple, graphical treatment of competition theory is sufficient to show that coexistence is most unlikely in a simple universe if the two species compete for a single resource (MacArthur, 1972; Colinvaux, 1973). Theory and laboratory experiment both verify Gause's axiom, but the problems with the definition of the niche in the real world have limited the practical application of the concept. The theory has been much elaborated in recent years and the major results from equilibrium competition theory have been summarized as follows by MacArthur (1972) and Levins (1979).

(1) Species must use the environment differently in order to coexist. This is the exclusion principle of Gause.
(2) The way in which different species use the environment may be assessed by measuring the different types of resources used. The statistical distribution of resources used gives a measure of niche breadth and the overlap of niches between species. Resource is used here in terms of a factor which, in the broadest sense, is consumed by the organisms and which leads to greater growth rates when its availability is increased (Tilman, 1980).
(3) The number of coexisting species cannot exceed the number of resources.
(4) The number of coexisting species in a community depends on the relationship between niche overlap and niche breadth.
(5) A group of competing species may more easily eliminate an invading species than a single competitor. This statement really concerns the way in which niches are packed together along a resource gradient.

Equilibrium theory considers the world in terms of closed systems. Interactions take place in constant environments, in 'closed boxes' at equilibrium. Thus there is no concept of history or time and no migration or dispersal. The multivariate nature of the natural environment has precluded any rigorous description of the niche of any species so that the axiom of Gause has considerable conceptual value but little practical value. The axiom may only be tested in the most simple universe. MacArthur (1972) pointed out two weaknesses in the generalization of the basic statement to n species; first, no one really knows what a resource really is: it must be defined in relation to the species in question. This can come close to being a circular argument. Secondly, resources may be an interactive function of both the environment and the species themselves. In this case it should be possible to have more species than apparent resources. Levins (1979) has recently voiced an objection to this practice of defining the environment and the organism separately and suggested that a more useful approach would be to stress the mutual dependence of organism and environment.

2.4 The equilibrium theory of community structure

Niche theory has now developed from the axiom of Gause to a more complex problem. If two species with similar niche requirements cannot coexist, how similar is similar? This question has been tackled by assuming a particular relationship between the competition coefficients of the Lotka–Volterra equations and the parameters of niche overlap. Once again the usual solution is dominated by the convenience of the mathematics rather than reality. The basic conclusion is that the separation of the niche centres (d) must be about 1.0–1.4 times the standard deviation of the niche dimensions (w) if competition is to be avoided (May, 1973). May also reviewed examples of field data which appeared to support this conclusion. Diffuse competition, or the interaction of a community of organisms with an invading competitor (MacArthur, 1972), also depends on the d/w ratio and the availability of 'empty' niches. Equilibrium theory predicts that resource gradients will become packed with species until $d/w=1.0$ at which time invasion will be effectively resisted. It must be remembered that the theoretical models assume a single resource gradient, competitive equilibrium and a constant environment. Despite these restrictions the idea of diffuse competition leads to the concept of guilds of coevolved species which tend to occur together and which resist competition as a group (Diamond, 1973, 1975). Thus the equilibrium theory of competitive interactions and the niche leads directly to a discussion of the dynamics of community structure and associations between species. The whole concept of the community, as the name implies, assumes that the interactions between the organisms are important and that there is some kind of joint identity. In short a community is presumed to be more than a collection of species which happen to be in the same place at the same time.

Perhaps the best known piece of equilibrium theory for community structure is the Island Biogeography theory of MacArthur and Wilson (1967). This relates the size, and position of islands in relation to a source of immigrants, to the overall species diversity. The model asserts that there is a regular relationship between species diversity and island area and that the rates of immigration and extinction are monotonic functions of the number of species present. Rates of immigration are assumed to decrease with increasing numbers of species already present (presumably due to diffuse competition) while rates of extinction increase for similar reasons (Simberloff, 1972). The equilibrium between immigration and extinction for islands of different sizes leads directly to a relationship between species diversity and area which has been derived from sampling data in many types of community (Connor and McCoy, 1979).

According to equilibrium theory (and following Darwin) the observed diversity of living organisms has arisen as a result of niche specialization and niche diversification. Such processes are assumed to have occurred as a result of competition, but there are three possible mechanisms by which the present observed diversity of organisms in communities may have arisen. The first mechanism requires that the observed niche dimensions are the result of current competition. Thus the organisms are occupying realized niches which are presently limited in their dimensions by interspecific competition. This requires also that the organisms have some flexibility in their resource utilization and that the final outcome of 'who does what' depends on whatever other organisms may be present. The second possible mechanism requires that the observed niche diversification resulted from competition in the past. This assumes that competition is now minimal and that the situation has arisen as a result of mechanisms which minimize competition over evolutionary time scales. This assumes that both the fundamental and realized niches of the organisms are restricted by genetic factors and that, consequently, the resource utilization of the organisms is highly specialized. The third possible explanation is that there never has been any competition at all and that the organisms merely evolved separate strategies by utilizing different resources in a multivariate environment. The worrying thing about all this is that it is not possible to rule out any of these explanations.

In a multivariate environment it is not possible to obtain a rigorous definition of the niche of any species so that exact tests of the various hypotheses are not possible. Merely collecting data which supports one or other of the explanations does not constitute a valid test. Abrams (1976) wrote a convincing criticism of the theory of niche diversification and attacked the equilibrium theory of niche packing on a number of grounds. The d/w ratio is an explicit function of the mathematical relationship which is used to link the competition coefficients of the Lotka–Volterra equations to the parameters which determine the dimensions of the niche. If different mathematical relationships are used then the critical ratio of d/w can vary significantly. What is there to say which formula describes the precise relationship of a pair of competing species or that all the interactions of competing species can be described by one expression?

The d/w relationship requires that competitive equilibrium has been achieved, that competition is occurring on a single resource axis and that the environment is not varying in time. Abrams (1976) showed that species are arranged on many resource axes and that they are not arranged linearly. Thus the theory of the d/w ratio is inapplicable and the suggested linear arrangement of species on a single resource axis arises from inadequate information. Natural environments are certainly not constant in time. Data on the size differences of congeners, on apparent constant niche overlap in different environments and on the apparent patterns in actual resource utilization data may also be criticised. Abrams points out the first is only indirect evidence, the second may merely indicate that the environment is not limiting niche overlap and in the third case there is always the possibility that there is another significant dimension that has been overlooked. If Morowitz (1980) is correct in his assertion that there are between four and six significant dimensions to the niche, then the niches of the species in a community should be spread out in four to six dimensional space. There are many apparent cases of niche overlap which may be explained by the existence of differences on dimensions which were not studied.

2.5 Criticisms of equilibrium theory

The fundamental problem with the foregoing equilibrium theory however, is that it is impossible to test if the observed community structure really is a result of competition and coevolution. This point has been made before but has not slowed the elaboration of equilibrium theory. The first shots in the war were fired over twenty years ago (Hairston *et al.*, 1960; Murdoch, 1966; Birch and Ehrlich, 1967; Ehrlich and Birch, 1967). To invoke coevolution is to confuse proximate and ultimate explanations and, furthermore, it is almost impossible to demonstrate the existence of coevolution in the field (Connell, 1975, 1980). Connell concluded that it was only on rare occasions that competition could be unequivocally demonstrated in the field and he refused to be persuaded by those who would invoke 'the ghost of competition past'.

There has been some considerable debate of late about the validity of these equilibrium views of community structure. As might be expected the two poles are occupied by those who believe that competitive interactions are of great importance (MacArthur, 1960; Whittaker, 1965; McNaughton and Wolf, 1970; Diamond, 1973, 1975; Grant and Abbott, 1980; Grant, 1981; Hendrickson, 1981) and those who believe that communities are generated by random and individual colonizing events (Simberloff, 1970, 1978, 1983; Shapiro, 1978; Connor and Simberloff, 1979; Strong *et al.*, 1979; Strong and Simberloff, 1981). It comes down to a debate between those who favour a competitive or a neutral model (Caswell, 1976). The arguments of the proponents of competition and coevolution are bolstered by much 'invocation of Chairman Darwin's wisdom' (Shapiro, 1978) while the random assemblage school insists that it is impossible to distinguish between natural communities and those generated at random by a computer. Strong *et al.* (1979) and Connor and Simberloff (1979) postulated a null hypothesis (that island communities are constructed by random colonization events) and then attempted to refute the hypothesis. Their conclusion was that it was not possible to do so. They also showed that many of Diamond's (1975) rules for the assembly of bird communities could be dismissed as 'either tautological, trivial, or a pattern expected were species distributed at random' (Connor and Simberloff, 1979). The evidence points to a weak competition effect, if any, though the battle rages in the pages of the scientific journals and the matter is far from closed.

The evidence in favour of weak competition does not imply that competition never occurs; it is simply that in the real world the environment fluctuates and competition may be important only occasionally when ecological 'crunches' occur (Wiens, 1977). Schaffer (1981) has recently proposed a solution to the controversy by showing that the usual models of population dynamics are abstractions from a much more complex natural situation. Schaffer suggests 'that the resolution of the dispute may depend on our ability to classify subsystems of species (i.e. guilds) with regard to the extent to which their internal organization is influenced by variation in the larger communities in which they are imbedded'; i.e. Schaffer uses a hierarchy of holons to classify the important interactions. Schaffer also shows that the degree to which the abstracted growth equations accurately describe the species of interest depends to a large degree on the time scale of the species relative to that of the species which are omitted. Thus species operating on different time scales are uncoupled and the hierarchy may be decomposed into discrete holons. In a world where fluctuations are the norm it is possible to avoid competition by exploiting the temporal variability in a resource in a different way to other species (Levins, 1979). Once again it is seen that adequate description and classification of the time scales of biological and environmental fluctuations is an important task. As a result of his analysis of the usual form of community models Schaffer (1981) was prompted to refer to the principles of community ecology as the 'late' principles.

Allen and Starr (1982) point out that the use of the species as a taxonomic holon may not be the appropriate ecological holon. The apparent lack of niche diversification between species may be due to both the fact that the correct niche dimensions were not measured and the fact that the taxonomic holon does not correspond to any significant ecological entity. Cell size, growth rate or sinking velocity may be a better ecological criterion. Structural and functional boundaries between holons may differ significantly. Schaffer's models of communities which are based on time scales of processes are hierarchical and are in accord with the holon concepts.

There can be no doubt that natural selection has occurred over time and that evolution has resulted in the present range of species. The precise role and importance of competition will, however, depend on which model of speciation is favoured. Equilibrium theory takes one extreme viewpoint; that competition is acting, was acting, and always will be acting. Perhaps a more reasonable position is to assume that competition is something that contributes to the generation of diversity over evolutionary time scales but that it may not be necessary to invoke competition as a mechanism to explain the maintenance of diversity at ecological time scales (Williamson, 1981b). Thus ecological questions are not necessarily answerable in evolutionary terms. Evolution reflects the long term effect of the means and variances of all the environmental components and at such time scales we are looking at the long term survival and adaptation of a sequence of populations which together make up the species (Schindel, 1982). Thus, if we relax the most stringent requirements of equilibrium theory whilst retaining some of the basic concepts, we may find a middle ground which is more realistic. We cannot always distinguish between coexistence (which implies some kind of stable equilibrium or coevolution), cohabitation (which implies sharing of habitat or resources, but not necessarily an equilibrium) and cooccurrence (which merely implies that species are found together). We certainly cannot use theories of coexistence to account for cooccurrence. Connell (1980) has shown that for coevolution to occur, long term coexistence is necessary. In other words coevolution takes time and cooccurrence must remain in effect while the process continues. In communities with rapidly changing species composition, such as phytoplankton, the brief periods of cooccurrence may not be sufficient to ensure coevolution. There is no reason to assume that all communities may be formed by the same processes. In a changing world organisms with different life spans may perceive the world to be different, and, depending on the interaction between the time scales of change and growth, competition may influence some more heavily than others.

2.6 Non-equilibrium theory

Non-equilibrium theory represents the other extreme point of view from equilibrium theory: that competitive interactions are minimal and that community structure is dominated by the vagaries of the environment. A non-equilibrium model requires that environmental disturbances occur with sufficient frequency that the course of competitive exclusion is disrupted. Such arguments have time as an explicit component; a component which is completely lacking in equilibrium theory. The time component must be relative to the growth rate and hence the speed of competitive exclusion for any group of organisms, thus time is not measured in absolute terms but in terms of organism size and growth rate (Stearns, 1976). This point was first made by Hutchinson (1941, 1953, 1961, 1967) in a series of papers in which he used phytoplankton as an example. In 1941 Hutchinson 'put forward the idea that diversity of phytoplankton was explicable primarily by a permanent failure to achieve equilibrium as the relevant external factors changed' (Hutchinson, 1961). Competitive exclusion will be largely avoided if the time between disturbances is somewhat less than the time it takes for competitive exclusion to occur. This is the intermediate disturbance hypothesis of Connell (1978). Figure 2.3 is taken from Hutchinson

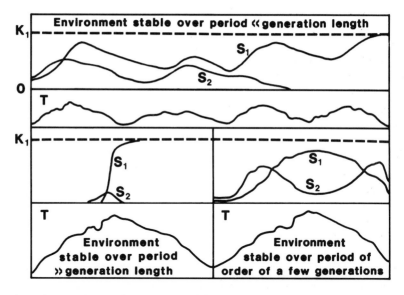

Fig. 2.3 The interaction of environmental fluctuations and the course of competitive exclusion. K_1, the supposed carrying capacity; S_1 and S_2, the abundances of the competing species; T, temperature. Reproduced from Hutchinson (1953).

(1953) in which a fluctuating coexistence is envisaged between two species when the environment is only stable for a period of a few generations. Huston (1979) obtained the same result using a series of Lotka–Volterra equations. He showed that periodic population reductions could prevent the approach to competitive exclusion and could lead to increased diversity

(Fig. 2.4). Non-equilibrium theory is also explicitly a theory of open systems: or in our terminology, small holons at the base of the hierarchy. Caswell (1978) modelled the non-equilibrium dynamics of populations in a system where each patch was a closed system (in that predation caused the extinction of the prey) but migration between patches was allowed. Long term persistence of species resulted in a form that looked rather like equilibrium coexistence. Huston and Caswell's non-equilibrium theory has as its primary assumptions:

(1) Most natural communities exist in a state of non-equilibrium where forms of density independent mortality and fluctuations in the physical and biotic environment prevent competitive exclusion.
(2) In a community of competing species different species have different intrinsic rates of increase. Thus a given periodicity of disturbance will affect different species in different ways depending on the relationship between growth rates and the frequency of disturbance.
(3) Ecological communities are open systems with species playing 'hop scotch' between patches. Extinction or exclusion is therefore only a local event for the local population.

These basic assumptions should be compared to the first three major results of equilibrium theory. The differences between the two theories are clear in that result (1) above, (the different use of resources) now is modified to include the use of the same resource at different times. Result (2) (the measurement of niche dimensions by the measurement of resources used) must be modified to include the role of fluctuations in the resource over time. Result (3) (the relationship between the number of resources and the number of species) becomes more complex as the fluctuation in the resource becomes a resource in itself (Levins, 1979). Levins' treatment stressed the relationship between the organism and the required resources, particularly the covariance of the two in time, and pointed out that, in a fluctuating environment, if 'the resource utilization functions were non-linear then some complex properties ensued and persistent coexistence was possible even when no stable equilibrium existed'. Levins (1979) also drew attention to the fact that for a temporally fluctuating non-linear function, the average of the function over time was not necessarily identical to the long term integral of the function. What all this means in simple language is that we must take the time course of variation into account and that simple averages are not a sufficient measure of what is going on.

There is considerable evidence that, at the species level, many types of community are not at competitive equilibrium (Wiens, 1976, 1977). Some examples include: temperate forests (Auclair and Goff, 1971; Bormann and Likens, 1979; Shugart and West, 1981); tropical forests (Connell, 1978); herbaceous plant communities (Gleason, 1926; Grime, 1973; Grubb, 1977;

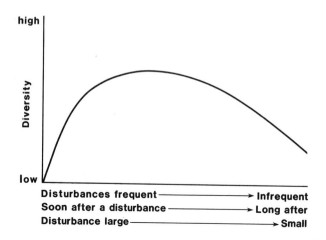

Fig. 2.4 The interaction of disturbance and diversity. After Connell (1978).

Fox, 1979, 1981); saxicolous lichens (Armesto and Contreras, 1981); coral reefs (Connell, 1978; Porter *et al.*, 1981; Knowlton *et al.*, 1981); marine benthic communities (Tunnicliffe, 1981); coral reef fish (Sale, 1977); decapod crustaceans (Abele, 1976) and phytoplankton (Hutchinson, 1941, 1953, 1961, 1967, Richerson *et al.*, 1970, Harris, 1982, 1983). Significantly Connell (1978) used an intermediate disturbance hypothesis to account for the very high diversity of tropical forests and coral reefs. These two very diverse communities require very great niche diversification according to the equilibrium theory and it has not been possible to demonstrate the necessary diversification. Time scales are important. Tropical forests are disturbed by processes operating over vastly longer time scales than those which influence phytoplankton populations, but the generation times of the two types of organisms are vastly different also. In relation to the generation time of the organism it is necessary to know whether specific environmental fluctuations are fast or slow so that their effect can be classified (Stearns, 1976, Schaffer, 1981).

While widespread acceptance of the viewpoint has come only recently, non-equilibrium views of community structure have almost as long a pedigree as the dominant, equilibrium viewpoint. Gleason (1917, 1926), writing at much the same time as Clements, wrote of the individual concept of the plant community and stressed the influence of environmental factors on the complement of species found at each point. Note the use of the term 'individual concept'. When the role of internal interactions is subjugated to the influence of external influences there is no need to assume a dominant role for the interactions between the organisms, and the community becomes less of a community and more of a loose association of coexisting or cooccurring species. This viewpoint concentrates on the lower (species) holons at the expense of the community level properties (Allen and Starr, 1982). Gleason's view was precisely this; that environmental factors were more important than the biotic interactions between species and he suggested that after a period of seed dispersal 'every plant association tends to contain every species of the vicinity which can grow in the available environment' (Gleason, 1926). A recent report by Silander and Antonovics (1982) shows that competitive interactions between plants cannot be totally ignored however, as selective removal of certain species from coastal plant communities revealed a range of specific, diffuse, reciprocal and non-reciprocal effects. Thus perturbation experiments have revealed that at least some potentially important interactions exist in nature.

2.7 Some ideas from non-equilibrium thermodynamics

A number of years ago von Bertalanffy (1952) distinguished between true equilibria in closed systems and the steady state of open systems. Open systems show continuous fluxes of energy and materials across their boundaries; properties characteristic of life itself. It is therefore difficult to describe living systems in terms of true equilibria. Von Bertalanffy viewed the world as an hierarchy of interacting open systems and in doing so set the stage for the ideas of Allen and Starr (1982). Allen and Starr's view of the world is basically that of von Bertalanffy (1952) with the terminology redefined in more modern terms. As defined above, Allen and Starr's holons have characteristic scales at each level in the hierarchy and the holon boundaries determine the fluxes of energy and materials and the nature of the interactive messages which pass between the holon and its environment. The environment of the holon consists of either all holons above it in the hierarchy (if the hierarchy is fully nested), or the higher holons plus other influences external to the hierarchy (if the hierarchy is not fully nested). Ecological systems seem to be rarely fully nested and this explains the controversy between the equilibrium and non-equilibrium approaches.

Recent developments in physics and chemistry provide a striking parallel to the developments in ecology. The development of non-equilibrium thermodynamics has led to a new understanding of irreversible processes (Prigogine, 1978) and has shown that non-equilibrium may be a source of order. Prigogine calls these new dynamics states 'dissipative structures' (Johnson, 1981). Von Bertalanffy (1952) noted many years ago that living systems were open, non-equilibrium systems which consumed energy and which were thus equivalent to dissipative structures. Living organisms are highly ordered systems which minimize entropy production and, as Johnson (1981) pointed out 'energy is actively acquired, time delayed and processed as it proceeds to a state of least dissipation'. This amounts to another definition of a holon in the terms of Allen and Starr (1982). Johnson (1981) sees the basic unit of ecology as the ecodeme, a group of individuals that together form a dissipative structure. The major developments of non-equilibrium thermodynamics which are of interest here include the properties of dissipative structures, the role of fluctuations and the relationships between microscopic chaos and macroscopic order (i.e. the relationships between middle and large number systems).

Prigogine (1978) showed that for states close to equilibrium, non-equilibrium systems will settle down into a state of minimum entropy production and when boundary conditions prevent the system from reaching equilibrium the system settles down into a state of least dissipation. This is only true for small perturbations close to equilibrium. Far from the equilibrium point fluctuations and perturbations become amplified to the point where a macroscopic order of a new kind emerges and dissipative structures grow into ordered fluctuations stabilized by energy exchanges with the outside world. Prigogine (1978) showed formally how a set of chemical reactions can, in the presence of diffusion within the reaction vessel, produce chemical waves, ordered fluctuations and non-uniform steady states (Turing behaviour). Recently Powell and Richerson (1985) have developed the two- and three-species competition models described above into a more general model of interspecific interactions in natural waters by introducing growth, death, nutrient recycling and horizontal diffusion. The result is a Turing instability with complex fluctuations and waves of abundance which allow coexistence where simple equilibrium theory would predict exclusion. These instabilities produce a type of long range order through which the system operates.

According to Prigogine (1978) three aspects of dissipative structures are always linked; the function, the space–time structure and the fluctuations which trigger the instabilities. These relationships introduce a component of history as the time sequence of events is crucial. Developments in catastrophe theory (Thom, 1972) are highly relevant to this discussion. Johnson (1981) interprets the structure of the ecosystem as a series of linked dissipative structures or 'shock absorbers'. Note that this type of explanation explicitly requires non-equilibrium conditions and the existence of fluctuations – features of considerable realism and generality. According to Johnson, each ecodeme (or holon) will be dependent on a specific energy fluctuation and the configuration of the ecosystem will depend on the component fluctuations with the total. The degree of interaction and dissipation will depend on the scales of fluctuation and the scales of response.

In a hierarchical context it is possible for low level (microscopic) processes to be essentially random, yet for there to be a high level (macroscopic) order. Prigogine (1978) provided a succinct explanation of how random molecular collisions may, in sufficient number, appear as Gas Laws and other 'large number' properties of systems. Non-equilibrium thermodynamics provides an explanation for the appearance of high level order from low level chaos. In ecological terms the interactions between populations may be so weak as to be almost non-existent but high level order may none the less emerge.

CHAPTER 3

Some basic physics

The physical processes of turbulent mixing are clearly of great importance to the ecology of phytoplankton populations (Margalef, 1978b). Physical processes are the basis of many of the perturbations which are so important for ecology of the organisms. In fact Margalef regards the external supply of turbulent energy to the water column as an energy supplement to the community. The spatial and temporal patchiness in the physical environment arises from the interaction of solar heating and wind mixing in all bodies of water from rain puddles to the oceans. As I indicated previously the differences between limnology and oceanography are only those of scale: granted there is salt in the oceanic water which modifies density structures, but the basic processes of the generation and dissipation of turbulent kinetic energy are the same. What is really different about large bodies of water as opposed to small ones is the presence of motions driven by the rotation of the Earth. The balance of solar heating and wind mixing in large bodies, when coupled with rotational effects, results in large scale horizontal water movements which add a whole new dimension to the ecology of organisms in these systems. For a full treatment of physical limnology and oceanography the reader is referred to the more specialized texts: it is my purpose here to provide some basic definitions and to outline some important features of the environment relevant to the organisms.

All bodies of water, of whatever size, are basically turbulent environments. The classical treatment of turbulence is of a field of nested eddies of varying size, where turbulent kinetic energy (TKE) cascades from the largest eddies to smaller and smaller eddies until viscous scales are reached, when the energy is finally dissipated as heat. It is assumed that the water flow is turbulent as eddies are generated by the influence of the wind and by the rotation of the Earth. Shear occurs when one eddy tears into another and this interaction between the eddies results in a cascade of eddies and energy from large to small (Ozmidov, 1965). As the old saw has it:

> Larger swirls have lesser swirls
> that feed on their velocity;
> and lesser swirls have lesser swirls,
> and so on to viscosity.

The wind induced motion in surface waters results in both horizontal and vertical motion and the turbulent diffusion of heat, salt and other substances. The turbulent diffusion results from the fact that as the water is moved by the wind dissolved and suspended materials are carried about by the water. It is possible to think of the process of turbulent diffusion in terms of eddy diffusion coefficients. The eddy diffusion coefficients say little about the basic physics involved, they are analogous to mass transfer coefficients which result from the transfer of materials by the eddies. The coefficients may be conveniently separated into vertical and horizontal components which may be treated independently. The reason for this lies in the fact that most water bodies are wide and shallow and the presence of density gradients in the vertical dimension produces a tendency for the major motions to be stretched out in the horizontal plane. The eddies which overturn in the vertical are smaller because of the restricting effects of the density gradients. The horizontal turbulent diffusion of materials is therefore much more rapid than the vertical turbulent diffusion of the same. The eddy diffusion coefficients are essentially empirical coefficients and they are usually defined in terms of the spread of patches of dye or other easily measured substances and are scale dependent because larger eddies move things about more rapidly than smaller eddies. As might be expected from the relative eddy sizes, coefficients of horizontal turbulent diffusion (K_h) are larger than the corresponding vertical coefficients (K_z). Observation of the spread of dye patches in the horizontal and vertical planes has led to the characterization of K_h and K_z and has shown that K_h is about two orders of magnitude larger than K_z (Boyce, 1974).

The results from the horizontal diffusion of dye patches have led to some well-defined relationships between the time scales and space scales of mixing in lakes and the oceans, as the scale dependent values of K_h appear to be consistent in many different waters (Bowden, 1970; Boyce, 1974). Figure 3.1 shows some data from the oceans which compares well with Great Lakes data (Murthy, 1973). These relationships are the basis of the time and space scales defined in Harris (1980a) and are the means by which time and space may be linked in the planktonic environment. Measurements of turbulent diffusion in the horizontal plane indicate that patches of the order of 1 km in size should be mixed at scales of about one day: this observation will become the basis of theoretical arguments about the patchiness of phytoplankton to be presented later. There is also a clear relationship between eddy size and basin dimensions. Small lakes cannot have eddies as large as those in large lakes and the oceans, so the scale dependence of the horizontal diffusion is cut off by the size of the basin. It can be argued (Boyce, 1974) that the largest possible eddy can only be about half the limiting basin dimension and still have room to swirl about. So there is a link between the morphometry of the basin and the mixing processes in the

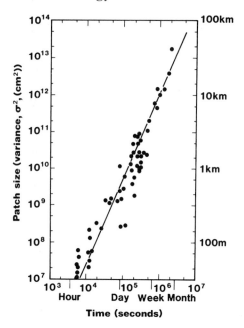

Fig. 3.1 The relationship between time and space scales as revealed by dye diffusion studies in lakes and the oceans. Reproduced from Bowden (1970).

basin. It is therefore to be expected that phytoplankton communities will be different in different sized water bodies.

The generation and dissipation of TKE in surface waters can be largely accounted for by physical theory but there are still a number of areas which cannot be rigorously defined. The basic processes of the generation of TKE by the wind and of the generation of buoyancy by solar heating are well understood and may be written down in terms of simple physical parameters. It has been suggested that God must have been a limnologist or oceanographer as the peculiar physical properties of water lead directly to the observed physics of lakes and the oceans, and without the particular relationship between density and temperature exhibited by water it would be very difficult for life to exist in surface waters.

3.1 Scales of turbulent kinetic energy generation

When the wind blows over water, work is done on the surface, motion is induced and TKE is generated. TKE therefore arises both from the stirring effect of the wind and from shear effects caused by the movement of the water. TKE is thus fed into the water at a number of scales. Knowledge of the scales of input of TKE is not good because of the interactions between basin size, wind stress, internal motions and rotational effects. Any summary of the temporal and spatial scales must be highly schematic. In the oceans (Fig. 3.2) there appear to be three major scales of TKE input (Monin *et al.*, 1977). The largest scales (1000 km) correspond to ocean basin scales, the scales of atmospheric weather patterns and the scales of ocean currents. The medium scales (10 km) correspond to internal inertial motions and tidal oscillations. Small scale inputs (10 m) correspond to wind and wave interactions. At spatial scales between these TKE inputs the TKE cascades down from one peak to the other with a predicted slope of − 5/3 (Fig. 3.2), giving the slope of the spectrum of cascading TKE decay. Woods (1980) has argued that the energy spectrum in lakes and the oceans is, in reality, a function of wave motions and that the peaks in the energy spectrum are determined by wave motions and breaking waves at a variety of scales. The largest possible scales are constrained by basin size so the full spectrum is only observed in the oceans and the energy spectrum of lakes of different size will differ accordingly.

The TKE generated by the stirring effect is the first source of TKE to become important when the wind begins to blow over a body of water. As the wind rises, work is done and stirring begins. More slowly a mass downwind motion of the surface water is induced which causes TKE to be generated by shear as masses of water slide over one another. If the water body was stratified when the wind began to blow (i.e. there was a warm surface layer overlying a colder, denser, bottom layer) further motions are now induced as the two layer system begins to oscillate and internal waves are generated at the interface between the two layers. The two layer assumption is necessary to make the theory tractible and, making that assumption, it is possible to quantify the effects of the internal motions and write down the amount of TKE so produced (Sherman *et al.*, 1978). In large water bodies, where horizontal motions are induced by the effects of the rotation of the Earth, then a further source of TKE arises from the fact that the internal motions become organized into identifiable currents which flow around the basin. Time scales are important in this whole discussion of TKE generation as the effects of a given wind event build with time until all the water in the basin is in motion and then decay away as the wind drops and the motions are dissipated by friction in the basin. Spatial scales must also be considered. The effects of wind upon a small basin differ from the effects of the same wind on a larger basin in that the fetch over the water is smaller and the scope for internal motions is restricted. There is therefore a strong morphometric effect and the same climate, when applied to lakes of different sizes, produces quite different stratifications in the water (Ford and Stefan, 1980) and influences the ecology of the phytoplankton accordingly. While there is an obvious connection between the scales of TKE input (the energy cascade), for the purposes of this discussion it will be simpler to discuss the scales separately.

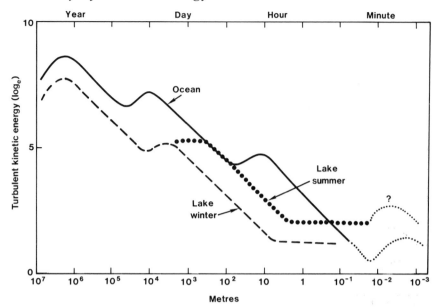

Fig. 3.2 A diagrammatic representation of the TKE spectrum in lakes and the oceans. Data from Monin et al (1977) and Harris (1983). For details see text. The dotted bump at small scales is the presumed patchiness resulting from biological regeneration of nutrients at small scales by microheterotrophs.

3.2 The physics of surface waters

This section is concerned with the derivation of a number of important measures of physical conditions in surface waters. The phytoplankton are very sensitive to changing conditions in surface waters and respond to fluctuations in the input of TKE in a number of ways. The measures of TKE derived below will be used at a number of points in the book so a reasonably full derivation is given. It is important to see not only how some of these physical measures are derived but also to discuss the scales at which different processes operate.

3.2.1 SMALL SCALE EVENTS
Measures of turbulence in surface waters

When the wind blows over the surface of the water TKE is generated because the water is stirred and waves are produced (Sherman *et al.*, 1978). The usual assumption is that the wind profile over the water is logarithmic (Phillips, 1966). This means that the wind speed increases with height above the surface in such a way that a plot of velocity (U) against the logarithm of the height (z) produces a straight line. Under these circumstances the shear stress is constant with height and related to the slope of the line. The fact that the wind speed at any height is given to turbulent gusts and lulls means that the wind speed measured at any given height can only be an average figure.

The surface stress, T_o, is a function of the square of the wind speed at some reference height. T_o is closely proportional to U^2 but not exactly so. The reason for the lack of exact proportionality lies in the fact that, over water, the nature of the surface changes with the wind speed, so the surface roughness, length (z_o), increases as the wind speed increases. As:

$$U_z = a(\ln z - \ln z_0)\qquad\qquad 3.1$$

a, the slope of the profile, defines the velocity scale for the wind which is normally written as U^*/k, where U^* is the friction velocity for the wind and k is von Karman's constant. Hence T_o is proportional to the square of U^* not U_z. The normal wind profile may thus be written as:

$$U_z = U^*/k \quad \ln z/z_0\qquad\qquad 3.2$$

and by this definition, T_o may be found from (Monteith, 1973):

$$T_o = \varrho_a U^{*2}\qquad\qquad 3.3$$

Rather than using the formulation for finding T_o which includes z_o it is more common to use an empirical relationship which includes a drag coefficient as follows:

$$T_o = \varrho_a C_{10} U^2\qquad\qquad 3.4$$

where T_o is the surface stress, ϱ_a is air density, C_{10} is the drag coefficient referenced to the velocity U at 10 m above the ground. The value of C_{10} may be found in a number of references but is generally between 0.001 and 0.002 (Deacon *et al.*, 1956; Smith, 1970; Large and Pond, 1981). The magnitude of the drag coefficient appears to be a weak function of the wind speed in that the value of C_{10} rises slightly as the wind speed rises; presumably because the surface roughness increases also.

The coupling of a velocity scale for the water to that of the air is accomplished by defining a turbulent friction velocity for the water as follows:

$$W^* = T_o \varrho_w\qquad\qquad 3.5$$

where ϱ_w is water density. W^* can be thought of as a surface current velocity. Thus we may convert the measured wind velocity into a velocity scale for the generation of TKE in surface waters. The shear stress (W^{*2}) can be found from the time average of the u' and v' velocity components in surface waters, which scales as:

$$\overline{u'v'} = W^{*2}\qquad\qquad 3.6$$

and is the primary source of TKE input to surface waters (Denman and Gargett, 1983; Sherman *et al.*, 1978).

The well mixed surface layer of water may also be treated as a logarithmic boundary layer just like the air (Jones and Kenney, 1977). Under these circumstances the important velocity scale is W^* and the important length scale is the depth z. The velocity gradient then is:

$$du/dz = W^*/k_z \qquad\qquad 3.7$$

If the surface water layers are neutrally stable (no density gradient) then the rate of dissipation of TKE (assuming the shear mechanism produces an energy cascade as above) can be found from:

$$e = W^{*2}du/dz = W^{*3}/k_z \qquad\qquad 3.8$$

(Dillon *et al.*, 1981; Oakey and Elliott, 1982). Thus the rate of dissipation of TKE scales as the cube of W^*. In unstratified water (the ocean) the thickness of the mixed layer should be approximately equal to the thickness of the Ekman layer (Monin and Yaglom, 1971; Blackadar and Tennekes, 1968) as follows:

$$L_e = 0.4W^*/f \qquad\qquad 3.9$$

where L_e is the length scale of the turbulent layer thickness and f is the local value of the Coriolis parameter. For application to the ecology of phytoplankton it is necessary to calculate the vertical velocity of the cells and the time taken by an algal cell to circulate in the surface mixed layer (Denman and Gargett, 1983). This can be done by assuming, for the time being, that the depth of the mixed layer is equal to the Ekman turbulent layer. That being so a reasonable estimate of the turbulent velocity in the largest eddies is $u(t) = 2.0W^*$ (Denman and Gargett, 1983) and the time scale for vertical movement around the eddy is $T_n = L_e/u(t)$. Given the definition of L_e, above, then $T_n = 0.2/f$. At 45 °N, latitude $f = 10^{-4}$ s^{-1} and with $W^* = 0.01$ m s^{-1} (medium wind speeds) the vertical length scale $L_e = 40$ m and the time taken for vertical movement around the eddy is 33 minutes (Denman and Gargett, 1983).

We now consider what happens in stratified water columns.

The energy balance of the water surface

Figure 3.3 gives a summary of the terms of the energy balance of the water surface. The total energy balance at the water surface may be expressed as follows:

$$H = R_n - G \qquad\qquad 3.10$$

where H is the flux of sensible and latent heat, R_n is the net radiation and G is the storage term, the term of interest as changes in the amount of stored

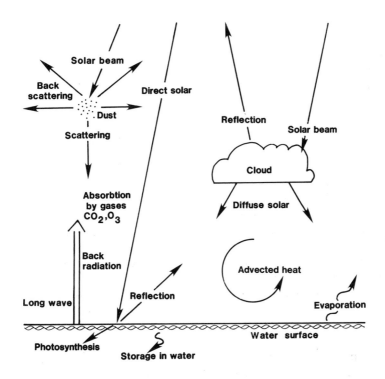

Fig. 3.3 The energy balance of the water surface.

heat influence the temperature in surface waters. This equation basically says that the amount of energy supplied by R_n must be balanced by loss due to sensible and latent heat fluxes or by storage in the water.

The components of H, the fluxes of sensible and latent heat, may be quantified in a number of ways that have been known to meteorologists for some time. The problem has been tackled by assuming that the transfer of energy may be expressed in terms of a flux of sensible heat (warm water directly warms the air flowing over it) and a flux of latent heat due to the evaporation of the water and the consequent extraction of heat from the surface (Phillips, 1978; Schertzer, 1978). The usual approach (Montieth, 1973) is to take a turbulent transfer approach and to make some assumptions about the vertical transfer of momentum, heat and water vapour over the surface. This approach has been well tested over a number of surfaces; indeed a water surface is about the easiest surface to work with. As noted above when the wind blows over the water the ideal wind profile over the surface is logarithmic. If this assumption holds then the shear stress is constant with height and it is possible to derive an 'air resistance' term that effectively describes the rate at which water vapour and heat are transferred away from the surface.

The rate of evaporation of water is assumed to be a function of the net radiation, the difference in the concentration of water vapour between the water surface (saturated) and the air, and the turbulent transfer of the vapour away from the surface. The sensible heat flux is likewise assumed to be a function of wind speed and the temperature difference between the surface and the air. Thus the heat flux into the water may be calculated from measurements of wind speed, temperature, relative humidity and R_n.

By taking the turbulent transfer approach and by measuring R_n it is possible to obtain estimates of the storage term in the heat balance equation. It is also possible to estimate the terms of the heat balance by empirical methods and a number of such approaches exist (Dake, 1972; Schertzer, 1978; Phillips, 1978). This storage term G may be used to calculate either the net flux of buoyancy across the surface or the temperature rise in a layer of given depth.

In the absence of wind mixing the expected temperature profile in a lake would be an exponential decrease of temperature with depth. This results from the exponential decrease in irradiance with depth underwater. Beer's Law states that:

$$I_z = I_o e^{-\eta z} \hspace{3cm} 3.11$$

where the irradiance at depth z is determined by the surface irradiance I_o and the attenuation coefficient for downwelling irradiance η. (η is dependent on the wavelength of the downwelling irradiance.) This equation essentially states that most of the energy stored in the water will be stored near the surface as most of the incoming short wave radiation will be absorbed near the surface. Because of the quadratic relationship between temperature and density the buoyancy generated by the storage of heat will tend to be maximal close to the surface where the temperature rise is highest. This will effectively stabilize the water column. Wind mixing serves to redistribute the stored heat downwards and to produce a surface mixed layer of uniform temperature overlying a sharp density gradient – the thermocline. Thus wind mixing opposes the effects of buoyancy. The equilibrium between the fluxes of heat and TKE is not reached immediately and the time dependence of the set up of the surface mixed layer has been modelled by Denman (1973). It may take a number of hours for a new equilibrium to be reached when the wind rises or the heating terms change.

Measures of turbulence in mixed layers with buoyancy

We now have the essentials of a number of measures of the turbulence structure of surface waters. The wind profile gives W^* and the heat budget gives G, the storage term. Basically all that follows are a series of ratios, in a variety of forms, of the stirring effect of the wind and of the stratifying effects of buoyancy generation.

A simple measure of stability in surface waters is the Brunt–Väisälä frequency defined as:

$$N^2 = -g/\varrho \; d\varrho/dz \qquad\qquad 3.12$$

of period

$$T_b = 2\pi/N \text{ (Phillips, 1966)} \qquad\qquad 3.13$$

which may be thought of as the period of oscillation of a parcel of water displaced from rest in a stratified fluid. If the parcel of water is lifted, it will sink to return to the position of neutral buoyancy, overshoot and fall below the position of rest. Now, being lighter than the surrounding water, it will rise and will oscillate up and down with frequency N. The stronger the stratification, the faster the oscillation. Frequencies of oscillation range from a minute or so to many hours, both in freshwater and in the ocean (Knauss, 1978). The Brunt–Väisälä frequency has a drawback in that it only takes into account the stability due to density changes and does not include any TKE component. The fact that it may be calculated from density profiles alone is, however, a distinct advantage.

The appropriate length scale for the largest vertical eddies in a stratified fluid may be found from a buoyancy scale:

$$L_b = e^{1/2} N^{-3/2} \qquad\qquad 3.14$$

(Dougherty, 1961; Lumley, 1964; Ozmidov, 1965; Denman and Gargett, 1983). When $z \gg L_b$ the buoyancy forces inhibit turbulence and stop the eddies turning over. At $z \ll L_b$ turbulence overrides buoyancy forces. The more rigorous form of this ratio is a Monin–Obukhov length defined as follows:

$$L_b = \frac{W^{*3}}{k a_v \; g \; G/\varrho \; cp} \text{ (Phillips, 1966)} \qquad\qquad 3.15$$

where a_v is the thermal expansion coefficient of water, cp the specific heat and g is the acceleration due to gravity. This length scale also defines the depth at which the stirring effects of the wind are balanced by the generation of buoyancy from G. As this is essentially an instantaneous flux ratio there is no time scale involved here. L_b may be thought of as the depth of the mixed layer at equilibrium; or the mixed layer depth if the fluxes were continued for a length of time sufficient to achieve equilibrium. Equilibrium is rarely achieved however because of the episodic nature of wind events, the diurnal components of the heat flux and the time required to achieve equilibrium. L_b is difficult to determine as the fluxes of W^* and G are difficult to determine over the water. Denman and Gargett (1983) used the simple form of the above equation to calculate the depth of mixing in the ocean and to obtain time scale estimates for circulation around the eddies in the presence of stability. They showed that with the smallest measurable values of N the depths and circulation times were comparable to the unstratified calculations above. In the presence of stability the vertical excursions were reduced as were the velocities. (Table 3.1.)

The fluxes of W^* and G do not achieve equilibrium immediately, so how do we estimate the time required to achieve equilibrium between the fluxes? One way to do this is to think of the entrainment of the stable fluid by the turbulent front as it moves downwards (Phillips, 1977). The turbulent front obtains its energy from the stirring effect of the wind and it eats down into the stable layers below at a rate (W_e) proportional to W^* and to the inverse of the mixed layer Richardson number defined as follows:

$$R_i = \frac{g\Delta\varrho h}{\varrho W^{*2}} \text{ (Phillips, 1977)} \qquad 3.16$$

where $\Delta\varrho$ is the density difference over the depth h. As R_i increases the stability of the layer increases and the rate of entrainment decreases. A number of experiments have shown that W_e/W^* is inversely proportional to R_i (Turner and Kraus, 1967; Phillips, 1977; Kullenberg, 1978) and that W_e has a magnitude of approximately 0.01 cm s^{-1}. The set up time of the thermocline must therefore be about 100 s cm^{-1} or 27 hours for a 10 m thermocline. Thus the equilibrium between the fluxes of heat and TKE will only be achieved in about one day. The diurnal components of the wind stress and of the solar flux therefore ensure that such equilibrium is rarely achieved.

A number of forms of the mixed layer Richardson number exist and all are analogous in that they are ratios of buoyancy forces to wind induced TKE inputs. In this respect the Richardson number is superior to measurements of N^2 as isothermal water or water of uniform density is not necessarily mixed in the absence of a source of TKE. Thus a large value of L_b does not necessarily mean that mixing is occurring; it merely means that mixing and entrainment will not be unduly hampered by the presence of strong density gradients if TKE is generated by the wind.

Local zones of active turbulent mixing may be identified by the calculation of a gradient Richardson number defined as:

$$R_i = g(d\varrho/dz)/\varrho(du/dz)^2 \qquad 3.17$$

so that:

$$N^2 = R_i(du/dz)^2 \text{ (Thorpe, 1977)} \qquad 3.18$$

This formulation includes the shear component from the velocity gradient in the water and it is known that when $R_i < 0.25$ active turbulent mixing will occur. When $R_i > 0.25$ the layers of water distinguished by their different densities will slide by one another without turbulence.

Estimates of K_z have proved to be difficult to obtain as measurements of shear must be made in surface waters. This has always been complicated by the presence of surface waves and (until recently) by a lack of good equipment. Clearly, K_z must be a function of wind stress and the stability in surface waters and a number of formulations have been proposed (Kullenberg, 1978). On dimensional grounds Denman and Gargett (1983) proposed:

$$K_z = 0.25 \, e \, N^{-2} \qquad\qquad 3.19$$

and Table 3.1 shows some results from that paper calculated from this relationship. In surface mixed layers the value of K_z varied from $10^{-3}-10^{-5}$ m^2 s^{-1}, the estimate of $2 L_b$ (the depth of the mixed layer) varied from 1–14 m and the time required for a particle to reach 5 m by turbulent diffusion varied from 4 minutes (with no stratification) to a few days. These are the fundamental time and space scales which govern the ecology of phytoplankton.

It is pertinent to discuss the subject of Langmuir cells here because they have often been invoked as important mechanisms for moving phytoplankton up and down in the surface layers of lakes and the oceans (Harris and Lott, 1973a,b; Harris, 1973a,b). When the wind blows over water at speeds in excess of about 4 m s^{-1} a series of streaks may be observed on the surface. Langmuir (1932) was the first to make note of these streaks when crossing the Sargasso Sea on a cruise ship. Floating objects, such as Sargasso weed, line up and accumulate in the streaks. Measurements of temperature and vertical velocities under the streaks indicate the presence of downwellings with weak return upwellings between the streaks (Harris and Lott, 1973b). It is the downwellings which leave the floating objects lined up on the surface. The common picture of Langmuir cells is of a series of roll vortices (Wetzel, 1975) aligned downwind, thus forming a series of linear downwellings and upwellings. Physical limnologists and oceanographers have always had problems with Langmuir cells as people who think of turbulence tend to think more of turbulent diffusion coefficients rather than of organized motions and velocities. Phytoplankton ecologists have tended to be more interested in the latter because of their interest in questions about the motion of the cells. Denman and Gargett (1983) reviewed the evidence for Langmuir cells and came to an interesting conclusion. Langmuir cells scales as the largest possible vertical eddies. Thus there is no need to invoke a special mechanism to account for the presence of the cells and the dismissal of Langmuir cells by physicists as 'just more turbulence' becomes understandable. We have already seen that there is a cascade of turbulent motions in the mixed layer, Langmuir cells are merely the largest eddies and

Table 3.1 Estimates of N, K_z, the scale of the largest eddies and the time taken to mix 5 m for a range of conditions (from Denman and Gargett, 1983)

N (cph)	N (s^{-1})	*$2\pi L_O$ (m)	K_z ($m^2 s^{-1}$)	†t^s	Remarks
0	—	—	—	4 min	U_{10} = 8 m s⁻¹
2	3.5×10^{-3}	1	4.1×10^{-5}	3.5 days	Upper 20 m, low winds 5 m s⁻¹
10	1.7×10^{-2}	0.3	8.6×10^{-6}	17 days	Base of mixed layer, low winds
20	3.5×10^{-2}	0.1	4.1×10^{-6}	35 days	Thermocline, low winds
2	3.5×10^{-3}	14	4.1×10^{-3}	50 min	Upper 20 m, high winds 15 m s⁻¹
10	1.7×10^{-2}	1	1.7×10^{-4}	20 hours	Base of mixed layer, high winds
20	3.5×10^{-2}	0.3	2×10^{-5}	7 days	Thermocline, high winds
11	2×10^{-2}	0.4	1.9×10^{-5}	8 days	Mixing layer (10–15 m), winds 5 m s⁻¹
11	2×10^{-2}	4	1.9×10^{-3}	2 hours	Mixing layer, winds 15 m s⁻¹
—	—	—	8×10^{-3}	25 min	Upper 10 m, Loch Ness Winds 8–10 m s⁻¹ (average over 3 days)

* Characteristic vertical scale for largest eddies.
† Time to mix 5m.

will exist in the presence of all the other, smaller scale eddies. The most rapid vertical circulation times over mixed depths of the order of 10 m are about half an hour if we assume likely values for the wind speeds and density gradients.

In the thermocline (Table 3.1), the values of K_z were small because of the reduced turbulence and the increased density gradient (Hesselein and Quay, 1973) so that the time required to move 5 m by turbulent diffusion was of the order of a week to a month. As might be expected there is a clear relationship between K_z and N^2; as the water column becomes more stable the vertical eddy diffusivity is reduced (Lerman, 1971; Quay *et al.*, 1980). It means, in summary, that phytoplankton in the mixed layer, experiencing moderately windy conditions, will be circulated around the mixed layer in a time of 30 minutes to a few hours. These time scales will turn up again when we consider the time scales of a number of physiological responses. At the same time as cells are being circulated in the mixed layer, any cells sitting in the thermocline will be experiencing a much more stable environment and will not be dispersed over periods of weeks. As we shall see many phytoplankton grow in this stable environment over the summer period. We now turn to a consideration of events at these scales.

3.2.2 MEDIUM SCALE (SEASONAL) EVENTS

The interplay of wind mixing and solar heating results in a surface mixed layer overlying a sharp density gradient, the thermocline. The knowledge of the physical processes which lead to the development of the seasonal thermocline can be used to write computer models which describe the thermal structure of lakes and the oceans. These models have been widely used to predict the behaviour of bodies of water under a variety of conditions. The early models concentrated on the equilibrium between the fluxes of heating and wind mixing (Turner and Kraus, 1967; Kraus and Turner, 1967; Huber *et al.*, 1972; Mellor and Durbin, 1975; Kraus, 1977; Sherman *et al.*, 1978) and used these fluxes to predict the depth of the thermocline. More recently the models have incorporated the effects of internal motions as well and have therefore included the generation of TKE by shear from internal waves (Spigel and Imberger, 1980). These waves are generated by the effect of wind stress on surface waters which causes the surface water to move downwind and to generate oscillations in the thermocline. Massive shear in the region of the thermocline causes the thermocline to be weakened and the predictions of thermal structures solely from surface fluxes to be in error. The prediction of thermal structures from surface fluxes of TKE and solar heating is now well advanced (Simpson and Dickey, 1981; Klein and Coantic, 1981) in both lakes and the ocean and the availability of such models will prove to be a useful tool in the understanding of phytoplankton ecology. It will become clear in later chapters that the physics of the mixed layer is of great importance for the growth and survival of phytoplankton populations. The ability to predict what might happen in lakes and reservoirs will prove to be a useful management tool.

It is the seasonal components of the heat budget (Schertzer, 1978; Phillips, 1978) and wind speeds which produce a seasonal thermocline in temperate latitudes. We have already identified the important features of lakes and the oceans which determine how the physical features of the water column evolve over the year. Wind speed, the fetch over the water and the basin dimensions determine the surface stress and hence the rate of TKE generation. The seasonal components of solar irradiance determine the rate of buoyancy generation. For a given basin, the mixed layer will be deeper in colder, windy weather as compared to warmer, calmer weather; and large basins will have deeper mixed layers than small basins. There is an interaction between the seasonal heat budget and the rate of generation of TKE in that cooling of the water surface may produce convective overturn (Sherman *et al.*, 1978; Spigel and Imberger, 1980). Water cooled at the surface will be more dense than the underlying water and will stream downwards until it finds water of similar density. Thus one of the reasons for autumnal overturn is the added TKE generated by surface cooling particularly at night. The more rigorous treatments of the dynamics of wind stress and buoyancy generation include terms for the generation of TKE from convective overturn (Spigel and Imberger, 1980).

The peculiar relationship between the density and temperature of water is responsible for the small density stratifications found in winter at temperatures less than the temperature of maximum density. The small reduction in density as water cools towards the freezing point means that any density difference will soon be overcome by TKE generated by the wind. Thus small lakes freeze before large ones and the temperature at which large lakes freeze will be lower than that at which small lakes freeze over. For example, the Great Lakes rarely freeze over completely in winter (and then only in very calm weather) and circulate at temperatures of about 2 °C. Smaller lakes and embayments freeze to a depth of as much as one metre. The fact that ice floats means that lakes and the oceans freeze from the top down. Were this not so life would have a much harder time surviving in winter under water. Ice cover is important for phytoplankton because it not only cuts down the underwater irradiance but, by cutting off the supply of TKE from wind stress, it also drastically reduces the vertical mixing. Consequently heavy phytoplankton tend to sink rapidly under ice.

In temperate latitudes, at the end of summer, the fluxes of heat and TKE reach a form of quasi-equilibrium where the thermocline comes to lie at a depth where the largest summer storms will not drive it deeper because of the dissipation of TKE with depth. For any group of lakes this thermocline depth is largely a function of fetch or lake diameter. Such morphometric links have been known for some time as Ragotzkie (1978) has shown that the depth of the thermocline in a number of Canadian and US lakes could be predicted with some accuracy from:

$$z_{th} = 4L^{1/2} \qquad\qquad 3.20$$

with L measured in km. For an irregular basin one way to approximate L is to use the square root of the area. Clearly, the fetch will depend on the basin shape and the wind direction, so a simple measure of L from lake area may fail in some circumstances but on average it should work adequately. Boyce (personal communication) has recently shown that there is also a link between z_{th} and the mean depth of a lake as it appears from an extensive analysis of data from a number of North American and European lakes that a thermocline will not form at depths greater than 80% of the mean depth of the lake. Thermoclines appear to be unstable at depths greater than 80% of the mean depth and mixing goes right to the bottom. This relationship seems to make a boundary between stratified and mixed lakes.

We might reasonably expect, therefore, to see distinct differences in the density stratification of lakes and the oceans at different latitudes at different times of the year (Woods, 1980). The effects of basin size scales right across the boundary between lakes and the oceans as long as we remember the fact that large water bodies develop quite different motions from small bodies and that salt affects density as well as temperature in the oceans. In lakes, the effects of altitude parallel the effects of latitude. In a recent important paper Spigel and Imberger (1980), discuss the dynamics of the mixed layer of lakes in terms of the dominant processes which occur. They classify the types of process in terms of time scales and of mixed layer Richardson numbers so that it is possible to quantify the effects of a wind event on a basin of a given size. Only small to medium sized basins can be treated, as the theory will only account for the stirring effects of the wind and for the effects of internal motions. Rotational effects cannot, at present, be accounted for. As we have seen, the rate of erosion of the thermocline is inversely proportional to R_i. At the same time as it is eroding, the interface between the warm surface water and the colder bottom water is tilting as surface water is blown downwind. Spigel and Imberger (1980) have shown that it is possible to classify the types of mixing to be expected in the presence of both stirring effects and the energy incorporated in internal motions. We will now consider the effects of the interaction of wind stress and stratification on the production of internal motions.

Internal motions in small to medium sized basins

The effect of wind stress is to cause a set-up in the surface water and a set of internal standing waves. The downwind motion of the surface water causes the mixed layer to deepen at the downwind end of the basin and causes the interface between the upper and lower layers to set-up with an equilibrium tilt of $1/R_i$ and a characteristic period of oscillation. It is possible to calculate the speed of the internal wave and for reasonable values of the density gradient and of basin dimensions the velocity is about 30 cm s^{-1}. Thus the internal wave period is about twice the basin length times this velocity. The motion of the internal wave produces shear across the interface. This shear weakens the interface and contributes to the overall TKE budget. It does not appear to be a major contributor to the overall TKE budget, however (Spigel, 1980). Spigel and Imberger (1980) show that, while the shear component may be episodic as the internal waves oscillate, in the presence of internal damping, stirring is by far the most important component which contributes to the deepening of the mixed layer.

Basin morphometry again interacts with physical processes as the length and depth of the basin control the internal wave periods. Spigel and Imberger (1980) classify the internal dynamics of lakes in terms of the aspect ratio of the lake – functions of length, depth and the depth of the surface layer. Together with R_i these terms can be used to describe what will happen in response to a wind event. Figure 3.4 shows the broad outline presented by Spigel and Imberger. If $R_i < 1$ then the response of the lake is very rapid horizontal and vertical mixing in less than a day. Any thermal gradient which existed before the wind event is rapidly destroyed by

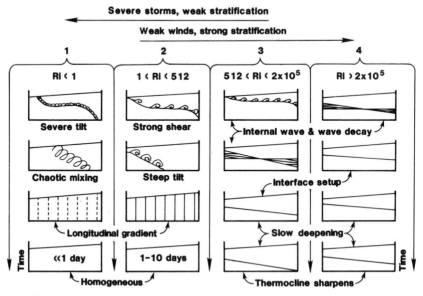

Fig. 3.4 The interaction of R_i and the progress of mixing in surface waters. The quoted values of R_i are applicable to a water body about 5 km by 8 km (Hamilton Harbour, Lake Ontario). From Spigel and Imberger (1980).

entrainment and severe tilting of the interface. The interface breaks down in large breaking internal waves. With $1 < R_i < 500$ or so for a lake about 5 km across and 20 m deep, strong shear and a steep tilt destroy the thermocline in 1–10 days. These conditions lead to active vertical mixing and a rapid deepening of the mixed layer. With $500 < R_i < 200\,000$ less erosion of the strong thermal gradient occurs and more internal wave activity appears. Under these conditions the thermocline deepens a little and becomes sharper and vertical mixing is restricted to the epilimnion. When $R_i > 200\,000$ there is very little deepening of the thermocline and the interface oscillates about the equilibrium position.

3.2.3 LARGE SCALE EVENTS

The reason that Spigel and Imberger's (1980) model cannot be applied to large basins lies in the fact that rotational effects become important and horizontal motions become very significant. The fact that the Earth is rotating throws the water about in the basins and causes swirling eddies and currents to appear. In small basins the largest eddy size is restricted, but in the oceans, massive horizontal motions such as the Gulf Stream may appear. The fact that the Gulf Stream flows up the western edge of the North Atlantic is a direct result of the rotation of the Earth; there is a similar western boundary current in the Pacific, the Kuroshio. Recent studies, particularly satellite observations which have given us synoptic observations for the first time, have revealed that the oceans are composed of eddies of varying sizes and that the major current systems eddy and flow leaving massive swirls, rings and fronts. Satellite images have given us our first really good look at the circulation patterns of the oceans because, once the features under study become a number of km across, it takes so long to move the ship around to sample the ocean that the ocean itself moves while you are doing it. This makes the interpretation of ship data difficult.

Rotational effects become important in basins of more than about 5 km across (Mortimer, 1974). This may be demonstrated by first calculating the Coriolis effect:

$$f = 2\Omega \sin \phi \qquad\qquad 3.21$$

where Ω is the angular velocity of the Earth's rotation and ϕ is the latitude. f is a frequency of period $2\pi/f$ called the inertial period. In middle latitudes (the Great Lakes) f is about 17.5 h. The Coriolis force, fu, caused by the Earth's rotation acts at right angles to the N/S axis of rotation and acts to the right in the Northern hemisphere and to the left in the Southern hemisphere. Water set in motion in a large body will, in the absence of friction, perform an inertial circle of radius:

$$r = u/f \qquad\qquad 3.22$$

and come back to the starting point. If we assume that $u = 10$ cm s^{-1}, then for 20 °N latitude, $r = 2$ km and for 40 °N, $r =$ approximately 1 km. For these rotational motions to become important the basin must be wider than 5r, and for such motions to become dominant the basin must be wider than 20r. Thus in lakes the size of the Great Lakes and in the oceans, such motions are dominant and of great importance. If the basin size is coupled with the correct magnitude of R_i, then the response of the basin to a wind stress is massive internal wave motion. This is particularly well shown in the behaviour of the Great Lakes in summer where the basin size and thermal stratification interact to produce characteristic types of internal waves. Rotational effects may, of course, be found in the circulation patterns of many large lakes (Emery and Csanady, 1973).

The physics of large lakes

The circulation patterns of large water bodies such as the Great Lakes are the result of the balance of four forces: wind stress, the Coriolis force, pressure forces and friction forces (Boyce, 1974). Together these forces produce both surface and internal oscillations in the water column. The time required to reach equilibrium between all these forces may be as long as days or even weeks (Boyce, 1974) and is certainly equivalent to the time between wind events in the basin (200 h or so – see below). Thus the physical structure of these lakes is rarely, if ever, at equilibrium. The lakes are always in the process of responding to a past wind stress event and the fact that the wind does not blow constantly, but comes in events associated with atmospheric weather patterns, is of great importance; both for physical processes and the organisms. As we have seen, the velocity of the internal waves which are induced by the wind are less than the surface waves and the response times are correspondingly longer. The response time of the internal waves has been estimated to be of the order of 50–200 h (Mortimer, 1974). This response time interacts with the frequency of the wind stress events to produce almost continuous internal wave activity in the lakes in summer. The 200 h forcing scale will be seen to be very important for the ecology of the organisms as it interacts strongly with the generation times and growth rates of the phytoplankton.

   This theoretical assessment of the situation is borne out by observation (Boyce, 1974; Csanady, 1975, 1978; Mortimer, 1974, 1979). Mortimer (1974, 1975, 1979) showed that two types of internal waves were present in large lake basins – Poincaré waves across the whole basin and shore-trapped Kelvin waves. Mortimer (1974, 1975) presented the basic theory and showed that the internal wave motions could be broken down and analysed as the two distinct wave patterns. The rotating Kelvin waves are an important feature of many large lakes and Mortimer's definitive work was done in Lake Michigan and Lac Léman (Mortimer, 1974). The internal waves may be of sufficient magnitude to cause upwellings of hypolimnetic water on the upwind or 'left-hand' shore (Csanady, 1978). The asymmetry is caused by the influence of the Coriolis force on the lake-wide circulation. The result of the Coriolis force is to cause the 'right-hand' or downwelling shore to be the warm shore, and the 'left-hand' or upwelling shore to be the cold shore. The interaction of wind stress with the internal wave motion causes the incidence of upwellings to be sporadic and coupled to the presence of internal Kelvin waves and the shore parallel currents (Csanady, 1972).

One outstanding feature of the spring circulation of the Great Lakes of great biological and physical interest is the 'thermal bar' (Csanady, 1971, 1974, 1975, 1978; Mortimer, 1974; Rogers, 1965, 1966, 1968). This is a feature of the early summer thermocline in which the density gradient becomes tilted to form a vertical 'bar' which prevents offshore transport of the spring runoff from the basin (Fig. 3.5). This causes the water quality of coastal regions to differ markedly from the mid-lake region. Surveys have revealed that algal biomass becomes concentrated in this region in early summer as the supply of nutrients is increased, the water is warmer than in mid-lake and the mixing depth is reduced close to the shore. The 'thermal bar' is maintained by the counter-clockwise circulation pattern in the lake (Mortimer 1974) and Csanady (1971, 1974) has shown that the 'bar' may be either lens shaped or wedge shaped depending on the direction of the shore parallel currents in relation to the shore. The 'bar' is, in reality, a type of front where a density gradient breaks the surface. The front is maintained in a tilted position by the rotation of the Earth and the influence of that rotation on the coastal jets. Similar fronts are of great importance in the oceans and we shall see that the generating mechanisms of oceanic fronts are not all that different from those in the Great Lakes.

The persistent shore parallel currents – or coastal jets – are a particular result of the effects of the interaction of the four forces listed above on the stratified lake. The presence of the coastal jets produces a coastal zone in which the currents are concentrated. The coastal jets may be treated theoretically as solutions to circulation models or as solutions to Kelvin wave models (Csanady, 1972, 1975, 1978; Mortimer, 1974, 1975, 1979). The presence of both coastal jets and Kelvin waves leads to some very complex currents and current reversals in the coastal zone that have great biological significance (Boyce, 1974). The reversals in the coastal currents occur about every five to ten days (Blanton, 1974; Boyce, 1974, 1977; Csanady, 1978) and are associated with the passage of the internal waves beneath the surface. The net effect of the coastal jets is to concentrate the velocity components parallel to the shore and to minimize the offshore transport of materials. Thus pollutants and organisms are concentrated in the coastal zone until a current reversal occurs, at which time the offshore transport is very rapid (Boyce, 1974).

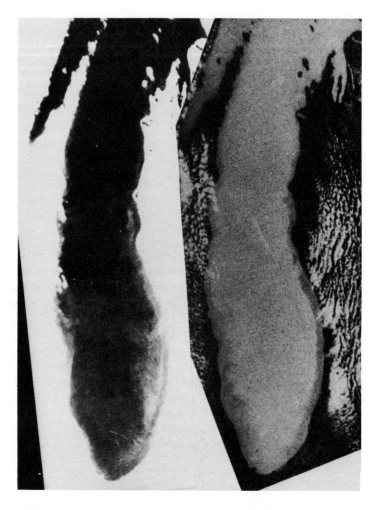

Fig. 3.5 The Lake Michigan thermal bar. Enlarged images from two channels of the Coastal Zone Colour Scanner on Nimbus-7 satellite, orbit no. 2895, 21 May 1979, 2m passage starting 16h 48m 40s GMT. (a) Channel no. 3, screen 0.54–0.56 pm, lighter bands near shore display higher colour intensity in the green, maximum in the region of the temperature front. (b) Channel no. 6 infra-red 10.5–12.5 pm, darker bands near the shore display water surface temperature, with the thermal bar presumed to be at the outer edge. (Original negative kindly provided by Clifford Mortimer.)

The coastal zone of the Great Lakes, which is dominated by large excursions of the thermocline and shore parallel currents, is about 8 km wide. The motions in this zone are concentrated at low frequencies. Blanton

(1974) found that most of the energy in this zone was concentrated at periods in excess of three days with a very sharp transition to the offshore zone where the energy was concentrated at high frequencies of the order of the inertial frequency (Birchfield and Davidson, 1967; Malone, 1968). Thus the coastal zone is dominated by low frequency motions of great significance for the biota (Boyce, 1974; Harris, 1982) and the reason for the shift in the dominant frequency of motion is associated with the typical internal wave motion which occurs in the two zones.

The circulation patterns of the oceans are a function of the same four forces as cause the circulation of the Great Lakes. Density structures are a little more complicated because of the presence of the salt but, overall, the physics is the same. The oceans generate motions of a different type from the Great Lakes because of basin size; eddies may be larger and a crucial extra component is introduced – the tidal component. Recent advances in remote sensing by satellites and in the design of physical experiments in the oceans has revealed the presence of large eddies in the oceans at scales of tens to hundreds of kilometres. These 'mesoscale' eddies appear to be ubiquitous and are of great biological importance. Thus in the oceans there exists the full spectrum of turbulence (Fig. 3.2) from basin scale currents like the Gulf Stream, to mesoscale eddies, rings and fronts, down to surface waves. Monin *et al.* (1977) have attempted to discuss the physics of the oceans by dissecting it into an hierarchy of scales. The subject of ocean variability may be treated as a problem of the coupling between ocean and atmosphere so that there is variability at scales from decades to seconds.

Fronts and rings

Because of the horizontal extent of the oceans and the importance of rotational effects, vertical motions and density gradients become displayed at the surface. As with the 'thermal bar' of the Great Lakes the tilting of the density stratifications which is maintained by Coriolis and pressure forces, results in fronts at the sea surface. Thus in basins of sufficient size vertical and horizontal patterns are coupled by the slope of the lines of equal density. In small basins internal motions are rarely of sufficient magnitude to be manifested at the surface, but as the basin gets larger the aspect ratio of the basin increases (Spigel and Imberger, 1980) so that sporadic upwelling occurs as the thermocline oscillates and breaks the surface. In ocean basins there is sufficient rotational energy to ensure that the density gradients are permanently tilted and appear at the surface as sharp density gradients or fronts. Thus as the Gulf Stream swings north-east up the east coast of North America it is marked by very sharp fronts and the meandering stream of warm water is easily seen in satellite images. The Gulf Stream meanders around as though a small boy were directing a hose. As it meanders, some of the larger loops become pinched off to form 'rings' or more or less circular eddies, again defined by sharp fronts which are maintained by the rotation of the ring.

All the major ocean currents shed these rings and there have been observations of such phenomena along the Gulf Stream (Saunders, 1971; Richardson *et al.*, 1978; Haliwell and Mooers, 1979), the Kuroshio (Cheney, 1977) and the East Australia current (Andrews and Scully-Power, 1976). As Richardson *et al.* (1978) pointed out 'the formation of rings provides the dominant mechanism by which mass, momentum, vorticity, energy, and chemical and biological material are transported northwards across the Gulf Stream into the slope water and southward into the Sargasso sea'. The biological significance of the rings lies in the fact that they are maintained as discrete entities for up to two years and represent whole ecosystems suspended in what is essentially foreign territory. Rings of warm water from the tropical ocean become suspended in colder slope water and rings of colder water are pushed southward into the warmer water of the Sargasso Sea. The warm core (anticyclonic) rings rotate clockwise on the north side of the Gulf Stream and move slowly southwest until they merge once more with the main current. The cold core (cyclonic) rings rotate counterclockwise and lie on the south side of the current. The warm core rings average about 100 km in diameter and move at an average speed of 6 cm s^{-1} (Haliwell and Mooers, 1979). The survey of Richardson *et al.* (1978) carried out in 1975 revealed that nine cold core rings and three warm core rings were present at one time. The warm core rings lose energy and heat by interaction with the atmosphere and by turbulent mixing (shear) at the walls (Saunders, 1971). Hence they have a tendency to shrink over their life span and may decrease in diameter by as much as 30 km (Haliwell and Mooers, 1979).

The other feature clearly visible in the satellite image of the western Atlantic was the shelf-break front along the east coast of the continent. This front is a consistent feature of the western North Atlantic (Haliwell and Mooers, 1979; Mooers *et al.*, 1978) and occurs at the point where the continental shelf breaks sharply downwards into deep ocean water. Basically, such fronts occur where warmer, more saline, deep ocean water meets colder, fresher water on the shelf. At this point the density stratification is tilted and breaks the surface. At scales of a day the variability in the front is coupled with energy input from the diurnal and semi-diurnal tides (Mooers *et al.*, 1978) and at scales of two to ten days (50–250 h) wind forcing is important. It appears that the front oscillates in much the same way as a tilted thermocline would after storms. Pingree (Simpson and Pingree, 1978) has identified a particular type of tidally induced front which occurs in areas of the continental shelf. These fronts may be seen particularly well around the British Isles where the water becomes shallower at the approaches to the Channel and in the Celtic Sea. Pingree calculates a ratio which is very similar to the Monin–Obukhov length (above) as it is the ratio of the rate of buoyancy generation by the surface heat flux to the rate of energy dissipation by the tidal stream. Thus:

$$R \text{ (Pingree's ratio)} = \left[\frac{ga_vGh}{cp\varrho} \middle/ cd \, U^3 \right] \simeq h/U^3 \qquad 3.23$$

where cd is a drag coefficient. Pingree identified a critical value of $\log_{10}$ (h/cd U^3) of 1.5 so that when the ratio is plotted on a chart of the British Isles the positions of the fronts are identifiable. At ratios of less than 1.5 the energy from the tidal stream is sufficient to ensure that the water is vertically mixed, whereas at $R > 1.5$ the sea will be stratified. The position of these shallow sea fronts has been confirmed by satellite infrared images (Fig. 3.6). Many of these frontal features are not easy to find from ships, but they may be seen quite easily in satellite IR images. Apparently the tidal, shallow sea front off the southwestern approaches to the Channel was never found by all the research cruises from the Marine Biological Laboratory in Plymouth, even though one of their master stations was quite near the average position of the front. This shows how difficult it is to find some of these dynamic features of the ocean unless you know what you are looking for. Shelf-break fronts and tidal fronts provide environments akin to seasonal thermoclines in that they exist for time periods much longer than the generation times of the phytoplankton and thus provide environments where biomass may accumulate and competitive exclusion may occur.

Upwellings

Upwellings are regions of the oceans and of large lakes where deep water is brought to the surface. Upwellings occur when the isopycnals become tilted as a result of wind stress or internal wave activity and are usually associated with decreased temperatures at the surface. Usually this deep, cold, water has its origin below the thermocline and carries with it supplies of nutrients thus increasing the productivity at the surface. Perhaps the best example of this type of large scale interaction between the atmosphere and the oceans is the Peruvian upwelling. The physical events underlying the appearance of the upwelling couple the activity of internal waves and of fronts to events of great biological significance. The Peruvian upwelling was one of the great ocean fisheries of the world before a series of events in the early 1970s led to the decline of the fishery (Idyll, 1973). The high productivity of the upwelling regions is driven by the appearance at the surface of colder, nutrient rich, deep ocean water. Occasional events, known as 'El Nino', occur when excessive amounts of warm Pacific ocean water flood over the top of the upwelling and close it off. Thus the onset of El Nino is signalled by a sharp increase in the sea surface temperature (Fig. 3.7). This warm Pacific water is

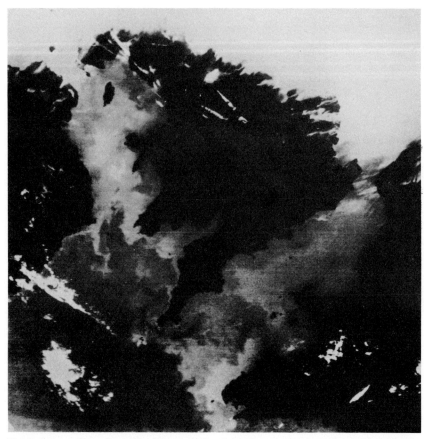

Fig. 3.6 An infra-red satellite image of the Celtic Sea, Irish Sea and English Channel, August 1976. (From Simpson and Pingree 1978.)

poor in nutrients, the productivity of the area is drastically reduced, the fishery collapses (Idyll, 1973) and the multitudes of sea birds starve (Boersma, 1978). The mechanism of the upwelling can now be explained by a Kelvin wave model. As the waves are always present, including El Nino events, the source of the waves must be remote from the coast of Peru.

The generating mechanisms for the whole sequence of events are complex but one explanation suggests that the origin of El Nino lies in the atmospheric circulations of the Pacific and in the strength of the southeast trade winds in the central Pacific (Wyrtki, 1975; Wyrtki *et al.*, 1976). When the southeast trades are strong, water is moved away from the coast of Peru allowing the Kelvin waves to break the surface and for upwelling to occur. The wave motions in the Peruvian coastal zone are particularly well displayed in the temperature signal of Fig. 3.7. When the trade winds relax

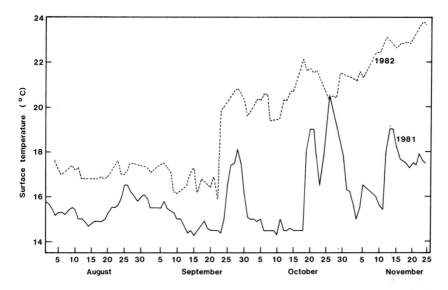

Fig. 3.7 The daily record of surface temperatures at Paita, Peru for the 1982 El Nino. Reproduced from Barber *et al* (1983).

water sloshes back across the equatorial Pacific as a massive internal wave and floods the upwelling with a layer of warm water. Thus the events which lead to El Nino events are remote in space and time from the actual events and local winds do not influence the sequence of events (Barnett, 1977). The eastward movement of warm water may be seen as a rise in sea level on the eastern side of the Pacific which may be detected from Alaska to the coast of Chile (Enfield and Allen, 1980). The spread of the warm water poleward may be modelled as a shore trapped Kelvin wave (Hurlburt *et al.*, 1976) and this solution is consistent with the analysis of the sea level data (Enfield and Allen, 1980) which revealed a poleward propagation of a wave at a speed of about 180 km d^{-1}. Thus equatorial events are reflected in poleward propagation of massive internal waves which influence the whole coastline of North and South America. The wave motion may be seen in current data from the Peruvian upwelling area (Smith, 1978; Brockmann *et al.*, 1980).

Thus energy imparted to the Pacific basin by the zonal trade winds, when coupled with the circulation of water in the basin, produces a series of internal and shore trapped waves analogous to the situation in the Great Lakes. The phytoplankton of the upwelling area respond to the changes in the physical structure and the nutrient status of the water column (Margalef, 1978a). Subtle changes in the hydrographic conditions appear to influence the phytoplankton populations particularly strongly (Margalef, 1978b; Blasco *et al.*, 1980).

CHAPTER 4

The chemical environment

The chemistry of natural waters has been reviewed in a number of major texts (Hutchinson, 1957b; Davis and DeWiest, 1966; American Chemical Society, 1971; Allen and Kramer, 1972; Broecker, 1974; Holland, 1978; Stumm and Morgan, 1981; Drever, 1982) and the reader is referred to such texts for the complete treatment of the topics to be outlined here. There has been much emphasis of late on the global biogeochemical cycles of nutrients and other elements and the influence of human interference on such cycles (Woodwell and Pecan, 1973; Stumm, 1977; Trudinger and Swaine, 1979). Events in natural waters must be seen as part of a much larger, global picture. The ionic composition of natural waters is as much a function of the solution chemistry of the lithosphere as it is a function of chemical equilibria *in situ*. The ionic composition of the oceans may be thought of as resulting from a massive titration, with the acids provided by gases from volcanic activity and the bases provided by the rocks (Stumm and Morgan, 1981). Similarly the ionic composition of lakes results from the dissolution of rocks and other materials in the basin. In this case the dominant mechanism of chemical weathering is carbon dioxide dissolved in the groundwater. The quality of water in a lake is therefore heavily influenced by events in the drainage basin and must be seen in the light of soil type, land use and agricultural practice in that basin. In all that follows I will concentrate on events in surface waters where the phytoplankton grow. The distribution of nutrients in surface waters is, however, very much influenced by regeneration in deep water and events at the sediment interface and, where pertinent, such events will be discussed here. The reader is referred to the texts listed above for details of the full cycles of elements within lake and ocean basins.

Phytoplankton are largely photoautotrophs; that is they use light energy to build macromolecules from simple precursors. The major nutrients of concern are consequently those which are important for all plants; carbon (C), nitrogen (N) and phosphorus (P). In various combinations C, N, and P, together with hydrogen (H) and oxygen (O), form the basis of macromolecular synthesis and energy metabolism in cells. Sulphur (S) and silicon (Si) can also be added to this list as Si is the major component of the cell walls of diatoms and S is an essential component of protein. These elements are called major nutrients because they are required for growth in relatively large amounts. Minor nutrients, on the other hand, are only required in trace amounts and include a large number of elements such as copper, manganese, magnesium, iron and zinc. A number of these are metals which limit growth if absent, are essential in trace amounts but are toxic in larger concentrations. Most minor nutrients are essential components of enzyme function.

4.1 The Redfield ratio

The most widely quoted stoichiometry for the major elements in plankton is (by atoms): 106 C: 263 H: 110 O: 16 N: 1 P: 0.7 S (Stumm and Morgan, 1981; Trudinger *et al.*, 1979). Rapidly growing phytoplankton exhibit uptake ratios in accordance with the Redfield ratio (Redfield, 1958) of 106 C: 16 N: 1 P and this stoichiometric ratio is frequently observed in ocean plankton (Redfield *et al.*, 1963). In the western Atlantic (Richards, 1958) the uptake ratios of Si were about the same as that of N leading to uptake ratios of 16 Si: 15 N: 1 P. The precise value of this ratio will depend on the species composition of the populations in question as only diatoms require Si for cell wall development. The Redfield ratio figures largely in the arguments which follow in this book as it has been used as evidence for the presence of rapidly growing phytoplankton in natural waters (Goldman *et al.*, 1979). There is some indication of variability in the elemental ratios of natural phytoplankton populations as, while the Redfield ratio by weight is 42 C: 7 N: 1 P, Boyd and Lawrence (1967) recorded a number of freshwater examples where the ratios of the major elements by weight were 116 C: 9.9 N: 1 P. Systematic variability in the elemental ratios of natural populations and the evidence for the interaction of such ratios with growth rates will be examined in a later chapter. The cycles of the major and minor nutrients are closely tied up with biological processes in the aquatic environment and the atomic ratios in the biota will reflect the way in which nutrients are processed in the water body. Organic forms of all the elements are an important part of the cycles as they represent nutrients tied up in macromolecules and in living organisms. In some cases such particulate organic pools are the major fraction of the total nutrient in surface waters.

All phytoplankton actively concentrate C,H,N,P and S together with minor nutrients, and the uptake of many of these elements is under physiological control. I shall discuss the metabolism of these elements in a later chapter so I will be concerned here with the geological, chemical and physical factors which regulate the availability of nutrients in surface waters. The distribution of a given nutrient in surface waters is a function of geochemical, physical and biological processes. Physical factors influence the 'patchiness' of nutrients in the horizontal and the vertical planes, largely through vertical redistribution of materials in the basin (Chapter 5) (Lerman, 1971). Biological processes such as uptake and recycling superimpose a second order pattern on the basic geochemical and physical environment. The methodological problems alluded to in the preface also begin to appear here as the measurement of dissolved and particulate elemental concentrations together with rates of uptake and regeneration provide some tricky problems in analysis, the use of isotopes and the interpretation of results.

The concept of time scales may be invoked to make some very important distinctions: the distinction between conservative and nonconservative substances and that between the concentrations and the fluxes of nutrients. The distribution of elements in basins may be treated as a 'box model' problem (Broecker, 1974; Imboden and Lerman, 1978) where the surface and deep waters of the basin may be regarded as completely mixed boxes. The important variables are the residence time of water in the basin, the rates of transfer of elements between water masses and the turnover times of elements. The turnover time of an element is defined as the pool size of the element in the box divided by the rate of flux through that pool. This, of course, assumes that every atom has the same chance of travelling through the pool and that the pool is completely mixed. The same argument may be used to define a residence time for water in a basin, by dividing the volume by the flow through the basin. The residence time of water in the oceans is about 4000 y (Garrels *et al.*, 1975) whereas the residence times of lake waters range from weeks to, at most, a few hundred years (190 y for Lake Superior).

The relationships between the rates of turnover (T_r) and the rates of horizontal (T_{mix},h) and vertical (T_{mix},z) mixing determine the spatial and temporal distribution of elements in the basin (Imboden and Lerman, 1978). If the turnover time of an element in surface waters is much less than the mixing time or the residence time of the water in the basin then there is a good chance that the element will become unevenly distributed depending on the sources and sinks in the system. Because of the differences between the rates of horizontal and vertical diffusion (Chapter 3) we are much more likely to see uneven vertical distributions than uneven horizontal distributions (Lerman, 1971; Imboden and Lerman, 1978). A conservative substance is therefore one which has a residence time much longer than the basin mixing time and consequently shows an even distribution all through the waters in the basin. A nonconservative substance has a relatively short turnover time as compared to the basin mixing time and therefore shows uneven distributions in space and time. As we are primarily concerned with the applied ecology of phytoplankton then we are also primarily concerned with those geological and chemical and biological processes which most directly influence the distribution and abundance of major and minor nutrients of importance to the organisms. We may use the mixing times and water residence times together with the generation times of the algae as useful time scales to organize our arguments.

4.2 The major ions

We know from isotope studies in the ocean (Broecker, 1974) that the mixing time for ocean water is about 1600 years. The major ions of sea water (Table 4.1) are conservative in that the turnover times of elements

Table 4.1 The composition of sea water (major ions). From Stumm and Morgan, 1981.

Ion	Concentration $(g. \, kg^{-1})$	Residence time in the oceans log_{10} years
Na^+	10.77	7.7
Mg^{3+}	1.29	7.0
Ca^{2+}	0.4121	5.9
K^+	0.399	6.7
Sr^{2+}	0.0079	6.6
Cl^-	19.354	7.9
SO_4^{2-}	2.712	6.9
HCO_3^-	0.1424	4.9
Br^-	0.0673	8.0
F^-	0.0013	5.7
B	0.0045	7.0

such as sodium, potassium and magnesium are of the order of 10–100 million years. As this is much longer than the ocean mixing time the concentration of the major ions in sea water does not vary by more than a few percent from place to place. Most of the minor elements in sea water show similarly long turnover times (Stumm and Morgan, 1981) so they may be regarded as conservative also. The notable exceptions appear to be aluminium and iron which have turnover times of the order of 100 years. Overall, the long turnover times of most of the major ions in sea water mean that the composition of sea water has remained constant for many hundreds of millions of years. The ions in sea water originate from rock weathering and runoff from the land but the ionic composition of the oceans cannot be obtained by merely concentrating river water. The overall composition must therefore result from the balance between a series of input and removal processes. The removal processes are many and varied and are still not completely understood. The recent discovery of hydrothermal activity on ocean ridge crests (Edmond *et al.*, 1979; Edmond and Von Damm, 1983) solved some problems but created others. Drever (1982) discusses the various removal processes in detail.

Stumm and Morgan (1981) have shown that the ionic composition of freshwaters may be predicted reasonably well from an equilibrium model of rock weathering by carbon dioxide in solution. Davis and DeWiest (1966) presented a summary of the ionic composition of potable water in the United States in the form of a series of cumulative frequency curves (Fig. 4.1). These curves show that calcium (Ca^{2+}) and bicarbonate (HCO_3^-)

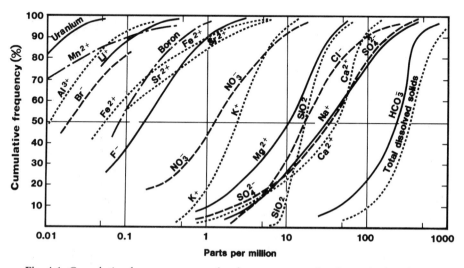

Fig. 4.1 Cumulative frequency curves for the occurrence of major and minor ions in potable water in the U.S.A. From Davis and deWiest (1966).

ions are the dominant ionic species as would be expected from a weathering model based on the solution of carbon dioxide in groundwaters. The fact that most of the cumulative frequency curves in Fig. 4.1 are parallel within broad limits indicates that, as in the ocean, the major ions in freshwater act conservatively and that there are a series of consistent ratios between them. The other way to examine the composition of freshwaters is obtain a large set of water samples and to plot the concentrations of the major ions against one another. This was done in some detail by Holland (1978) using data from Livingstone (1963) and other sources. The most significant correlations were those between total dissolved solids (TDS) and HCO_3^- and between TDS and Ca^{2+} as these are the major constituents of freshwaters. Margalef *et al.* (1982) obtained a similar result from a survey of Spanish reservoirs. With exceptions, the other major ions show significant correlations and consistent ratios also (Table 4.2). The correlation between Ca^{2+}

Table 4.2a Correlations between total disolved solids (TDS) and the major ions of freshwater. Data selected over two ranges – the concentration ranges are given as mg l^{-1}.

	Concentration range of ion	r^2	N
TDS	HCO$_3$ (0–50)	0.504	162
(0–75 mg l^{-1})	SO$_4$ (0–20)	0.018	160
	Cl (0–20)	0.068	159
	Ca (0–17)	0.421	160
	Mg (0–11)	0.191	162
	NaK (0–25)	0.133	159
	SiO$_2$ (0–15)	0.103	149
TDS	HCO$_3$ (0–180)	0.466	258
(75–250 mg l^{-1})	SO$_4$ (0–130)	0.133	259
	Cl (0–100)	0.037	258
	Ca (0–50)	0.489	256
	Mg (0–14)	0.382	256
	NaK (0–50)	0.059	254
	SiO$_2$ (0–25)	0.002	225

Table 4.2b Correlations between the concentrations of the major ions of freshwaters r^2 values, 171 cases.

					HCO$_3$			
h	HCO$_3^-$	SO$_4^{2-}$	Cl$^-$	Ca^{2+}	Mg^{2+}	Na$^+$/K$^+$	SiO$_2$	*TDS
HCO$_3^-$	1	0.143	0.018	0.417	0.382	0.296	0.278	0.589
SO$_4^{2-}$		1	0.052	0.487	0.539	0.293	0.057	0.578
Cl$^-$			1	0.181	0.127	0.568	0.027	0.330
Ca^{2+}				1	0.742	0.258	0.117	0.711
Mg^{2+}					1	0.257	0.096	0.654
Na$^+$/K$^+$						1	0.198	0.732
SiO$_2$							1	0.256
*TDS								1

*Total dissolved solids.

and HCO$_3^-$ indicates clearly that many of the freshwaters contain a concentration of dissolved CO$_2$ in excess of the concentration in equilibrium with the partial pressure of CO$_2$ in air again pointing to the importance of CO$_2$ generated by decomposition in soils.

Holland (1978) discussed the transport of the major ions in the river systems of the world and showed that there was an inverse relationship between TDS and runoff (Fig. 4.2). All dissolved components showed decreased concentrations in the rivers with the highest runoff but the relationships were not linear. Holland (1978) discussed data from all over

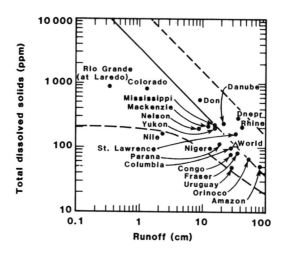

Fig. 4.2 The total dissolved solids in major rivers as a function of mean annual runoff. From Holland (1978).

the world in some detail. Clearly the rivers with the highest runoff figures (mostly in South America) have the lowest concentrations of TDS. The rate of erosion of the continental land masses is a complex function of total rainfall and its seasonality, of mean annual temperature, of elevation differences and of the soil and rock types in the basin. The mean elevations of the various continents do, however, account for a good deal of the variation between the major river systems of the world (Holland, 1978). The low TDS figures for the Amazon basin can be explained by the very heavy rainfall on the low portion of the basin (Gibbs, 1967). The provenance of the major ions in average river water is summarized in Table 4.3 and these data

Table 4.3 The provenance of average river water (meq kg^{-1}). (From Holland, 1978 and Stumm and Morgan, 1981.)

Source	HCO_3^-	SO_4^{2-}	Cl^-	Ca^{2+}	Mg^{2+}	Na^+	K^+
Atmosphere	0.58	0.09	0.06	0.01	≤ 0.01	0.05	≤ 0.01
Weathering or solution of							
Silicates	0	0	0	0.14	0.20	0.10	0.05
Carbonates	0.31	0	0	0.50	0.13	0	0
Sulphates	0	0.07	0	0.07	0	0	0
Sulphides	0	0.07	0	0	0	0	0
Chlorides	0	0	0.16	0.03	≤ 0.01	0.11	0.01
Organic Carbon	0.07	0	0	0	0	0	0
Sum	0.96	0.23	0.22	0.75	0.35	0.26	0.07

compare well with data for a number of lakes and rivers quoted in Hutchinson (1957b). The major ions in sea water and in freshwaters may therefore be treated as essentially conservative substances as their turnover times are longer than the residence or mixing times in the basins. One reason for the long turnover times is the fact that none of the major ions is required by the phytoplankton as a major nutrient.

Talling and Talling (1965) presented data on the chemical composition of African lake waters which are of interest because of the large range of total ionic concentrations encountered. Talling and Talling used conductivity as a measure of total ionic concentration, and used this scale to arrange the lake waters in a sequence of increasing concentration. The interest in African lake waters comes from the wide range of saline lakes caused by accumulation and evaporation in closed basins or by inflows rich in solutes (Beadle, 1981). Drainage from alkaline lavas is especially important. Talling and Talling (1965) found a strong correlation between conductivity and the concentrations of the major anions ($HCO_3 + CO_3$, Cl, SO_4) and between alkalinity and pH. Bicarbonate was almost always the principal anion. The distribution of the major cations was more complicated because the solubility products of the divalent cations were exceeded in the waters of higher ionic strength leading to precipitation of calcium and magnesium carbonates and the dominance of sodium in the waters of highest ionic strength. Sodium accounted for 95% of the cations in the most saline waters.

Reactions between inorganic ions in solution may usefully be regarded as being at equilibrium as, while there is a large spread of reaction rates (Stumm and Morgan, 1981), the rates of reaction in most cases are such as to bring about steady state distributions of reactants in times very much less than those that would be perceived by the cells. For most of the reactions involving the major nutrient elements even the slowest rates of reaction are such as to bring about steady state distributions of reactants in a matter of seconds. For this reason Stumm and Morgan (1981), successfully used equilibrium theory to describe the chemistry of natural waters. The analysis has been carried to its logical conclusion by the construction of large computer routines (such as REDEQL) which compute the equilibrium composition of natural waters (see Morel and Morgan, 1972) from a given composition of metals, ligands, acids and bases. These computer routines give details of the equilibrium distributions of free metal ions and complexes and give important information for the understanding of the chemistry and possible metabolic effects of micronutrients. Thus, for an element such as copper, it is possible to predict how much of the total metal is bound to ligands of various types and how much is present as the free metal ion. In many cases the toxicity of a metal in solution depends on the form in which it exists in the water (Stumm and Morgan, 1981).

4.2.1 THE DISTRIBUTION OF PHYTOPLANKTON IN RELATION TO THE MAJOR IONS

The relationship between the major, essentially conservative, ions and the distribution of phytoplankton has not been widely studied, probably because the relationships between phytoplankton and the minor, nonconservative, ions are stronger. Much of the work on phytoplankton and the major ions has been essentially empirical – 'large number' approach based on statistical correlations. As will be seen in Chapter 11 this is not necessarily poor science.

Lund (1954–55) was one of the first to investigate the relationships between the distribution of phytoplankton and the concentrations of major ions. He sampled 108 tarns in the English Lake district and showed that the standing crop of phytoplankton increased as the concentration of major ions increased. Jarnefelt (1956) obtained a similar result from a large survey of Finnish lakes. The relationship is a weak one and appears to depend on the broad correlation between all ions in waters of changing total ionic strength (Fig. 4.1). Waters with higher ionic strength tend to have higher concentrations of nutrients as well (Talling and Talling, 1965) though the correlation between conservative and nonconservative ions is often poor. The relationships between total ionic strength and overall productivity have recently been exploited by those interested in fish yields in lakes (Ryder, 1965, 1982), as total ionic strength is one component of the morphoedaphic index (MEI) used as an empirical predictor. The components of the MEI such as total dissolved solids and mean lake depth are only broad predictors of fish production for the reasons mentioned above and for other reasons discussed below in a chapter on the use of empirical state predictors in limnology (Chapter 11).

There have been a small number of investigations concerned with the relationships between the concentrations of major ions and the species composition of phytoplankton in surface waters. Talling and Talling (1965) noted that species of *Melosira* were dominant in African lakes of low ionic strength whereas *Spirulina*, a blue–green alga, was dominant in the high alkalinity waters. Many of these more alkaline and saline lakes are also highly productive (Hecky and Fee, 1981; Melack, 1981). One confounding factor in these lakes was the correlation between ionic strength and pH so that the distribution of *Melosira* was as likely due to pH effects as to ionic strength. Many diatoms are unable to grow at high pH (Jaworski *et al.*, 1981). Recently, Gasse *et al.* (1983) have classified the diatom communities of a broad range of African lakes and have shown that the species composition of the lakes could be correlated with a number of water quality parameters including temperature, alkalinity, conductivity, pH and the ratio of monovalent to divalent cations. The major conclusion of the work was that the diatom assemblages were clearly linked to the composition of major ions (see also, Hecky and Kilham, 1973).

The use of the monovalent to divalent ion ratio as a predictor of species composition has been investigated by Shoesmith and Brook (1983) and Brugam and Patterson (1983) who have shown that the ratio is a useful predictor of species composition in populations of desmids as well as diatoms. Margalef and Mir (1979) and Margalef *et al.* (1982) studied the phytoplankton populations in a number of Spanish reservoirs and showed that the assemblages of species in surface waters could be correlated with the total mineral content of the waters. Species of diatoms, green algae and blue–green algae occurred in characteristic clusters in different water types. In the Great Lakes there has been a sustained increase in the total ionic strength of the waters of the Lower lakes due to urban runoff and the use of salt on the roads in winter. This has led to the appearance of a number of new species of phytoplankton including a species of *Coscinodiscus*, a genus characteristic of brackish and marine waters (Harris and Vollenweider, 1982).

The concentration of major ions in the water is therefore an imprecise indicator of algal biomass and productivity. The distributions of individual species can sometimes be correlated with the concentrations of the major ions but only over broad ranges. As all the major ions have conservative distributions in natural waters the dynamics of growth and of the seasonal patterns of abundance are more closely correlated with the spatial and temporal fluxes of nonconservative nutrient elements in surface waters.

4.3 The major nutrients

The major nutrient elements (C,N,P,S and Si) cannot be regarded as conservative substances as the turnover times of the pools of these elements in surface waters are usually much less than the mixing and residence times of water in the basin. The spatial and temporal distribution of these elements is influenced by the pool size of the soluble nutrient and by physical and biological processes such as uptake, growth, grazing and sedimentation within the water column. As $T_r < T_{mix}$, $h < T_{mix}$, z; physical processes determine the horizontal and vertical distributions of the elements (Lerman, 1971; Imboden and Lerman, 1978) in surface waters and because T_{mix}, $h < T_{mix}$, z ($K_h > K_z$, Chapter 3) the vertical patchiness is usually greater than the horizontal patchiness. Data on the global cycles of N and P clearly shows that the turnover time of the pools of these nutrients in the oceans is very rapid (Table 4.4). The residence time of water in the surface mixed layer of the ocean is about 10 y (Garrels *et al.*, 1975) so these elements show strong variations in space and time, and N and P are severely depleted in ocean surface waters.

Table 4.4 Turnover times for C, N and P in plankton and surface waters.

	Pool size (g)	Flux (g/y)	Turnover time (y)	Reference
Carbon				
Marine plants	1×10^{15}	2×10^{16}	$\simeq 0.05$	(1, 2)
Ocean total biomass	17×10^{15}		0.85	(1)
Ocean surface water	1.2×10^{18}		10	(1)
Deep ocean water	3.5×10^{19}		1500	(1)
Carbonate rocks	4.8×10^{22}		6×10^{7}	(1)
Sedimentary rocks	1.2×10^{22}		2.3×10^{8}	(1)
Nitrogen				
Ocean plankton	3.2×10^{14}	9.6×10^{15}	0.03	(1, 3)
Marine plants	8.0×10^{14}	10×10^{15}	0.08	(6)
N^2 fixation		1×10^{13}		
Marine biosphere	1.2×10^{15}	10×10^{15}	0.12	(7)
Atmosphere	4×10^{21}		4.4×10^{7}	(4)
Sediments	1×10^{21}		4×10^{8}	(4)
N_2 fixation (plankton)		4.8×10^{12}		(8)
Phosphorus				
Ocean biota	1.38×10^{14}	1×10^{15}	0.138	(4)
Ocean biota	1.28×10^{14}	1×10^{15}	0.128	(5)
Ocean biota	1.28×10^{14}	1.3×10^{15}	0.098	(3)
Freshwater biota	1×10^{12}	1×10^{13}	0.1	(5)
Ocean surface	2.7×10^{15}	6×10^{13}	45	(4, 5)
Freshwater	9×10^{13}	1.5×10^{13}	6	(5)
Soil	1.6×10^{17}	1.5×10^{14}	1000	(5)
Rock	1.1×10^{25}	1.2×10^{13}	$\simeq 10^{12}$	(5)

Data from the following sources: (1) Kester and Pytkowicz (1977); (2) Machta (1973); (3) Harrison (1980); (4) Garrels *et al* (1975); (5) Pierrou (1976); (6) Delwiche (1970); (7) Holland (1978); (8) Capone and Carpenter (1982).

As the N:P ratio in 'average' seawater is the same as the Redfield ratio, the precise form of nutrient limitation will depend on the relative turnover times of N and P. C, on the other hand, because of the large concentration of bicarbonate in surface waters, is only slightly depleted as compared to deep water (Broecker, 1974). The atomic ratios of C:N:P in 'average' seawater are 1017 C: 15 N: 1 P so the C is in great excess as compared to the supply of N and P (Redfield *et al.*, 1963). C in the oceans therefore acts as a more or less conservative element and the turnover time of 10 y in surface waters equals the turnover time of the water (Table 4.4). As we shall see, C may be limiting in dilute freshwaters. S may be regarded as a conservative element in sea water as it occurs in great excess and has a turnover time of the same order of magnitude as the major ions.

N and P in surface waters of lakes also show short turnover times. In most freshwater systems P has the shortest turnover times; much shorter than either the length of the seasonal stratification or the residence time of water in the basin. N is much more mobile in soils than P so the supply of N to freshwaters is, on the whole, greater than the supply of P. Tropical lakes tend to be relatively poor in N as the supply of N from tropical soils is low (Talling and Talling, 1965). The relative excess of N in temperate waters often results in N turnover times that equal the turnover time of water in the basin. The turnover times of the pools of C, N and P in the biota of surface waters may be as short as 0.1 y (Table 4.4) but it should be remembered that the turnover times of the soluble pools will be much more rapid as, in the case of N and P, the soluble nutrient pool is very small and the majority of the nutrient is tied up in the biota. As in the oceans, S is rarely limiting in freshwater as it is available in large amounts from atmospheric deposition or from runoff. In the more dilute freshwaters, however, S may be limiting algal production as physical, chemical and biological factors influence the vertical distribution of S in the lakes (Hutchinson, 1957b) and the element is clearly not acting as a conservative element. Si is known to be a limiting factor for the growth of diatoms in many lakes as the pool size of soluble Si falls to practically unmeasurable values during the period of the summer stratification (Lund, 1954; Hutchinson, 1957b). This indicates very clearly that the rate of uptake of soluble Si can greatly exceed the rate of regeneration in the water column.

4.3.1 CARBON

(a) *Ionic species*

C has a central role in determining events in many waters because of the importance of the equilibria between CO_2, bicarbonate and carbonate which determines the acidity or alkalinity of natural waters and because C is the element required in the greatest quantities by photosynthesizing organisms. Furthermore the relationships between the concentrations and fluxes of C in natural waters illustrate very well the problems with the interpretation of kinetic data and show how controversy may arise if these relationships are confused.

In sea water the total concentration of carbonic species equals 2.3×10^{-3} M with total CO_2 (TCO_2) defined as:

$$TCO_2 = [CO_2aq] + [H_2CO_3] + [HCO_3^-] + [CO_3^{2-}] \qquad 4.1$$

In freshwater the concentration may vary from 10^{-5} M to 10^{-2} M. A full discussion of the CO_2 equilibria may be found in Stumm and Morgan (1981) who devoted an entire chapter to the subject. In a closed system the equilibrium concentrations of the solute components can be described completely by a set of six equations and these may be used to determine the distribution of the solute species at equilibrium. Natural waters are, of course, not closed systems and are subject to the addition or removal of CO_2 by chemical and biological processes. There is an equilibrium between the CO_2 in the atmosphere and that in the oceans (Plass, 1972; Broecker, 1973) and considerable amounts of CO_2 are stored in the oceans as CO_2 is about 200 times more soluble in water than oxygen. Talling (1973, 1976) has given a detailed presentation of the limnological applications of the CO_2 equilibria and has discussed the analysis of freshwaters by Gran titration. The Gran titration gives information on TCO_2 as well as the speciation of the carbonic ions in solution (Stumm and Morgan, 1981). Rather than entering into a detailed discussion of the equilibrium chemistry of the CO_2 equilibrium it will be sufficient here to simply state the ecologically significant results and properties of the system.

(1) As bicarbonate is present in sea water at a concentration of 10^{-3} M and is the commonest ion in freshwater (Fig. 4.1) it, and the other members of the CO_2 equilibrium system, are the major determinants of the pH of natural waters. CO_2, HCO_3^- and CO_3^{2-} in solution effectively buffer pH changes.

(2) The presence of bicarbonate as the dominant ion ensures a slightly alkaline pH. The pH of average sea water is 7.9 and the pH of most freshwaters falls between 6.0 and 9.0.

(3) As the pH rises the proportion of free CO_2 in TCO_2 falls and the proportion of carbonate rises. There are, therefore, two considerations for the availability of carbon for photosynthesis; the value of TCO_2 and the portion of TCO_2 in available forms (Harris, 1978).

(4) Many natural waters contain more CO_2 than would be expected from air equilibrium alone. Two thirds of TCO_2 in natural waters comes from the atmosphere, either directly, or by photosynthesis and the decomposition of organic matter in soils (Holland, 1978). There is thus a net flux of CO_2 from many natural waters to the air.

(5) The excess of CO_2 in solution above the air equilibrium value means that the carbonate will be dissolved by this 'aggressive' CO_2 and furthermore, when CO_2 is lost on exposure to air, the carbonate may be deposited again as calcite or marl (Wetzel, 1975). Holland (1978) estimates that, on average, one third of TCO_2 comes from the dissolution of carbonate in rocks. Marl deposition may be accelerated by photosynthetic carbon uptake.

pH and the distribution of phytoplankton

There has been much interest of late in the relationship between pH and the distribution of phytoplankton in natural waters. Part of this interest arises from the pollution problems associated with acid rain and the acidification of natural waters. The remains of diatom populations preserved in lake sediments have been used to attempt to reconstruct the history of acidification. This has been done in a number of publications (Renberg and Hellberg, 1982; Huttunen and Meriläinen, 1983; Gasse and Tekaia, 1983; Flower and Battarbee, 1983) and there are now accepted transfer functions for obtaining pH from the characteristics of the diatom community. Many diatom species are sensitive to C availability and to pH (Jaworski *et al.*, 1981) through a C limitation of photosynthesis. The transfer functions which predict pH from the diatom assemblage are however dependent on relatively few species in the overall assemblage and most diatoms are not very sensitive to major changes in the pH of the water.

Moss (1972, 1973a,b,c) carried out a series of experiments to determine the effects of water quality on the growth of phytoplankton and showed that species characteristic of eutrophic (nutrient rich) water were tolerant of higher pH than those of oligotrophic (nutrient poor) water. Talling (1976) showed that different species of phytoplankton have differing abilities to utilize C at high pH, when most of the C in the water is in the form of bicarbonate or carbonate (cf. review by Harris, 1978). Species typical of late summer blooms in eutrophic water (dinoflagellates and blue–green algae) tended to be able to utilize C at higher pH values than other species.

The ability of blue–green algae to grow at high pH appears to be correlated with their ability to produce late summer blooms in eutrophic waters. It is possible that the growth of the blooms arises from competitive interactions between the phytoplankton as the enhanced ability of the blue–greens to utilize C at high pH will give them an advantage over other species. It has been suggested that it is possible to control the growth of blue–greens by reducing the pH of lake waters (Brock, 1973; Shapiro, 1973a). Quite how or why this works is unclear and the use of acid to reduce the pH of natural waters has generated some controversy (Goldman, 1973; Shapiro, 1973b). The results from field experiments are not always easy to interpret (Shapiro *et al.*, 1982) and the expected shifts in species composition do not always occur at the expected pH values. If the pH of natural waters is rapidly decreased by the addition of acid then all sorts of chemical equilibria will be affected and it is not clear whether the cessation of growth by the blue–green algae is due to the physiological effects of altered C availability, to toxic effects due to changed metal solubilities or to pathogenic effects from viruses (Shapiro *et al.*, 1982). The blue–green algae tend to be replaced by green algae at lower pH but the replacement is not due to competition, as the green algae do not begin to grow until after the death of the blue–greens.

(b) *Vertical distributions*

The distribution of TCO_2 in lakes and the oceans is tied up with biogeochemical events in both the drainage basin and in the water body itself. CO_2 produced by decomposition of organic matter in soils dissolves in the ground water and, if the drainage basin includes carbonate rocks, this 'aggressive' CO_2 will dissolve carbonates and produce hard water lakes (Wetzel, 1975). Thus the TCO_2 content of natural waters is very much a function of the geology of the basin. Within the water body itself the distribution of TCO_2 is conservative unless photosynthesis and sedimentation remove C from surface waters and deposit it on the sediment surface or in deep waters. There is a close relationship between production in surface waters and oxygen depletion in bottom waters or at depth below the thermocline, which brings up the importance of the flux of nutrients from the surface to deep water brought about by the settling of particles. Ohle, (1934, 1952, 1956) was the first to systematically study the relationships between photosynthetic production and the uptake of CO_2 in surface waters, and the release of CO_2 by the decomposition of organic material in bottom waters and the sediments.

Oligotrophic (nutrient poor) waters tend to show fairly even vertical distributions of TCO_2 as there is little production of organic matter at the surface, low algal biomass and little sedimentation of material from surface waters. Eutrophic (nutrient rich) waters, on the other hand, show strong vertical differentiation in TCO_2, as much C is removed from surface waters by photosynthetic production of organic C, which sediments out. In bottom waters, this particulate organic C may decompose, use up oxygen and regenerate the CO_2. Wetzel (1975) discusses the relationships between the fluxes of organic C and inorganic C in lakes at some length and examines the consequences of this 'displaced metabolism'. The sedimentation of organic matter into deep water and onto the sediment surface has a profound effect not only on the distribution of C within the basin but also on the biogeochemical cycles of other elements.

The decomposition of the organic matter consumes oxygen and, in the ocean, Redfield and his co-workers showed that the rate of oxygen depletion in deep waters was a direct function of organic production in surface waters. In deep basins where oxygen is depleted by the degradation of biological material, C,N and P increase in the water in the ratio 106 C: 15 N: 1 P and oxygen disappears in the ratio of about 276 O: 1 P (Redfield *et al.*, 1963; Richards, 1965). The requirement of 276 O atoms for the complete oxidation of biological material containing one P atom indicates the reduced nature of the C in living material as, in the living state, the ratio is only 110 O: 1 P. Redfield and his co-workers were the first to point out the importance of biological processes in controlling the concentrations of the major nutrients in ocean waters by showing that on many occasions the changes in C,O,N and P were such as to indicate the synthesis and degradation of biological material. The flux of C to the deep oceans is further complicated by the flux of calcite which precipitates out of surface waters and may dissolve again at depth (Broecker, 1974). Calcite precipitation may also occur in some of the Great Lakes and these calcite 'whitings' are clearly visible from space (Strong and Eadie, 1978).

(c) *Fluxes and regeneration*

As the concentration of TCO_2 in many waters is of the order of 10^{-3} M and surface water may equilibrate with the atmosphere, it may be supposed that carbon is rarely limiting to algal growth in the oceans and in lakes. This impression is reinforced by the realization that the pH of these waters is regulated by the buffer capacity of the CO_2 equilibrium. Through the CO_2 equilibrium system a photosynthetic organism has access to practically the whole of the TCO_2 pool, as well as to CO_2 from the atmosphere. The rate of the reaction between bicarbonate and CO_2 is slow by chemical standards but is sufficiently rapid to supply the rate of photosynthesis, and this has led to statements that C is not a limiting factor for photosynthesis and growth in surface waters (Goldman *et al.*, 1972, 1974). This is not an entirely academic question, as part of the controversy about the effects of municipal and industrial nutrient additions on algal growth in lakes, concerned whether or not it was added P or C which was stimulating algal growth and causing nuisance blooms. Early publications seemed to show that atmospheric CO_2 was an insignificant source of CO_2 for phytoplankton growth (King, 1970; Kuentzel, 1969). Municipal wastewater contains elevated concentrations of C, N and P in solution, so those wishing to play down the role of P in stimulating the increase in algal biomass wished to show that it was the increased C in surface waters which led to increased algal biomass.

Whole lake eutrophication experiments in dilute freshwaters with low TCO_2 demonstrated that C would never be limiting for algal biomass, as atmospheric CO_2 could always supply the requirements for algal growth (Schindler *et al.*, 1972). These whole lake eutrophication experiments also demonstrated that strong pH rises occurred as a result of photosynthetic activity during the day. Only in the more dilute freshwaters will TCO_2 be sufficiently small to allow the pH to rise when photosynthetic activity strips CO_2 and bicarbonate from the water. The report that the invasion of CO_2 from the atmosphere was sufficient to fully support algal biomass increases produced some controversy (Verduin, 1975; Schindler, 1975).

Truck loads of P were added to lakes and massive algal blooms resulted, apparently confirming that C was not limiting algal biomass. The constancy in the TCO_2 in the lake after fertilization and constant relationship between TP and algal biomass indicated that increased TP in the lakes was the reason for the increased biomass. So, does low TCO_2 in lake waters ever limit anything? Not biomass, clearly, but despite assertions to the contrary, there is good evidence that C may limit rate processes in surface waters at high pH (Jaworski *et al.*, 1981; Talling, 1976). In the waters of the Canadian lakes there was a strong diurnal excursion in the pH of surface waters as C was depleted during the day by photosynthesis and was replaced at night by respiration and by invasion from the air (Schindler, 1971; Schindler *et al.*, 1971). Photosynthesis may become limited by lack of C if, as the pH rises because of C demand, the concentration of CO_2 falls to the CO_2 compensation point of the algae in question. Photosynthesis now becomes limited by virtue of the fact that the algae cannot use bicarbonate and because the concentration of CO_2 is sufficiently low to reduce C uptake to żero (Jaworski *et al.*, 1981). The fact that the speed of the reaction from bicarbonate to CO_2 will supply the needs of photosynthesis is of no avail if photosynthesis itself is inhibited. Some algae are restricted to utilization of CO_2 for photosynthesis (Harris, 1978) so the ability of certain species to use bicarbonate directly (Talling, 1976; Harris, 1978) is a crucial asset which ensures growth and survival in waters of low TCO_2 but elevated N and P. While the overall biomass remains a function of P not C, the species composition of that biomass will be influenced by the availability of C in the water and the ability of the different species to use it.

So what was limited in those Canadian shield lakes? Rate processes during the day were clearly influenced by the lack of C in surface waters. All the photosynthesis took place early in the morning before the C was depleted (Schindler, 1971; Schindler *et al.*, 1971). The algal biomass was, however, apparently unaffected by the daytime lack of C. The depletion of CO_2 in surface waters is not sufficient evidence of C limitation. Limits to biomass are not the same as limits to uptake processes, such as photosynthesis and limits to uptake processes may not be limits to growth. The relationships between uptake, growth and biomass will be discussed in Chapter 6.

4.3.2 NITROGEN

(a) *Ionic species*

The dynamics of the cycling of nitrogen in the oceans and in lakes is complex because of the existence of four dissolved forms of inorganic N which can only be interconverted by bacterial action in the environment or within living cells. In surface waters the pools of dissolved inorganic nitrogen (DIN, dissolved N_2 gas, NH_4^+, NO_2^-, NO_3^-) are therefore very much dependent on biological uptake and regeneration. McCarthy (1980) has recently reviewed the availability of nitrogen in surface waters and has pointed out that interest in the availability of nitrogen for phytoplankton growth in the oceans and in lakes has led to a number of excellent reviews of the subject (Vaccaro, 1965; Brezonic, 1972; Goering, 1972; Martin and Goff, 1973; Hutchinson, 1973; Dugdale, 1976).

About 95% of the N in the oceans is present as molecular N_2 and only two thirds of the remainder is NO_3^- (Martin and Goff, 1972). Profiles of the concentration of NO_3^- in the oceans show marked depletion and practically undetectable concentrations in many surface waters (McCarthy, 1980) and an increase to about 30 μg − atoms l^{-1} below 1000 m (Dugdale, 1976). Historically, the work of Redfield (cf. reviews by Redfield, 1958; Redfield *et al.*, 1963) is of some importance as these papers point out the fact that the deep ocean water contains N and P in the atomic ratio of 15:1. It may be concluded that the concentrations of these nutrients result from the degradation of biological material synthesized in the surface waters.

In the deep ocean the supply of NO_3^- to surface waters depends on vertical advection of deep water or eddy diffusion across the thermocline. As was shown in the last chapter eddy diffusion across thermoclines is slow compared to direct vertical advection in the form of upwelling (Table 3.1). In the central areas of the subtropical ocean therefore, the surface concentrations of NO_3^- are practically unmeasurable (McCarthy, 1980) and NO_3^- only appears in significant concentrations in surface waters in areas of upwelling such as the coast of Peru. Fogg (1982) has recently reviewed the cycling of N in oceanic waters and has concluded that the estimated total of N returned to surface waters is currently less than the estimated losses. There is no need to assume that the N cycle of the oceans is at steady state over geological time scales. Losses due to denitrification in anoxic bottom waters will be critically dependent on productivity in surface waters and on physical conditions.

Because of the rapidity of nitrification in natural waters and the ubiquitous presence of oxygen in surface waters, NO_2^- is present in oceanic water at much lower concentrations than NO_3^- and profiles of NO_2^- may only show a deep water maximum associated with the oxygen minimum (McCarthy, 1980). This NO_2^- is presumed to be produced by the excretion of NO_2^- from phytoplankton at the deep chlorophyll maximum. A similar NO_2^- maximum occurs in the Great Lakes and a similar coincidence of a deep chlorophyll maximum and a deep NO_2^- maximum is found (Mortonson and Brooks, 1980). Studies indicate that the NO_2^- maximum arises from the partial reduction of NO_3^- by phytoplankton in the deep chlorophyll maximum (French *et al.*, 1983). The mechanism of NO_2^- excretion from phytoplankton will be examined in a later chapter. Ammonium nitrogen is similarly rare in oxygenated waters as it is also rapidly nitrified. NH_4^+ accumulates in anoxic waters where the process of decomposition and nitrification is stopped by an absence of oxygen. Effluents from municipal waste water treatment plants are particularly rich in NH_4^+.

In oligotrophic freshwaters NO_3^- is the dominant form of DIN because of the rapidity of nitrification (Hutchinson, 1957b) and the even distribution of oxygen throughout the basin. In eutrophic waters, however, both NO_2^- and NH_4^+ may become abundant as decomposition processes in the sediments and in deep water deplete the oxygen and release reduced forms of N from organic material. In a classic piece of work, Mortimer (1941–1942) examined the release of NH_4^+ from the sediments of an English lake as the water became anoxic and showed that the redox conditions of the sediment surface were a crucial factor in determining the rates of nitrification in his system. Turnover of temperate lakes in Spring and in Autumn leads to oxygenated conditions in deep water and to nitrification of any NH_4^+ produced during the summer stratification. The rate of nitrification may be temperature limited in winter so that NH_4^+ may build up in lake waters particularly if reduced forms of DIN enter lakes in the effluents from municipal waste water treatment plants and industry. The oxygen demand of nitrification may then be considerable in Spring when warming of the water occurs. In Hamilton Harbour (Harris *et al.*, 1980a) the oxygen demand from the nitrification of as much as 3 mg l^{-1} of NH_4^+ that builds up over the winter, was sufficient to remove all the dissolved oxygen from bottom waters in less than three weeks in May and June.

The fixation of N_2 gas directly into organic N is a process that can only be carried out by a small number of different types of organisms. All nitrogen fixers are prokaryotes, either bacteria or blue–green algae (cyanobacteria). Nitrogen fixing blue–green algae are common in freshwater, rare in the oceans and apparently absent in estuaries (McCarthy, 1980). Since that was written in 1980 the status of the oceanic blue–green algae has been revised with the discovery that very small blue–greens are common in many of the world oceans. The almost ubiquitous occurrence of unicellular blue–green algae in the oceans will not affect the estimates of N_2 fixation as they lack heterocysts, the specialized structures required for N_2 fixation (Fay *et al.*, 1968) and none appear to show evidence of aerobic nitrogen fixing activity (Stewart *et al.*, 1978). Garrels *et al.* (1975) estimated the flux of N into the world ocean by N_2 fixation as about 0.1% of the total photosynthetic N flux (Table 4.4). Fogg (1982) estimated that the total N flux due to N fixation was 0.2% of the total photosynthetic flux. Capone and Carpenter (1982) estimated the flux of nitrogen due to N fixation to be 0.3% of the total flux due to primary productivity in the sea. They interpret this result to mean that N is not limiting in the ocean, otherwise the rate of N fixation would be higher. The abundance of N fixing blue–greens in many fresh-water systems means that such organisms are a significant component of the N cycle in such environments (Horne and Fogg, 1970) and may contribute more than 2–3 times the N flux from other sources at certain periods (Granhall and Lundgren, 1971).

The conventional wisdom is that N is limiting in the oceans and P is limiting in freshwater. As we have seen, the definition of 'limiting' depends on a number of assumptions about the temporal and spatial components of the processes in question. The composition of deep ocean water is certainly the same as that of rapidly growing phytoplankton so the deep upwelled water contains the correct elemental ratio for growth. Precisely which element is limiting will depend on the relative turnover times of the pools of DIN and DIP, not their concentrations. Because of the fact that DIN must be released from macromolecules by digestion or decomposition before it can be made available for further algal growth, and because it must undergo changes in oxidation state before metabolism, it may be expected that the velocity of DIN cycling may be less than that for DIP.

The sources of N for algal growth are threefold: DIN recycled within surface waters; DIN recycled in deep waters and brought to the surface by upwelling; and DIN brought in by rivers and groundwater (Fogg, 1982). In the central subtropical oceans the algal growth and productivity are clearly mostly dependent on the first mechanism. The presence of a continuous thermocline prevents massive upwelling and the remoteness from coastal inputs reduces the impact of river waters containing N. In coastal oceanic waters all three sources of N are significant with upwelling and river inputs certainly dominating the *in situ* regeneration of N at certain times of the year. As the overall productivity of the water body increases the dependence on nutrients regenerated within surface waters seems to be less (Eppley and Peterson, 1979; Harris, 1980a). In inland waters the sources of N are also threefold but, again, the three processes differ in importance depending on temporal and spatial components which are largely determined by basin size. In most freshwaters the concentration of DIN is large compared to ocean water and the importance of *in situ* regeneration is consequently reduced. The high concentration of DIN in freshwaters is largely due to the mobility of DIN in soils (Vollenweider, 1968) and the relative importance of river and groundwater inputs (water residence times are small). DIN inputs from soils are largely in the form of NO_3^- as a result of nitrification. Inputs of NH_4^+ may be significant if municipal or industrial wastes are discharged into lakes in a poorly treated form.

(b) *Vertical distributions*

The flux of organic material into deep water may be considerable in eutrophic waters, and may be sufficient to produce oxygen depletion either at depth in the water column or at the sediment interface. If oxygen concentrations are reduced, denitrification may result in the reduction of NO_3^- and NO_2^- to N_2 gas and the loss of N from the system. There are some remarkable parallels between lakes and the oceans in that the deep water oxygen minimum in the ocean is associated with the sedimentation and decomposition of organic materials produced in surface waters. Indeed the magnitude of the downward flux of N may be used as a measure of production in surface waters (Knauer *et al.*, 1979). Thus oceanic oxygen profiles may show a minimum around 1000 m or so where organic material sedimenting out of surface waters is decomposed and the major nutrients are released in accord with the Redfield ratio. Deep lakes may show a similar pattern but the process may go further towards anoxia because the processes of production and decomposition are more rapid and the water volume and oxygen stock smaller. As organic material settles out of surface water, oxygen is consumed below the thermocline and denitrification removes DIN from the water column as N_2 gas. The concentration of DIN may rise again below the DIN minimum if NH_4^+ is released from the sediment interface. Because of the link between anoxia and denitrification, the process becomes increasingly important in eutrophic lakes and anoxic oceanic basins. As we shall see, denitrification becomes important in the overall cycling of N in eutrophic lakes.

(c) *Fluxes and regeneration: large scale processes*

In the oceans, the flux of nutrients from river inputs is only sufficient to satisfy a fraction of one percent of the demand from algal productivity (Harrison, 1980; Fogg, 1982). The data for N, P and Si indicates that at least 98% to 99% of the nutrient demand must be supplied by recycling *in situ*. A very important question is raised by this data and it concerns the temporal and spatial scales at which this recycling occurs. Are the majority of the nutrients recycled on a scale of whole basins and years or does the process occur much faster, over scales of microns and minutes?

We have already seen that Redfield observed a close link between the nutrient composition of deep ocean water and degraded biological material. When the deep ocean water upwells these nutrients become available to plankton in surface water. By this mechanism the recycling of nutrients occurs over large temporal and spatial scales in the oceans as the mixing time of ocean water is measured in thousands of years. The downwelling, sedimentation and upwelling sites may be separated by thousands of kilometres (Broecker, 1974). This large scale recycling mechanism leads to upwelling ecosystems in such areas as Peru and very high productivity. We have already discussed the link between physical and biological events in that system. Upwelling ecosystems are thus driven by nutrients which sediment out and are released at sites remote from the site of the upwelling. These upwelling events also occur in large lakes such as the Great Lakes (Chapter 3) because the same Kelvin wave model may be applied in both circumstances.

The oceans and very large lakes have another feature in common in that they both have coastal zones where nutrients may be available directly from runoff and from physical processes which serve to bring nutrients up from deep water. The high productivity of coastal regions results from this combination of external inputs and regeneration from within the basin. For example the high productivity in the region of the shelf-break front off the east coast of North America results from a combination of nutrient inputs from the land and nutrients made available from deep water through upwelling (Mooers *et al.*, 1978). High algal productivity in such areas is associated with high productivity in the other levels of the marine food chain because of the relationships between the particulate pools in the food chain. The Peruvian fishery is (or was) an important global resource (Idyll, 1973) and the important fishing grounds off the east coast of North America are associated with the shelf-break front. In areas such as the Grand Banks the important fishery is associated with the average position of the shelf-break front and somehow the demersal fish production occurs in the same place as the highest phytoplankton production. This can only be achieved by a coupling between planktonic and benthic events by sedimentation of material produced in surface waters.

An important feature of many planktonic systems is evident from these data also. In temperate oceans where there is a seasonal thermocline, NO_3^- pools may be restocked by spring overturn and depleted by algal growth in the summer (Cooper, 1933a,b; Ketchum *et al.*, 1958a,b). The proportion of nutrient supplied from *in situ* recycling is not constant over the year. In spring much of the nutrient supplies come from water column reserves; reserves built up during the winter as a result of reduced algal demand and deeper mixing. Ketchum and Corwin (1965) showed that during the spring bloom in the Gulf of Maine 86% of the P came from water column reserves, 12% from vertical mixing and 2% from regeneration. This very low figure for regeneration during blooms appears to be typical as the flux of regenerated nutrients from grazing is reduced as the growth rate of the algae has temporally outstripped the grazing rate. Such affairs cannot continue indefinitely. After the bloom when the grazers catch up with the growth of the phytoplankton the proportion of P regenerated *in situ* rose to 43%. This condition is more normal (Postma, 1971) as blooms occur infrequently. Thus the limitation of growth by nutrients in the water has spatial and temporal components which make generalization dangerous; what is true this month may not be true next month.

There is a real debate here, as the situation in many temperate waters is clearly not at steady state. Physical events may serve to redistribute the DIN at seasonal and storm event scales so that pulses of DIN are brought into the surface waters. The steady state arguments (Chapter 2) go like this: the apparent absence of inorganic N for most of the year in surface waters must mean that phytoplankton growth is slowed to values less than the maximum possible values. The Michaelis–Menten relationship between the concentration of DIN and growth rate dictates the result. The concentration of DIN is clearly not a sufficient parameter, however, as the flux of DIN in the surface waters of the ocean is a function of both regeneration and uptake.

Regeneration of DIN at seasonal scales makes a pulse of DIN available each Spring which may be rapidly depleted by phytoplankton growth. As this supply occurs only once each year, at time scales of many generations, the phytoplankton must eventually run out of DIN, as cellular storage mechanisms cannot tide the cells over periods of many months.

Regeneration of DIN at more frequent intervals may occur as the frequency of storm events and major frontal weather patterns is of the order of 100–200 h in temperate latitudes (Chapter 3). This provides a pulsed source of DIN at more frequent intervals by entrainment of deep water across the thermocline. Such pulses of DIN may be small compared to the supply made available each Spring, but the frequency of the pulses is such as to make it possible for the phytoplankton to rapidly take up and store essential nutrients so that rapid growth is maintained. The effect is something like a monthly pay cheque and a bank account. If personal finances are managed well it is possible to maintain a consistent life style despite large fluctuations in the size of the bank balance. The balance may even briefly go to zero at the end of each month without detriment to the holder of the account! Evidently there must be a close coupling between the time scales and magnitudes of the regenerative pulses of nutrient and the uptake rates, storage capacities and growth rates of the phytoplankton if continued rapid growth is to be achieved. The existence of cellular storage mechanisms will be dealt with in a later chapter.

(d) *Fluxes and regeneration: small scale processes*

Such large scale events, occurring at time scales of a few generation times, are not possible in small lakes as the internal motions in the basin are scale dependent. In these cases, and in the case of ocean productivity at sites with permanent thermoclines remote from the coastal and upwelling sites, the nutrient demands of algal productivity can only be satisfied by *in situ* recycling. This realization has led to the study of rates of nutrient regeneration *in situ* and the debate as to whether rates of regeneration within surface waters are sufficient to satisfy the requirements of rapid algal growth. Harrison's (1980) review looked at the sources of regenerated nutrients in the oceans and concluded that 20% came from deep ocean sources and 80% from *in situ* recycling within the surface waters. Fogg's (1982) estimates were 9% and 91%. Such values are global averages and must indicate that practically 100% is be regenerated *in situ* in the surface waters of the deep oceans. The importance of *in situ* nutrient regeneration was evident in early studies (Atkins, 1930; Cooper, 1933a,b) as the pool size of the major nutrients in surface waters was, on occasion, inadequate to account for the observed biomass and productivity of the algae. Later studies (Riley, 1956; Harris, 1959, Ketchum and Corwin, 1965) showed that, in coastal regions, at least 50% of the N and P was supplied from *in situ* regeneration on an annual basis and that the turnover times of 0.1–0.2 y for the total N pool were reasonable; this data is consistent with Table 4.4.

Jackson (personal communication) has indicated that the pool of dissolved organic nitrogen (DON) plays an important role in the regeneration of N in the water column in the ocean. He noted that if the DON pool is added to the DIN concentration in surface waters then the Redfield ratio of concentrations is preserved. Redfield *et al.* (1963) showed data where the concentration of DIN and DIP in surface waters were plotted against one another. That data indicated frequent cases where the DIP pool was significant but the DIN pool was unmeasurable so the slope of the line conformed to the Redfield ratio but the line did not go through the origin. Jackson (1982) showed that when pools of DON and dissolved organic phosphorus (DOP) were added to the graph the slope agreed with the Redfield ratio and the line passed through the origin. Jackson calculated that the DON pool need only show a turnover of about 3% per year to have a significant impact on the availability of N in surface waters of the ocean.

4.3.3 PHOSPHORUS

(a) *Inputs and vertical distributions*

As with C and N the flux of P to surface waters is a function of geochemical processes in the basin. P is quite immobile in soils because of the fact that PO_4 is bound into clay minerals and other soil components (Vollenweider, 1968). Two types of P inputs must therefore be considered: soluble P inputs to ground waters consisting of dissolved inorganic P (DIP) and dissolved organic P (DOP), and particulate P bound into various forms which may vary greatly in biological availability. Vollenweider (1968) has summarized the basin features which contribute to P inputs to lakes. In a summary of European data 45% of the N came from underground seepage, whereas 53% of the P came from municipal waste and urban runoff. Undisturbed terrestrial ecosystems conserve P, a conclusion supported by data from the Hubbard Brook experiment (Hobbie and Likens, 1973). With an annual input of 100 g P ha^{-1} in rainfall the undisturbed watershed in that experiment experienced a net gain of 87 g P ha^{-1} whereas the disturbed (cut) watershed exported 104 g P ha^{-1} more than it received in rainfall. The magnitude of the P loss in the forested watershed was small compared to the overall P flux of 1900 g P ha^{-1} in the annual leaf fall; data which indicates how strongly the P in the system was conserved.

P is to some extent a more simple element than N as the single ionic form (PO_4) is the ubiquitous form of the element found in both the organisms and in the environment. The metabolism of PO_4 will be discussed in a later chapter but it should be noted here that no oxidation and reduction steps are required for environmental cycling, metabolism and macromolecular synthesis. The most readily available form of P for algal growth in natural waters is DIP although the presence of phosphatases in phytoplankton means that DOP may be broken down and metabolized (Nalewajko and Lean, 1980). As with C and N, the availability of DIP in natural waters depends on some complex interactions between dissolved and particulate organic and inorganic P fractions. Lean (1973a,b) used the isotope ^{32}P to examine the forms of P in lake water and the rates of cycling between the various fractions. The vast majority of the total P (TP) was tied up in the particulate fraction and the DIP fraction was only a minute fraction of the total. In order to supply the needs of algal growth in surface waters it is necessary for the DIP pool to be turned over very rapidly. Evidence from ^{32}P kinetics in lake waters indicates that the turnover time of the DIP pool may frequently be measured in minutes (Rigler, 1966, 1973; Lean, 1973b; Nalewajko and Lean, 1980). The rapidity of the turnover of the DIP pool is facilitated by the fact that no redox changes are necessary for uptake and metabolism and by virtue of the fact that dead algal cells release most of their TP very rapidly.

DIP availability is determined by inputs from the basin and by biogeo-chemical processes in surface waters. In oligotrophic waters the sediment acts as a net sink for DIP as particulate P which sediments out of surface waters is incorporated into the sediment and remains buried. The mobility of P is closely tied to the redox conditions of the water and the sediment interface as P is readily bound by Fe^{3+} and Ca^{2+} ions and may coprecipi-tate with them in oxygenated water (Hutchinson, 1957b; Stumm and Morgan, 1981). Under anoxic conditions, however, the sediment becomes a net source of P as the Fe^{3+} is reduced and the bound P is released. Mortimer (1941–1942) showed a close relationship between the release of Fe and P into the water overlying anoxic sediments. Hutchinson (1957b) discussed the relationships between redox potential, Fe, S and P mobility in anoxic lake waters and sediments. The consequences of 'displaced metabolism' in eutrophic lakes are therefore considerable. A larger flux of particulate organic C, N and P formed by phytoplankton growth in surface waters will produce serious disturbances in the P cycling within the lake basin. This will be particularly true when it sediments into bottom waters in quantities sufficient to produce anoxia before the end of the summer stratification. Anoxia in deep waters will lead to a build up of PO_4 and NH_4^+ in bottom waters in summer which, when returned to surface waters by autumnal turnover, will serve to fertilize surface waters again.

(b) *Fluxes and regeneration*

Experiments with ^{32}P indicate that there is a pattern in the turnover rates of the DIP pool depending on the trophic state of the water and time of year.

Turnover times of the DIP pool are slower in winter and in eutrophic waters; i.e. in situations where the DIP pool is large or when the algal demand is reduced. Turnover times in such situation may be measured in hours or days (Nalewajko and Lean, 1980). The recent paper by White *et al.* (1982) examines the relationships between the turnover times of ^{32}P and a number of indices of algal biomass and trophic state. The conclusions of that work seem to be that the turnover times of ^{32}P are influenced by two major factors of supply and demand. The size of the DIP pool (supply) is positively correlated with ^{32}P turnover times (larger DIP pool, longer turnover times, Pomeroy, 1960) whereas the indices of algal biomass (demand) are negatively correlated with ^{32}P turnover (higher biomass, shorter turnover times).

In summer the turnover times of the DIP pool reduce to minutes or seconds as the demand increases and the pool size becomes smaller. As already noted, Ketchum and Corwin (1965) showed that the proportion of recycled DIP in coastal waters increased from 2% in Spring to 43% in Summer. Thus the same pattern as that for DIN emerges: a build up of DIP during periods of vertical mixing, followed by depletion of the DIP pool in Spring and Summer as algal growth proceeds. As with DIN also there must be close coupling between the processes of productivity and grazing if the DIP is to be rapidly and efficiently recycled in surface waters. Microplankton are apparently involved with the cycling of DIP in surface waters as they exhibit rapid uptake and loss of ^{32}P. The larger net plankton have high half saturation constants for ^{32}P uptake (Fig. 4.3) and so do not contribute greatly to the cycling of DIP when the pool size is small (Lean and White, 1983). In summer in oligotrophic waters the pool size of DIP is practically unmeasurable (Rigler, 1966; Lean, 1973a,b; Brown *et al.*, 1978; Tarapchak *et al.*, 1982) and cellular stores of DIP may be severely depleted as short term uptake experiments may show uptake of DIP far in excess of the apparent growth requirements of the cells (Lean and Pick, 1981). As the cellular stores of DIP are run down (the bank balance analogy again) growth may continue without external DIP supplies (Mackereth, 1953).

When we consider the availability of DIP in freshwater systems, precisely the same considerations apply to DIP in freshwater as apply to DIN in the oceans. Only the time scales and basin dimensions differ. DIP may be made available in three ways: as DIP from recycling within surface waters; as DIP from material decomposed in bottom waters and returned to the surface upon seasonal turnover or as a result of storm events; and as DIP from river and waste water inputs. The temporal and spatial components of these DIP inputs interact with the components of the algal demand to produce some complex patterns of change in the DIP pool size. In eutrophic waters the DIP input from external sources may exceed the requirements of algal growth, in which case a large pool of DIP may build up in surface waters. For example,

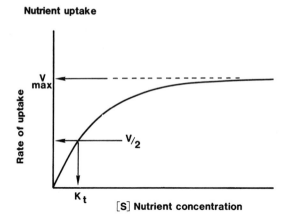

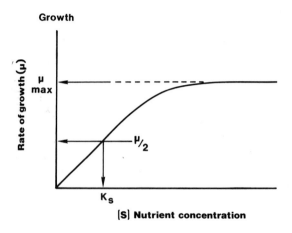

Fig. 4.3 Sketches of the Michaelis-Menten curves for the relationships between nutrient concentration and uptake and growth. The affinity constant for uptake K_t is usually smaller than the half saturation concentration for growth K_s.

in Hamilton Harbour which receives large external inputs of DIP from municipal waste water treatment plants the concentration of DIP rarely falls below 5 µg l^{-1} (Harris *et al.*, 1980a) and there is a continuous background concentration of DIP. In mesotrophic waters the seasonal regeneration of DIP at overturn may provide a large pool of DIP which is depleted during the summer stratification so the DIP concentration falls to unmeasurable values in summer. In oligotrophic waters the phytoplankton may be dependent on DIP recycled within surface waters for most of the year. The spatial and temporal components of the DIP pool are therefore dependent on the trophic state.

CHAPTER 5

Defining the scales of interest

We saw in Chapter 2 that equilibrium theory could only be applied to small closed universes that were, in all probability, rarely found in the real world. The world of the field ecologist is one of 'patches' in space and in time and, while some of the patches may be sufficiently distinct to be classed as closed systems for the purposes of theory, there is much migration between patches. Some of the patches are clearly the result of external physical and chemical influences on the populations of organisms and some of the patches arise from internal biological interactions such as grazing. How important are the two influences? What constitutes a patch and how are they formed? Are there any patches that are sufficiently closed and persistent for the competitive equilibrium to be achieved? I will now attempt to describe the 'patchiness' to be found in lakes and the oceans and to show how this patchiness influences the ecology of the species in those systems. In later chapters I will look at distinct spatial and temporal scales as they influence certain aspects of the ecology of the organisms, so our job here is to sketch the outline of the processes and to broadly define the scales of interest.

5.1 Patchiness in space and time

5.1.1 VERTICAL MIXING AND SEDIMENTATION

We have seen that the water column of lakes and the oceans is much better mixed in the horizontal than in the vertical. The presence of density stratifications results in the restriction of the size of eddies that turn over vertically and produces a rotating fluid in which the eddies are stretched in the horizontal (Denman and Gargett, 1982). Vertical mixing over scales of metres may take a day or so in surface waters (Table 3.1) whereas horizontal mixing over scales of kilometres takes about the same length of time (Fig. 3.1). In the region of the thermocline, vertical mixing over scales of metres may take as much as a month. The vertical dimension is therefore of great importance to phytoplankton as the environment may be stable for periods greatly in excess of the generation time and growth at different depths may

be possible. Because of the exponential attenuation of downwelling irradiance, the gradient of photosynthetically active irradiance is very steep in
the vertical, as are the gradients of a number of other important ecological
factors. Thus, because the populations may come to lie in a stable density
gradient, it is possible for populations to grow under different conditions
separated by small distances in space but long periods of time.

Profiles of *in vivo* fluorescence (IVF, a measure of algal biomass) often
reveal strong layering in the upper water column (Berman, 1972) and
microscale patchiness in the region of the thermocline (Derenbach *et al.*,
1979). This is, of course, only possible in stably stratified surface waters.
The vertical structure in the phytoplankton populations is clearly correlated
with the physical environment (Manzi *et al.*, 1977). Many species show
distinct depth preferences. This is particularly true of algae which buoyancy
regulate or swim to a preferred depth. Dinoflagellates (Harris *et al.*, 1979)
and blue-green algae (Reynolds and Walsby, 1975) both show depth distributions that are controlled by the vertical eddies and TKE inputs to the
water column. Both groups of phytoplankton form nuisance blooms and the
importance of the interaction between the occurrence of such blooms and
the structure of the water column will be documented later. Fluctuations in
the stability of the water column which result from the passage of atmospheric weather patterns are also important because of the way in which the
density gradients may be disrupted and restored over time. It will suffice,
here, to note that the vertical density structure of the water column is of
vital importance to the phytoplankton, as it controls a number of processes
associated with the growth and loss of the populations the mixed layer
(Walsby and Reynolds, 1980).

For the purposes of describing the spatial and temporal variability in the
planktonic environment a vertical scale of mixing and sedimentation has
been defined (Fig. 5.1) consistent with the values of K_z and with the time
scales of mixing. Rates of sedimentation of the phytoplankton are commonly about 0.5 m d^{-1} (Smayda, 1970), and the spread of observed sinking rates (Fig. 5.2) correlates well with the range of vertical mixing rates
calculated by Denman and Gargett (1983). The phytoplankton that sink
rapidly obviously require rapid vertical mixing if they are to stay in the
water column. This evident adaptation of the sinking rates of the phytoplankton to the physical properties of the environment of the mixed layer
should be expected if phytoplankton are to survive in the water column!
The vertical scales of interest are therefore rather small (Fig. 5.1) ranging
from about 10 m to about 10^{-2} m. As these scales encompass strong
environmental gradients it is clearly necessary to obtain good resolution
when sampling in the vertical. The use of fluorometers (Berman, 1972) and
beam transmissometers (Heaney and Talling, 1980) has greatly improved
the resolution of studies of vertical patchiness.

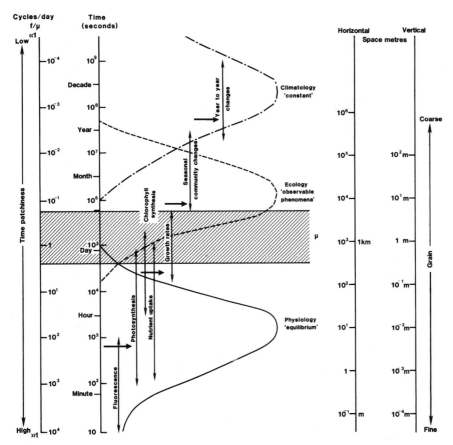

Fig. 5.1 The scales of phytoplankton ecology. Horizontal and vertical scales are determined by the respective diffusion coefficients K_h and K_z. The time scales for the algae are determined by the scales of growth (shaded band). The processes of importance at each scale are noted. In the past it was always assumed (wrongly) that physiological processes were at equilibrium and that climatological variability could be ignored.

5.1.2 HORIZONTAL MIXING

A similar scale for horizontal patchiness may also be defined based on the values of K_h and the fact that dye studies (Fig. 3.1) have shown that scales of 1 km and 1 day are equivalent (Harris, 1980a). The observation of patchiness in phytoplankton populations is not new (Hardy, 1935a, 1936) but the observations became much more systematic with the introduction of *in vivo* fluorescence (IVF) techniques. The measurement of IVF is easy but the interpretation of the resulting data is not (Harris, 1980c) as it depends to a large extent on the sensor in use and the method used. Those problems

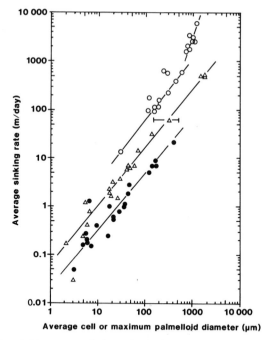

Fig. 5.2 Data for sinking rates of phytoplankton. Open circles, palmelloid stages of *Cyclococcolithus fragilis*; triangles, senescent cells; closed circles, actively growing cells. From Smayda (1970).

aside, what results is a long string of IVF data from a transect across the lake or ocean which may be analysed to determine the distribution of variance. Platt and his co-workers did most of the pioneering work in this area (Platt *et al.*, 1970; Platt, 1972) and introduced the use of spectral analysis as a means to summarize the variance structure of the data (Platt and Denman, 1975a; Fasham, 1978).

A large number of publications have resulted from this work (cf. reviews in Platt and Denman, 1980; Harris, 1980a) and the observed distributions of variance in the data have been supported by a theoretical model. The observed patterns of IVF distribution show that the variability increases as larger and larger areas of the oceans are sampled. It is therefore assumed that IVF (or algal biomass) is acting as a water mass marker and, over certain scales, is displaying the same eddy cascade structure as the water itself. A plot of the spectrum of variance in IVF should, if biological processes were unimportant, show the same −5/3 slope as a plot of the spectrum of TKE (Fig. 3.2). Theory predicts that a major difference between the distribution of TKE and algae should appear at scales of about 1 day or 1 km as the growth effects of the algae become evident. This result is obtained from the KISS model (Kierstead and Slobodkin, 1953; Skellam,

1951) which relates the rate of horizontal diffusion to the rate of algal growth and calculates a critical patch size below which the effects of diffusion will override any growth effects. The KISS model has a number of unrealistic assumptions (Harris, 1980a) and, like other mathematical formulations already discussed, is limited by the tractability of the mathematics, but the results generally appear to conform with observation. The model predicts a critical patch size of about 100 m below which diffusion should wipe out any growth effects. Beyond about 1 km however, the growth effects should become evident. Later spectral models by Platt and Denman (1975b), Denman and Platt (1976) and Denman *et al.* (1977) gave similar results.

Many IVF spectra do have a bend at about 1 km but many do not (Horwood, 1978) and slopes which differ widely from the theoretical norm have been observed. The spatial correlations between temperature and IVF (Denman, 1976) or current velocities have been examined (Powell *et al.*, 1975) and some significant results have been obtained. One of the problems with all this is the presence of vertical structure and internal waves which will tend to destroy any correlations between physical and biological variables. If the sensor is moving through internal waves the water masses will differ not because of horizontal patchiness but because of the vertical patchiness. The apparent horizontal correlations will suffer accordingly. In some cases a vertical separation of only a few metres is sufficient to destroy the horizontal correlations (Platt and Denman, 1980). As we have already seen, in the smaller basins of lakes the full spectrum of turbulence is constrained by the basin size. The cascade of turbulence may not, therefore, follow the $-5/3$ law in coastal regions or areas where the internal motions are constrained by the basin morphology (Palmer, 1973; Harris, 1983). In such areas the slope of the energy spectrum may be as steep as -3 (Fig. 5.3). In lakes therefore, the patchiness of the phytoplankton is a function of features such as river flows and nutrient inputs to the basin (Richerson *et al.*, 1978) as well as the cascade of turbulent eddies within the basin itself.

5.1.3 SPECTRA OF TURBULENT KINETIC ENERGY

The classification of the spectrum of horizontal and vertical motions into functionally discrete scales will depend on the extent to which the spectra may be decomposed into discrete units. The processes of vertical mixing act over quite different time and space scales from the processes of horizontal mixing so the two processes are functionally different from the point of view of the organisms.

The energy spectrum for Hamilton Harbour shown in Fig. 5.3 was determined by the variance in water column stability. Two features of note are evident: the rapid decrease in the variance at time scales below 5 h and the peak in the spectrum at 200 h. The steep decline in the energy spectrum

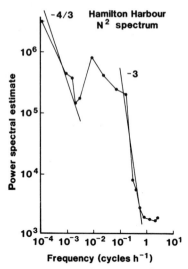

Fig. 5.3 The power spectrum of water column stability in Hamilton Harbour. Data from Harris (1983).

at scales below 5 h was due to the interaction of the basin morphometry with the turbulence field so that the truncation of the spectrum was a function of the size of the basin (Harris, 1982). The 200 h peak in the spectrum was due to the passage of atmospheric weather patterns and the disturbances caused by wind events. This 200 h peak appears to be characteristic of temperate latitudes (Boyce, 1974) and interacts strongly with the temperature structure of large lakes causing mixing and internal wave action. At seasonal scales the spectrum of N^2 shows a slope of about $-5/3$ with increasing variance at longer time periods. The spectrum of water column stability (Fig. 5.3) shows that the motions at a scale of minutes to hours are distinct from longer term motions and are, in terms of total energy, less important. The very sharp drop in the spectrum from scales of 10 h–5 h shows that the energy distribution may be decomposed into processes operating at three major scales – at seasonal scales, scales of a few days, and scales of minutes to hours.

As we have seen (Fig. 3.2) the energy spectra in all surface waters show a number of discrete peaks with smooth cascades of energy between. Much of the published work on horizontal patchiness falls into the region from metres to a few kilometres and hence tends to show smooth energy cascades whether physical or biological parameters are measured (Fortier and Legendre, 1979; Horwood, 1978). This appears to be mainly because the observations fall into the region between the small scale mixed layer dynamics and the mesoscale structures at scales of many kilometres. Not all energy spectra show the sharp drop at 10 h that exists in Hamilton Harbour. The most significant horizontal holons are the mesoscale structures at a scale of lake basins and of eddies, fronts and rings in the oceans. These structures persist for a number of generation times and hence show distinct patterns of species abundances and biomass. There is also a significant input of energy at scales associated with large scale frontal weather patterns.

In lakes the spectra differ from those in the oceans and the overall distribution of energy differs also. The coastal waters of the Great Lakes show a strong peak in the energy spectrum at a scale of a few days. Such motions are associated with the shore-trapped Kelvin waves. The energy spectra therefore show peaks at a scale of 4–8 days (Blanton, 1974). Lake basins appear to respond strongly to frontal weather patterns at scales of a few days, particularly if, as in the case of the Great Lakes, the wind forcing resonates with the scales of internal motions. Even small lake basins show large day to day fluctuations in stability in response to wind forcing and solar heating. Figure 5.4 shows the record of N^2 for Guelph Lake (Trimbee and Harris, 1983) over a period of 105 days in summer. Such variability at a scale of days has a profound effect on the phytoplankton populations in the lake.

The full energy spectrum for Guelph Lake (Fig. 5.5) shows a quite

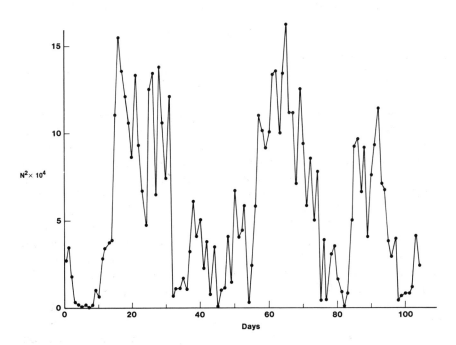

Fig. 5.4 The daily sequence of changes in water column stability in Guelph Lake, Ontario. Data from 105 days in the summer (Trimbee, 1983).

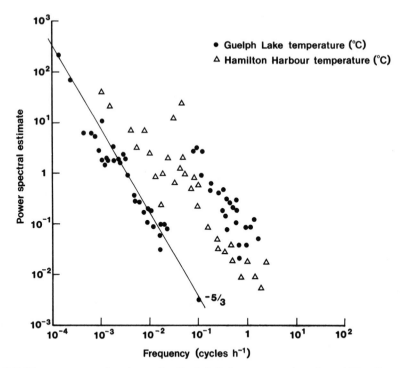

Fig. 5.5 The spectrum of variance for Guelph Lake compared to that of Hamilton Harbour. Note the large increase in variability at scales of 12–24 h in Guelph Lake and the greater variance in the Harbour at scales of 24–200 h.

different pattern from Hamilton Harbour and gives some indication of the differences that exist between lake basins, even those of similar size. Hamilton Harbour, as it is connected to Lake Ontario and is forced by the motions in the coastal zone of that lake, shows a flat energy spectrum from about 20 h to a few hundred hours. No strong diurnal signal is found. Guelph Lake, on the other hand, shows a very strong diurnal signal at certain times of the year with a diurnal variance nearly comparable with that of the seasonal signal. Guelph Lake is a closed basin which responds strongly to diurnal heating in summer. The physical processes in the two water bodies, particularly in summer, are therefore quite different and the species composition of the two are consequently totally different.

It is therefore possible to decompose the spectrum of physical variability in natural water bodies and to identify four basic physical scales. At the smallest scales (seconds and millimetres or less) there is the possibility of significant patchiness associated with grazing and nutrient regeneration (Jackson, 1980). At scales of minutes to hours and metres the dynamics of vertical mixing have a profound effect on the physiology of the phytoplankton in surface waters through control of such factors as the underwater light climate. At scales of days and a few kilometres the fluctuations in water column stability affect the population dynamics of the phytoplankton by influencing the growth rates of various species. The presence of a seasonal thermocline has a great effect on such scales of variance as the presence of a density gradient and introduces the possibility of high frequency motions. If the basin is large enough, mesoscale processes at scales of weeks result in patchy biomass distributions at scales of 100 km or more. At seasonal scales the formation and disruption of the seasonal thermocline is a major determinant of many chemical and biological processes.

These physical scales are separated into a hierarchy of temporal and spatial processes by discontinuities in scale of at least two orders of magnitude. Each lower process (eddy) is nested into the larger eddies and the connections take the form of the cascade of energy from large to small scales.

5.1.4 SPECTRA OF CHEMICAL FLUCTUATIONS

From what little data is available it is possible to make similar statements about the spectrum of variance in chemical (nutrient) factors. Observations in Hamilton Harbour (Harris, 1983) showed a spectrum of variation in oxygen and dissolved inorganic phosphorus which was very similar to the spectrum of variance in water column stability (Fig. 5.6). At seasonal scales

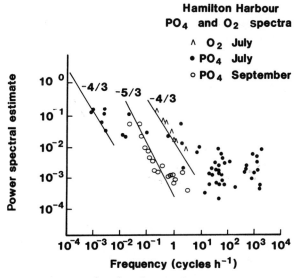

Fig. 5.6 The variance spectra for phosphate and oxygen in Hamilton Harbour.

the physical processes in the mixed layer serve to control the spatial and temporal distribution of nutrients. At a scale of days the presence of vertical density gradients induces greater variance in nutrient distributions. This reflects the greater degree of heterogeneity when density gradients are present. At these scales the distribution of nutrients should reflect a degree of feedback between the rates of supply and demand as populations may persist at various levels in the water column for a sufficient length of time to influence the nutrient supply. In Hamilton Harbour it does not appear that such feedback processes radically alter the spatial and temporal patchiness of the nutrients. Nutrient levels in the Harbour are very high, however, so that supply rates far exceed demand. In less eutrophic bodies of water there is a closer coupling of supply and demand and more biologically induced patchiness is to be expected.

Some spectra of variance in dissolved inorganic phosphorus are presented in Fig. 5.7. These data are from three lakes and display the variance in PO_4 at a scale of days. In Lake Ontario the overall variance is high but the mean PO_4 concentration is low. In Hamilton Harbour, on the other hand, the reverse is true whereas Guelph Lake is intermediate in both respects. The data from Lake Ontario shows the variance associated with the passage of the Kelvin waves in the coastal zone. The relationships between the mean and the variance of each resource raise some important questions about the role of biological processes in producing patchiness in the water column by the feedback between nutrient supply and demand. The nutrient and oxygen spectra in Hamilton Harbour (Fig. 5.6) show that the same major functional scales exist as were seen in the physical data. Patchiness is low at scales of minutes but rises rapidly and reaches a plateau at scales of one to a hundred hours. Temporal patchiness is greater when a density gradient exists (July) than when the water column is vertically mixed (September). The variance rises further to seasonal scales as would be expected. The day to day variability in both chemical and physical parameters in surface waters has a very great effect on the population dynamics and community structure of phytoplankton. A detailed discussion of these effects will appear below.

5.2 The biological response to variability in space and time

The problem with understanding the biological processes which occur in such a variable system as the surface mixed layer lies in the classification of the scales of variability. The biological responses to the scales of variance (Fig. 5.1) encompass a number of different processes which operate at different temporal and spatial scales but they may be simplified somewhat if they are viewed from the point of view of the phytoplankton. It is the way the organisms perceive the environmental fluctuations that is important. This aspect of ecology was recently discussed by Southwood (1977) who proposed a model of ecological strategies in relation to a habitat template (Fig. 5.8). Southwood's model exploits features of the environment as perceived by the organisms and relates these to general strategies for survival and reproduction. Thus the scales of spatial and temporal variability

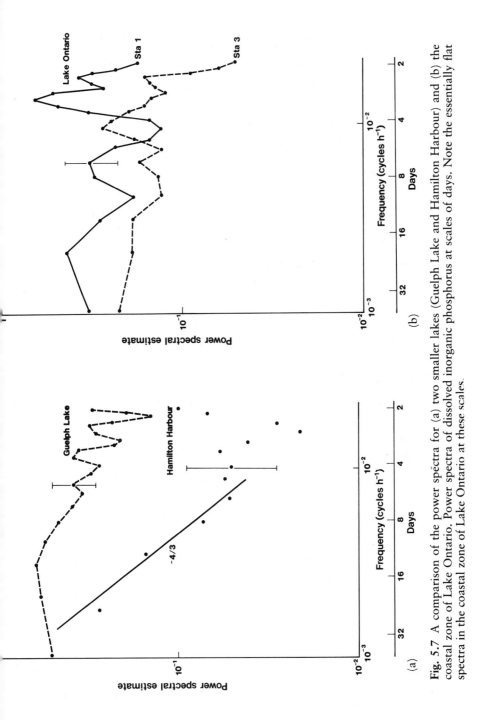

Fig. 5.7 A comparison of the power spectra for (a) two smaller lakes (Guelph Lake and Hamilton Harbour) and (b) the coastal zone of Lake Ontario. Power spectra of dissolved inorganic phosphorus at scales of days. Note the essentially flat spectra in the coastal zone of Lake Ontario at these scales.

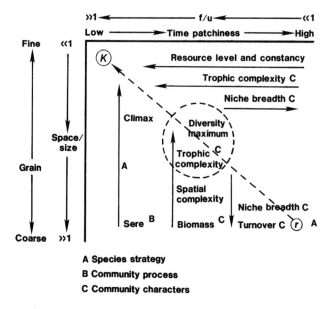

A Species strategy

B Community process

C Community characters

Fig. 5.8 A modification of Southwood's (1977) habitat template to suit the organisms of interest. The environmental fluctuations scale low patchiness, fine grain to high patchiness, coarse grain and the biological response changes from *r*- to *K*-strategies. The spectrum of possibilities runs from near equilibrium in the top left to near chaos in the bottom right.

are scaled to the size, dispersal abilities and generation times of the organisms.

The scheme shown in Fig. 5.8 explicitly includes the environmental variance that is an anathema to equilibrium theory and also restates many ecological problems in terms of what might be called 'tracking problems'. At equilibrium there is no temporal component so the past history of the environment is unimportant to the organisms. In a non-equilibrium, variable environment the past history of the environment is an important parameter (Levins, 1979) and the organisms have the problem of coping with the fluctuations as best they can. They may be said to be 'tracking' the fluctuations by physiological adjustments, vertical migrations or other means so that growth is maintained. Fluctuations in time are scaled by the generation time of the organisms as it may be argued that the generation time is a fundamental time unit for the organism. Fluctuations at time scales much less than the generation times of the organisms (vertical mixing, small scale eddies, turbulence) may be integrated by physiological mechanisms, while fluctuations at time scales much longer than the generation time (seasonal stratification cycles, frontal processes, interannual variability) may be avoided by perennation mechanisms or may be expressed as density independent catastrophes. Thus the time axis may also be related to Connell's (1978) intermediate disturbance hypothesis (Fig. 2.6). Environmental fluctuations at time scales of a few generation times (in particular the 200 h atmospheric time scale) will have a powerful effect on competition and community diversity (Huston, 1979).

The presence of smooth spectra of variance over time scales which are important to the organisms makes it almost impossible for the organisms to optimize their responses. In a statistical distribution of environments the best the organism can do is to filter, integrate and average the external fluctuations and to use the time weighted average response as a source of information. This is exactly what biological holons do so well (Koestler, 1967). The individual cells are both structural and functional holons which contain input hierarchies. 'Input hierarchies . . . are equipped with "filter"-type devices . . . which strip the input of noise, abstract and digest its relevant contents, according to that hierarchy's criteria of relevance' (Koestler, 1967). If optimal responses are out of the question in variable environments then it is possible for the organism to use heuristic methods to achieve its ends. If the organism is well adapted to the combinations of environmental fluctuations it experiences then the heuristic response may differ little from the optimal response. There is, however, a fundamental difference in the way in which the result is achieved. None of the environments discussed above is totally random. All environments contain noisy signals that may be used as a source of information.

The spatial scales of interest may be interpreted by the use of the concept of environmental grain. The distinction between fine and coarse grain environments (MacArthur and Levins, 1964; Levins, 1968; Pianka, 1978) depends on the size and mobility of the organism in relation to the patchiness of resources in the environment. Many phytoplankton bear flagellae, so limited motion is possible. Sournia (1982) indicated, in a review, that the average swimming speeds of phytoplankton are about an order of magnitude greater than the rate of sinking. In Sournia's classification nine of thirteen algal classes contain at least some flagellated members. As the rates of swimming therefore exceed the rate of mixing in many cases these phytoplankton have the ability to position themselves in the water column. Other phytoplankton are traditionally viewed as immobile: existing only at the mercy of turbulence. Such organisms are not strictly immobile however as they all sink. Sinking has often been thought of as a problem for phytoplankton as it results in losses from surface waters. If, on the other hand, we assume that the phytoplankton actively exploit the turbulence structure of their environment (Harris 1978, 1980a), then the interaction of vertical structures and sinking rates may be viewed as a necessary dispersal mechanism. Sinking moves the organism through the vertical environmental gradient in surface waters of lakes and the oceans. This may well be important as the vertical gradient is much steeper than any of the horizontal gradients the organisms may encounter.

Hutchinson (1967) showed that phytoplankton exhibit varieties of hydrodynamic form resistance which cause phytoplankton to tumble through the water as they sink. He concluded that this is a direct adaptation to ensure as much turbulence as possible near the cell surface, to facilitate nutrient uptake. The sinking motion, as well as helping to break down diffusion barriers on the cell surface, serves to move the cell through any environmental gradients and patches, so allowing the phytoplankton to exploit the patchiness of their environment. As differences in size lead to differences in the sinking rate (Fig. 5.2), different phytoplankton will perceive the environment differently; the 'graininess' of the environment depends on dispersal ability as well as size. We have, again, touched on the nub of an important set of arguments concerning the interaction between phytoplankton community structure and the mixing conditions in surface waters.

5.2.1 PHYSIOLOGICAL SCALES

Environmental fluctuations which impinge on the cells at time scales less than the generation time elicit a physiological response. This aspect of the ecology of phytoplankton was reviewed by Harris (1978). The environmental fluctuations which operate at such scales include small scale patchiness in nutrient fluxes and fluctuations in the underwater light field brought about by vertical mixing and turbulence. The small scale patchiness in the nutrient flux arises from grazing and the excretion of nutrients by the animals. There is some debate about the significance of such patchiness (Jackson, 1980) but there is known to be rapid uptake and storage of nutrients by phytoplankton at scales of seconds and minutes (Goldman and Gilbert., 1983; Gilbert and Goldman, 1981). The uptake and storage capacities of phytoplankton are considerable and the organisms appear to have evolved very great affinities for the major nutrients. The critical physiological parameters appear to be the half saturation coefficient for nutrient uptake (K_t), the maximum rate of uptake (V_{max}) and the cell quota of the nutrient (q). By having uptake capacities sufficient to replenish cellular quotas of nutrients in minutes, the cells can tide themselves over periods of nutrient starvation. By rapidly taking up and storing nutrients the cells have evolved the capacity to integrate and damp the external fluctuations in nutrient availability.

The time scales of vertical motion in the surface mixed layer (Table 3.1) interact strongly with the photosynthetic characteristics of the phytoplankton. At scales of seconds and minutes the fluctuations in the underwater light climate caused by vertical mixing are matched by the fluorescence responses of the cells (Harris, 1980c) and by photosynthetic responses (Harris and Piccinin, 1977). The phytoplankton filter and track the underwater light fluctuations (Harris, 1978) by physiological means at short time scales (minutes) and by structural means at longer time scales (hours). Chlorophyll synthesis and other structural changes take place if the cells experience a number of hours at low light. The fluorescence response of phytoplankton to the turbulence in surface waters was demonstrated by Harris (1984).

5.2.2 THE SCALES OF GROWTH RATES

The scales of physical and chemical variability in the water interact with discrete biological holons. From the point of view of population dynamics the environmental fluctuations at a scale of 50 h to 200 h are of the greatest importance as they interact with the growth rates and the population dynamics of the organisms. Such fluctuations are analogous to the perturbations which are of such importance to the coral reef systems. As we shall see the population dynamics of each species are non-equilibrium dynamics but, at the community level, the perturbations are essential for the maintenance of community diversity. Seasonal changes in thermocline depth determine the overall community structure at the next higher level in the hierarchy. Each level in the biological hierarchy transmits averaged, lagged and filtered information to the next higher level. Physiological information contributes to the determination of growth rate; growth rates determine population size and community structure.

Within any given community it is possible to identify a spectrum of reproductive strategies from so called 'opportunist' or r-strategists to the 'equilibrium' or K-strategies of others. Kilham and Kilham (1980) applied these ideas to phytoplankton. Here the r and K labels are derived from the equivalent symbols in the Lotka–Volterra equations. The concept of the r–K continuum was first introduced by MacArthur (1962) and MacArthur and Wilson (1967) and since then there have been a number of summaries of the concept, notably that of Stearns (1976) and Pianka (1978). The attributes of r- and K-species are summarized in Table 5.1. These attributes arose out of the 'Modern Synthesis' and the evolution of these traits was consistent with the accepted explanation of equilibrium ecology. Standard equilibrium theory of course assumes that the r species are rare in communities (Caswell, 1982) so that the opportunist species only last for a brief period of initial colonization. As may be seen from the table the attributes of r-strategists are that they put a large amount of energy into reproduction, they have short generation times and they tend to occur in unstable environments. The K-strategists, on the other hand, tend to grow and reproduce slowly and tend to be good competitors. Clearly one would expect r-strategists to occur in non-equilibrium environments whereas K-strategists would be expected to do well in equilibrium situations where competition was important.

Recently Caswell (1982) has reviewed the evolution of the r–K concept

Table 5.1 Some of the correlates of *r*- and *K*-selection (from Pianka, 1978)

	r-Selection	K-Selection
Climate	Variable and/or unpredictable; uncertain	Fairly constant and/or predictable; more certain
Mortality	Often catastrophic, non-directed, density independent	More directed, density dependent
Survivorship	Often Type III	Usually Types I and II
Population size	Variable in time, non-equilibrium; usually well below carrying capacity of environment; unsaturated communities or portions thereof; ecologic vacuums; recolonization each year	Fairly constant in time, equilibrium; at or near carrying capacity of the environment; saturated communities; no recolonization necessary
Intra- and interspecific competition	Variable, often lax	Usually keen
Selection favours	1 Rapid development 2 High maximal rate of increase, r_{max} 3 Early reproduction 4 Small body size 5 Single reproduction	1 Slower development 2 Greater competitive ability 3 Delayed reproduction 4 Larger body size 5 Repeated reproductions
Length of life	Short	Longer
Leads to	Productivity	Efficiency

and has shown how the idea became one of the bulwarks of equilibrium theory. But Caswell (1982) has written: 'Granting the premises of *r*- and *K*-selection, it is certainly legitimate to infer that equilibrium species will be *K*-selected. Since density independent growth cannot continue forever, it is also legitimate to infer that a species which is genuinely *r*-selected must be non-equilibrium. The problem is with the converse of these arguments: that *K*-selected species must be at equilibrium, and that non-equilibrium species must be *r*-selected'. Caswell goes on to show that, even in non-equilibrium environments, selection will tend to push species into the two strategies of predominantly *r* and predominantly *K* type life histories and that there will be an instability between the two strategies. This conclusion is buttressed by the reviews of Stearns (1976, 1977). Thus it is not necessary to invoke equilibrium theory to account for the observed differences in the life history strategies of natural populations and one explanation cannot be used to support the other. The presence of *K*-strategists does not imply equilibrium conditions (Stearns, 1977; Caswell, 1982). This will be seen to be important later when we consider the growth strategies of phytoplankton populations at different stages of the seasonal succession.

A spectrum of environmental fluctuations (a distribution of characteristic periodicities in time) should therefore result in a distribution of life history strategies and there is no reason why *r*- and *K*-strategists should not be found together (remember different time scales can uncouple species interactions). The distribution of perturbations implicit in the 'intermediate disturbance' hypothesis of Connell (1978) also predicts that in highly disturbed environments *r*-strategists should be favoured but in the relatively constant environments which exhibit little disturbance and reduced diversity (Fig. 2.6) *K*-strategists should be more common. This is consistent with the models of Huston (1979). The maximum diversity of the intermediate disturbance situation should see the coexistence of both types.

In order to make the case for the intermediate disturbance hypothesis it is necessary to demonstrate that phytoplankton growth rates are of the same order of magnitude as the strong 100–200 h environmental time scales (Fig. 5.3). The generation times of the organisms are not, of course, all identical; indeed there is a spread of values depending on temperature, light intensity, nutrient availability and cell size. It is possible, however, to identify a spectrum of growth rates amongst phytoplankton groups and to classify organisms on the basis of both rates and efficiencies (Kilham and Kilham, 1980). The literature on the growth rates of phytoplankton under defined conditions in culture is voluminous to say the least. I am, here, more concerned with realized rates in nature than with growth in culture.

The basic difference between culture conditions and natural conditions lies in the variability of the natural environment and the coexistence of many species. Many species which are common in the wild cannot be grown in culture and many species that can be grown in culture can only be grown in media with ionic strengths and nutrient ratios quite atypical of the natural state. Under optimal conditions, however, the growth rates of phytoplankton, both in the wild and in culture, should be similar. The major question is 'how often are conditions in nature optimal?' The high diversity of species present in surface waters allows species to be substituted so that conditions which are not conducive for the growth of one species may allow the rapid growth of others. Eppley's (1972) review of phytoplankton growth rates showed that the doubling times of phytoplankton in culture ranged from a few days to a minimum of three hours at temperatures in excess of 30°C (in ln units d^{-1} (k), rates equal to 0.1 d^{-1} to 5.5 d^{-1}). Few phytoplankton are capable of more than four doublings per day ($k = 2.8$ d^{-1}) and most of the data for realistic temperatures (0–25°C) show the expected growth rates to be rather less than this. The expected time scales of interest therefore lie in the range from six hours to a week or so – scales which interact strongly with the scales of environmental change.

The observed apparent maximum growth rates of phytoplankton in surface waters (in ln units d^{-1}) range from about 0.1 d^{-1} to 2.0 d^{-1} (Fig. 5.9 and Reynolds, 1984b). Much of the culture data quoted in Hoogenhout

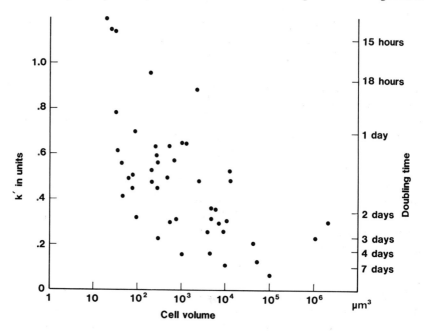

Fig. 5.9 A summary of observed natural growth rates as a function of cell size. For data sources see text.

and Amesz (1965) and Eppley (1972) falls into this range also. There is a clear effect of cell size with the smallest cyanobacteria and ciliates growing at about 1.5 d^{-1}. The larger net plankton (> 65 μm in diameter) grow at rates between 0.1 d^{-1} and 0.8 d^{-1}. Natural populations suffer from mortality and other losses so that *in situ* estimates must be minimum estimates. Nevertheless, the maximum rate of increase of natural populations, when corrected for grazing and sedimentation, must give rough estimates of k (Kalff and Knoechel, 1978; Knoechel and Kalff, 1978). By observing the growth of populations in large enclosures and accounting for all losses, Reynolds and his co-workers have managed to estimate accurately the growth rates of natural populations (Reynolds, 1982, 1983a,b; Reynolds *et al.*, 1982, 1983, 1984; Reynolds and Wiseman, 1982). Reynolds (1984b) has summarized this work. Talling (1955a) and Cannon *et al.* (1961) placed populations in bottles *in situ* (with obvious containment effects) while others (Elbrächter, 1973; Swift *et al.*, 1976; Heller 1977) used the frequency of dividing cells to estimate *in situ* growth rates. No technique is totally free from bias but the results summarized in Fig. 5.9 do appear to be broadly consistent. The decrease in growth rate with increasing size is consistent with the allometric trends displayed in Peters (1983) which show a general tendency for growth rates to decline as a power function of body mass.

The observed growth rates of natural populations are consistent with the idea of an *r* to *K* spectrum within the phytoplankton community. Small species grow rapidly and are lost rapidly (Sommer, 1981) and tend to occur early and late in the year. The larger, summer species grow more slowly and persist longer with lower loss rates (Tilzer, 1984). The data is also consistent with the coexistence of a broad range of species of different sizes, growing at different rates and responding to different characteristic periodicities in the spectrum of environmental perturbations.

5.3 Models of competition between phytoplankton

At scales of weeks there is evidence that in constant environments competition between phytoplankton can lead to the exclusion of some species and to dominance by others. In order to fit such events into a framework of scales of perturbation it is necessary to determine the time scales of such competitive interactions. This has been done by growing pairs of species in continuous culture and observing the course of events. Tilman has developed a series of models which describe the mechanism of competition between phytoplankton in culture (Titman, 1976; Tilman, 1977; Tilman *et al.*, 1981) and has attempted to extend the argument to natural phytoplankton populations and other types of organisms (Tilman, 1980). The models of competition which he has developed are derived from the equations for nutrient uptake and growth in continuous culture; cultures which are explicitly steady state systems.

The normal practice is to grow phytoplankton in culture vessels into which is continuously pumped a supply of nutrient medium. The nutrient medium is normally made up so that one element is limiting (commonly nitrogen, phosphorus or silicon) and the others are in abundant supply. If the limiting element is indeed limiting the growth of the organisms then, at steady state, the rate at which new medium is pumped into the culture will determine the growth rate up to a maximum rate of growth, after which the cells will be washed out of the culture. The work on continuous cultures had led to the development of a considerable body of information on the physiology and growth rates of phytoplankton under conditions of constant dilution, nutrient supply, light and temperature (Rhee, 1979).

The theory was first developed for populations of micro-organisms but has been widely adopted for use with phytoplankton. Under constant conditions, the growth rate of the organisms may be related to nutrient concentrations by the various forms of the Monod equation. There are a number of forms of the equation and relationships exist to define the growth rate in terms of both the external nutrient concentration in the medium (Monod formula) and the nutrient concentration inside the cell (Droop formula). At steady state, all the formulae are formally equivalent (Goldman, 1977a; Burmaster, 1979a; DiToro, 1980) as it matters little which parameters are determined and the rate of nutrient flux into the cells will be a direct measure of the rate of growth. Goldman (1977a) has stated the relationships between the relevant variables and has summarized the necessary assumptions.

Monod (1942) discovered that the following relationship described the uptake of limiting nutrient in continuous cultures of bacteria:

$$V = V_{max} \frac{S}{K_t + S} \qquad 5.1$$

where V is the rate of uptake of the limiting nutrient, V_{max} is the maximum uptake rate and S is the concentration of limiting nutrient. K_t is defined as the concentration of limiting nutrient at which $V = \frac{1}{2} V_{max}$. This equation describes a rectangular hyperbola (Fig. 4.5) for nutrient uptake versus concentration and in this form K_t is an affinity constant for uptake and transport into the cell. In continuous cultures:

$$D = \mu = \frac{1}{X} \frac{dX}{dt} = f(S) \qquad 5.2$$

where D is the dilution rate per unit time, μ is the specific growth rate per unit time, X is the biomass concentration and S is the concentration of the limiting nutrient. The usual form of $f(S)$ is the Monod equation:

$$\mu = f(S) = \bar{\mu} \frac{S}{K_s + S} \qquad 5.3$$

where $\bar{\mu}$ is the maximum specific growth rate, S is the external or residual concentration of limiting nutrient in the culture, and K_s is the half saturation constant for growth (i.e. S at $\frac{1}{2}\bar{\mu}$). The Droop equation relates the growth rate to the internal concentration of limiting nutrient as follows:

$$\mu = \hat{\mu} \frac{Q - kq}{Q} \qquad 5.4$$

where $\hat{\mu}$ is the maximum specific growth rate (not necessarily equal to $\bar{\mu}$), Q is the cell quota or the amount of limiting nutrient inside the cell and kq is the minimum cell quota for that nutrient. The derived nutrient uptake rate p, is equal to μQ and at steady state, p is a direct function of μ, depending on a yield coefficient which is determined by the elemental cellular ratios. At steady state, therefore, the fluxes of nutrient into the cells may be equated to the growth rate and the efficiency with which any given organism uses the nutrient supply may be measured by the values of K_t, K_s and the yield coefficient.

Taylor and Williams (1975) grew phytoplankton in continuous culture and essentially confirmed the major postulates of competitive theory (page 24) as the continuous culture environment is at equilibrium by definition. They also were the first to use a nutrient ratio approach. Titman (1976) (cf. also Tilman, 1977) grew *Asterionella formosa* (*A.f.*) and *Cyclotella meneghiniana* (*C.m.*) in continuous culture and determined the K_s values for growth under phosphate and silicate limitation. *A.f.* had a significantly lower K_s for phosphate than *C.m.*, whereas *C.m.* had a significantly lower K_s for silicate than *A.f.* These responses were measured by growing the species singly, in culture, with the single limiting nutrient. Titman showed that the result of competition between the two species could be predicted by the ratios of their K_s values. Thus he introduced the concept of nutrient ratios. The critical nutrient ratios were defined as follows: for phosphate, the K_s values were 0.04 μM for *A.f.* and 0.25 μM for *C.m.* and for silicate the K_s values were 3.9 μM for *A.f.* and 1.4 μM for *C.m.* From the growth equation (above) growth rates will be equal when:

$$S_1/(S_1 + K_1) = S_2/(S_2 + K_2) \qquad 5.5$$

$$S_1/S_2 = K_1/K_2 \qquad 5.6$$

where S_1 and S_2 are the concentrations of nutrient 1 and 2 respectively and K_1 and K_2 are the K_s values for species 1 and 2 respectively.

The boundary between phosphate and silicate limitation for *A.f.* should occur when the silicate to phosphate ratio equals $3.9/0.04 = 97$. When the ratio exceeds 97, *A.f.* should be limited by phosphate and for ratios less than the stated value the organism should be limited by silicate. Similarly the critical ratio for *C.m.* is $1.4/0.25 = 5.6$. In competition for phosphate and silicate these results predict that at phosphate to silicate ratios in excess of 97 both species are limited by phosphate but *A.f.* should outcompete *C.m.* as it uses phosphate more efficiently: it has the lower K_s. At ratios below 5.6 both species are limited by silicate and *C.m.* as it has the lower K_s. At ratios between 97 and 5.6 the two species should coexist as each is limited by a different resource, *A.f.* by silicate and *C.m.* by phosphate. Titman's (1976) culture data, summarized in that paper supported these conclusions.

Tilman (1977) presented much the same data as Titman (1976) and extended the analysis to the Droop (1974) model using the internal cell quotas rather than the K_s values for external nutrient concentrations. Of course, he obtained the same result because, as we have already seen, the

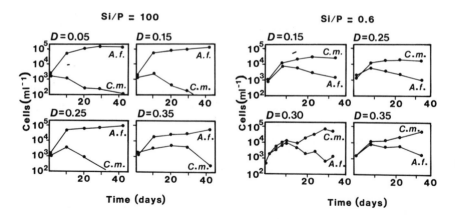

Fig. 5.10 The course of competitive exclusion in continuous culture. Competition between *Asterionella formosa* (*A.f.*) and *Cyclotella meneghiniana* (*C.m.*) at a range of dilution rates (*D* per day) and silica to phosphorus ratios (Si/P). Note that exclusion takes at least 40 days. From Tilman (1977).

two treatments are formally equivalent at steady state. Tilman's (1977) paper showed that the time required for competitive exclusion depended on the flow rate of the culture. As $D = \mu$ (at steady state) the growth rates are higher at higher flow rates so the time for competitive exclusion tends to be less at higher flow rates. Even so the time required for 99% domination by one species was of the order of 10–50 days (Fig. 5.10), with most of the values in the 20–40 day range. Thus we can assume that the full expression of competitive equilibrium was only reached after these time intervals.

Tilman *et al.* (1981) carried out much the same analysis for *A.f.* and *Synedra ulna* over a range of temperatures and showed that the effect of temperature on the growth kinetics of the two species could be included in the model and that competition over a temperature gradient could be effectively described. The times required for competitive exclusion were, if anything, longer than in the previous study requiring at least 50 days in many cases. This result parallels the earlier result of Taylor and Williams (1975) who observed a similar time scale for competitive exclusion.

Kalff and Knoechel (1978) discussed the competitive interactions between phytoplankton and pointed out that 100% dominance cannot be achieved with any finite growth rate. They suggested that the reason for the apparent coexistence of so many species lay in the fact that growth based competitive exclusion is never absolute. While it is possible for one species to achieve 99% dominance in the time scales discussed above, a 1% occurrence of the disadvantaged species still represents a considerable innoculum of individuals. Kalff and Knoechel showed that a recalculation of Tilman's (1976,

1977) data in terms of growth rate differentials led to the conclusion that the species succession should have been achieved in 25 days whereas it actually occurred in 11–15 days. This indicates that part of the observed change was not due to competition. As the outcome of the apparent interaction between the species is dependent on growth differentials, the loss terms are as important as the growth terms and they must both be taken into account. Loss rates from natural populations are therefore a crucial component of the seasonal cycles of species.

Thus in simple, constant environments theory and observation are consistent and Gause's axiom is confirmed. It should be remembered however that the culture environment is a closed universe and that no migration is possible between culture vessels. The outcome of competition is predictable when two species compete for a single resource or for a small number of resources in culture and, in the case of phytoplankton where the mechanism of competition for nutrients may be specified, a precise, mechanistic, explanation of the process may be given (Tilman, 1980, 1982). The process of competitive exclusion requires limiting resources and a constant environment for about 20–50 days. From the discussion of physical processes in Chapter 3 it is clear that the surface waters of lakes and the oceans are rarely stable for such periods and that competitive exclusion must therefore be a rare occurrence. This is not to say that competitive exclusion never occurs in nature, examples will be given below. Equilibrium conditions exist at one end of a broad spectrum of ecological possibilities.

CHAPTER 6

The measurement of productivity and growth rates

As we are discussing a group of small single-celled organisms floating free in surface waters the measurement of growth rates and nutrient fluxes presents some special problems. The enclosure of the water sample in a chamber or bottle encloses a fragment of the whole planktonic food chain so that any measurement of the flux of oxygen, carbon or nutrients is a measure of community processes including both autotrophs and heterotrophs. Any attempt to measure growth by increases in cell number must account for the possible presence of grazing organisms and other mortality terms. In many experiments attempts have been made to eliminate all but the photo-synthetic fluxes of oxygen and carbon by excluding the grazing animals and confining the phytoplankton in bottles. There is a long history of photo-synthesis measurements in limnology and oceanography as such measure-ments have been made for years under the guise of productivity measurements.

It is necessary to draw aside for a moment to make a very important distinction between changes in algal biomass and changes in the rate of turnover of that biomass. This distinction will be important at a number of subsequent points so it is well to properly explain the difference here. The increase in algal biomass over a particular period, usually a period of weeks or months may be referred to as 'growth' or 'new production' (Eppley and Peterson, 1979). It may be thought of as an exportable or harvestable quantity, so that areas of high production give not only high algal biomass after a fixed period of time, but also high rates of sedimentation from surface waters and even, on occasion, high fish yields (Fig. 6.1). Thus 'growth' is the increase in the pool size of, say, particulate organic C tied up in the phytoplankton. 'Specific growth' or 'regenerated production' refers to the rate of flux of C through that pool. Short term (hours) measurements of the uptake of ^{14}C or ^{15}N by phytoplankton in bottles which are usually referred to as productivity measurements are really measurements of 'specific growth' as it is the turnover of C or N in the particulate pool that is being measured.

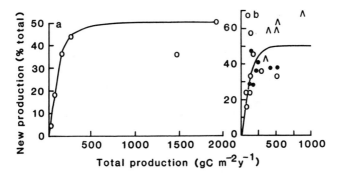

Fig. 6.1 The relationships between total production and new production as summarized by Eppley and Peterson (1979). When total production is small new production appears to be a tiny fraction of the total.

It is possible to have high 'specific growth' rates but low 'growth' rates if the particulate C pool turns over rapidly but does not get any bigger; due, for example, to grazing activity. We shall see later that oligotrophic (low nutrient) waters may show surprisingly high 'specific growth' but tend to show low 'growth'. This is due to the fact that most of the essential nutrient elements are rapidly recycled in surface waters and the coupling between algal growth and grazing is tight. Eutrophic (high nutrient) waters, on the other hand, tend to show low 'specific growth' and high 'growth' as the particulate C pool tied up in the phytoplankton grows in size and algal blooms appear. Nutrient recycling appears to be less efficient in such systems (Harris, 1980a) and the fate of much of the production is sedimentation onto the sediments, followed by decomposition. The distinction between 'specific growth' and 'growth' may be thought of as the difference between changes in the size of a wheel and changes in the rate at which the wheel is spinning (Harris, 1980a).

Since the earliest days, the rate of primary productivity in the sea and in lakes has been measured by resuspending bottles containing phytoplankton at the depth from which the algae were collected and observing the changes in oxygen and carbon over time. Traditionally the bottles are suspended *in situ* or are placed in an incubator which simulates the natural light and temperature as nearly as possible. A precise reconstruction of the natural conditions is very difficult to achieve in an incubator as the temporal fluctuations in the natural light regime and the natural spectral composition can only be simulated with difficulty. Notwithstanding these problems, and the fact that in *in situ* incubations the natural turbulence structure of the water column is lost, the extensive use of bottle incubations has characterized productivity measurements for years. The use of bottles in *in situ* experiments and in incubators has restricted the emphasis on the loss terms in Fig. 6.2 to respiration. Cell death, sinking and grazing have been largely neglected. As we shall see the effects of such neglect of grazing and nutrient regeneration in the natural state may have been considerable.

The earliest productivity measurements were made by measuring the changes in the concentration of oxygen in bottles suspended in the water using the Winkler technique (Gaarder and Gran, 1927; Jenkin, 1937). The introduction of the ^{14}C technique by Steeman-Nielsen (1952) increased the sensitivity of the methods available and made measurements in oligotrophic waters a possibility. As phytoplankton are predominantly photoautotrophs there is obviously a link between photosynthesis and growth, as photosynthesis is the major source of energy available to the cells. Thus there must be a link between the flux rate of O or C as measured by some sort of productivity measurement, and the rate of growth of the phytoplankton. The strength of the link depends on a number of methodological and physiological factors however. At steady state and assuming balanced growth the linkage is very tight, but under non-steady state conditions there are considerable discrepancies. In this chapter, I will investigate some of the relationships between the measurements of productivity and photosynthesis and the flux of nutrients into cells and the relationships of such measurements to estimates of the growth rate of the cells.

Productivity may be defined as a rate of production (mg C m^{-3} h^{-1}) usually averaged over some defined period of time. The time in question is usually determined by methodological and biological considerations such as the length of time required to obtain a reliable estimate of the rate. Production is 'the weight of new organic material formed over a period of time, plus any losses during that period' (Wetzel, 1975). Photosynthetic processes contribute to an increase in production while losses are due to such processes as respiration, death, sinking and grazing (Fig. 6.2). Gross productivity refers to the observed (net) rate of change in biomass plus all losses. Traditionally the gross rate of productivity has been obtained from the observed net productivity plus the respiratory losses, as the main loss process was assumed to be respiration (neglecting the other terms). The volumetric rates of productivity (mg C or O$_2$ m^{-3} h^{-1}) may be expressed as a 'specific growth' rate by dividing by a biomass estimate in the same units. The most commonly used biomass estimators are chlorophyll (the photosynthetic pigment) and carbon, so that either a turnover rate (h^{-1}) for the particulate C pool is obtained or a rate per unit chlorophyll (mg C or O$_2$ mg Chl^{-1} h^{-1}). Ratios of C:Chl (commonly 50–80) are used to convert from one biomass estimate to another. As we shall see these ratios are by no means constant and may, in fact, be used in themselves to estimate relative growth rates. The use of constant conversion ratios may be a significant source of error. For the purposes of this discussion, therefore, I

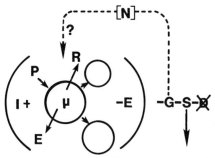

Fig. 6.2 A summary of the production and loss terms in bottle incubations. Within the bottles (brackets) immigration (I) and emigration (E) are contained while the cells may grow (μ). Photosynthesis (P), respiration (R) and excretion (E) are important aspects of cellular physiology. The balance of production may be lost through grazing (G), sedimentation (S) and death *in situ* (D). The process of grazing may lead to nutrient regeneration (N).

will define net and gross productivity as rates measured by the flux of elements, usually in bottle experiments. Such measurements produce estimates of 'specific growth' rates. 'Growth' will be defined as before as the increase in biomass as measured by cell counts, particulate C or some such. Relative growth rates are defined as rates of growth divided by the maximum growth rates of the organism.

Organisms which 'grow' by cell division and an increase in cell number rather than by an increase in body size represent a special case of a more general ecological relationship (Fig. 6.2). If we assume no significant changes in cell size in the population (a rash assumption that will suffice for the present) and we eliminate the mortality terms then the net productivity may be directly equated with the gross 'growth' in abundance. Population changes may be measured by changes in cell number, changes in particulate carbon or other elements, or changes in chlorophyll. Only at steady state will all the biomass estimates yield equal estimates of growth, as under fluctuating conditions of growth the elemental ratios of the cellular constituents vary as the cells cope with, and adapt to, the external conditions. In almost all cases the observed changes in the biomass of natural populations represent something less than net productivity ('net growth') as losses due to mortality and the effects of immigration and emigration cannot be estimated. Because net productivity measured by a C or O flux in a bottle does not usually include the mortality terms of Fig. 6.2, there may be a considerable discrepancy between the growth rate estimates so obtained and those obtained from changes in biomass or from population counts. This is why 'specific growth' as measured by bottle techniques rarely can be equated with 'growth' in the natural population.

Net and gross productivity are therefore defined herein as rates measured by the flux of elements, usually in bottle experiments. Specific growth rates are obtained from the turnover of the particulate pools of C, N or P whereas growth is a function of an increase in biomass. Relative growth rates are derived from the ratio of the observed growth rate divided by the maximum growth rate of the species or clone in question. As in the case of the turnover of the pools of DIP and DIN and the relationship between isotope studies and whole basin studies, there is a complex relationship between the flux of nutrients as measured by isotopes in bottles and the growth of biomass at seasonal scales. The precise relationship, as before, depends to a large extent on the degree of recycling ('new' vs. 'regenerated' production, Fig. 6.1, Eppley and Peterson, 1979) and precisely which terms of Fig. 6.2 are included and which are not. The interpretation of the results concerns questions of time scales of measurement and process. As in the case of the soluble pools, the flux rates over short periods may not be in steady state and the assumption of balanced growth will not hold. As we shall see this is not just a methodological problem, it is a highly significant feature of the ecology of the organisms.

Presumably we are interested in the measurement of productivity and growth for a number of reasons: firstly we are interested in absolute rate measurements for the comparison of lakes and oceanic areas and for the assessment of trophic state. Secondly, we require such measurements to improve our understanding of the processing of materials and of the cycling of energy in planktonic ecosystems. Thirdly, we are interested in the population dynamics and growth rates of the phytoplankton populations. These different interests require different techniques and different levels of sophistication in measurement. Perhaps one of the reasons for the current active debate about growth rates and methods of measurement, lies in the fact that an enormous number of uncritical bottle experiments have been performed over the years and that we now have a wealth of data that is consequently difficult to interpret. It is easier to measure the rate of ^{14}C, ^{15}N or ^{32}P uptake in a bottle than it is to decide what the result means. The oxygen and ^{14}C methods have been plagued by a number of problems with the methodology and with problems of interpreting the results in terms of the underlying physiological processes in the cells. It is not surprising that the results from the different methods are difficult to interpret. For example the inability of the ^{14}C method to measure respiration rates places a considerable constraint on the use of the method to measure community metabolism (Harris, 1978).

There have been a number of recent reviews of the methodology and of the interpretation of the results of productivity measurements (Berman and Eppley, 1974; Harris, 1978, 1980a,b; Knoechel and Kalff, 1978; Morris, 1980a, 1981; Peterson, 1980; Falkowski, 1980; Platt, 1981). Indeed the flurry of recent reviews and books is some indication of the speed of the development of ideas. I believe that our realization that something was wrong with our productivity measurements stemmed from four basic discoveries: (1) that most natural phytoplankton systems are not at steady state, at least at the time scales covered by bottle incubations (Harris, 1980a); (2) that the phytoplankton are integrating machines which track environmental fluctuations and are beautifully adapted to growth in fluctuating environments; (3) that growth is not just photosynthesis and (4) that photosynthesis is not just C-fixation or the release of oxygen. When added to the problems with the methods these four discoveries add up to a revolution in thought.

Cells growing at rates close to their maximum relative rates take up the major nutrient ions in a fixed ratio, the Redfield ratio (Goldman *et al.*, 1979). This is the C:N:P ratio for rapid, balanced growth and is usually assumed to be 106 C:16 N:1 P by atoms. Thus, if we assume balanced growth and no mortality, any measurement of the rate of uptake of C, N, or P should give a rate estimate similar to the rate of growth. In productivity measurements the flux of oxygen and carbon will again be balanced at steady state such that the photosynthetic quotient will be approximately 1.2 (Harris, 1978). Oxygen is evolved as one of the first steps in photosynthesis, whereas carbon fixation is a later event, an event which uses reducing power and energy derived from the electron transport system when exposed to light. Nitrogen is required for the synthesis of protein and cannot be utilized without the supply of C skeletons to complete the molecules. Phosphorus is an essential component of the energy metabolism of cells which involves the incorporation of DIP into ATP. Balanced growth combines the essential elements into cell metabolites of a variety of types but the overall stoichiometry remains constant.

One way to make the following discussion clearer is to distinguish between the energy and nutrient requirements of cells growing in the mixed layer. Phytoplankton obtain their energy from sunlight through photosynthesis and their nutrients from the water. Nutrient uptake requires energy, as does the subsequent metabolism of that nutrient to form macromolecules. Carbohydrates and lipids may be thought of as a form of energy storage in that ATP and reducing power may be released by the catabolism of these molecules. P and N are required for energy transfer and for the anabolic and catabolic activities within the cell. Thus stored carbohydrates may be thought of as energy stores which, together with stored N and P, are the main 'bank balance' of the cell. To continue the 'bank balance' analogy, the 'accounts' for stored nutrients and energy may fluctuate but if the cell is tracking the external fluctuations in light and nutrient supply efficiently the rate of growth should be unaffected. Growth may continue in the absence of external nutrients as the internal stores are depleted and if successful the organism should not go 'broke' too often. Biological 'overdrafts' are also possible, but only for short periods, as it is possible for the organism to keep growing by shrinking. This cannot go on for long as there is a minimum cell quota for each essential element. Once that minimum cell quota is reached, cell division and growth must cease. Under non-steady state conditions the uptake and storage of the major nutrients becomes a highly complex temporal sequence of events controlled by the integrative nature of cellular physiology and the priorities placed on the synthesis of cellular constituents.

Under non-steady state conditions the methods used must be able to resolve the true time scales of the biological processes (Harris, 1980a,b). The problems with the methods and the interpretation of data are concerned with time scales and the technology of detecting small changes in the concentrations of C, N and P in dilute solutions (Talling, 1971). I have already shown that the concentrations and turnover times of the soluble nutrient pools differ greatly in waters of differing trophic state. Thus the methodological problems differ in different water types. Anyone working in the field will know that the interpretation of measurements from natural populations is by no means a simple matter. If anything, it has been the application of advances in the areas of physics and physiology that has led to the biggest advances in the interpretation of productivity measurements: the basic methodology has remained unchanged since the early pioneers. Perhaps the greatest advances have been conceptual, with the realization that the surface mixed layer is not a constant environment and that the physiological processes of phytoplankton are adapted to a fluctuating environment. The time scales of the relevant processes in the mixed layer have turned out to be faster than we thought (Harris, 1980a).

6.1 The interpretation of kinetic measurements

The continuous culture work has produced a considerable body of information on the physiology and growth rates of phytoplankton under conditions of constant dilution, nutrient supply, light and temperature (Rhee, 1979) and much of the continuous culture work has been extrapolated to field conditions. Continuous culture work produces estimates of the half saturation constant for growth, whereas the measurement of nutrient uptake estimates a half saturation constant for uptake. The half saturation constant for uptake, or transport throughout the wall (K_t), measures the affinity of the alga for the nutrient or substance in question (Fig. 6.3). Thus the measurement of the flux of nutrients into the cells will only measure the growth rate at steady state. The measurement of nutrient fluxes into cells

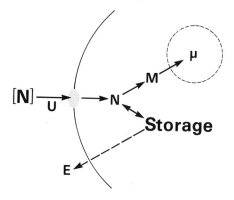

Fig. 6.3 The terms of the nutrient (N) dynamics of an algal cell. Nutrients are taken up (U) actively, and, once within the cell, may be metabolized (M), stored, excreted (E) or used for growth (μ).

has been an active area of research in recent years as the introduction of isotopes has made such measurements apparently very simple.

I believe that such measurements require care in their interpretation. The elemental fluxes of elements into algal cells are complex, even at steady state. A recent paper by Smith and Horner (1981) points out that improper or inadequate analysis has often been used by researchers studying multi-compartmented systems. They even go as far as to say '. . . marine bioscientists are largely ignorant of the body of theory required to successfully employ radioisotopes. . .' So what is the problem? The problem lies in the fact that when isotopes of the major nutrients are added to a suspension of algal cells the isotope is taken up by the cells and incorporated into a number of internal pools and pathways. Thus the specific activity (the fraction of the particular pool which is labelled) of these internal pools will change over time, even at steady state, until the specific activity of all the internal pools is the same as the specific activity of the external nutrient supply. If there is exchange between internal pools of the nutrient, or exchange between internal and external pools, then the labelling patterns over time may be highly complex (Smith and Horner, 1981).

6.1.1 ^{14}C UPTAKE

As an example, I will use the theory of ^{14}C uptake in photosynthesis measurements, as this is one of the commonest types of measurement carried out. It should be remembered that these arguments apply to all isotope uptake experiments, although the specific assumptions will differ from element to element. Li and Goldman (1981) phrased the arguments as follows. As already noted, at steady state, the specific growth rate and the rates of uptake measured by the fluxes of the individual elements are equivalent. Thus an estimate of growth from the dilution rate of the culture should equal the specific growth rate as measured by the uptake of carbon. Li and Goldman (1981) used μ for the growth rate estimated by the dilution rate of the culture and μ' for the growth rate estimated by the uptake of carbon. If the rate of carbon uptake is constant over the incubation period then:

$$\mu' = \frac{f}{a\text{CC}} \frac{d^{14}C}{dt} \qquad\qquad 6.1$$

where f is the isotope discrimination factor (usually assumed to be 1.05 for ^{14}C), CC is the total cellular carbon, $d^{14}C/dt$ is the net flux of labelled cellular carbon, and a is the specific activity of the ^{14}C in the aqueous medium surrounding the cells. The mass balance of unlabelled C is:

$$dC/dt = P - R \qquad\qquad 6.2$$

where P is the photosynthetic flux of carbon into the cells and R is the outward flux of carbon due to respiration and excretion. Both P and R must be constant in time. Similarly the mass balance of labelled carbon is:

$$d^{14}C/dt = (aP - bR)/f \qquad\qquad 6.3$$

where b is the specific activity of the ^{14}C in the biomass. From the definitions of μ and μ' (above) and setting R equal to R' (a specific loss rate), these equations reduce to a useful relationship, if it is assumed that recently fixed carbon is immediately available for respiration and that such 'hot' compounds are respired at the same rate as 'cold' carbon. If such be true then:

$$\mu' = \mu + R'(1 - b/a) \qquad\qquad 6.4$$

For this to be true then $d^{14}C/dt$ must be constant in time and the time of the incubation must be sufficiently short that a and the total cellular carbon do not change significantly. Over such time periods b will be much smaller than a, implying complete isotopic disequilibrium.

Thus at steady state the ^{14}C method will measure gross photosynthesis over short time periods and net photosynthesis over time periods long enough to achieve isotopic equilibrium (Hobson et al., 1976). Dring and Jewson (1982) have recently extended this analysis and modelled the time course of ^{14}C incorporation into cells. Their analysis was based on the fact that the average ^{14}C incubation lasts at least four hours because of the requirement of obtaining a sufficient number of counts in the algae. Their analysis showed that over a period of four hours the ^{14}C uptake rate will not be a simple function of either gross or net photosynthesis but will vary in a complex manner depending on the $P:R$ ratio of the cells. This discrepancy between the estimated rate of photosynthesis (from ^{14}C uptake) and the true rate of photosynthesis becomes greater when the $P:R$ ratio is almost unity. The variation in the ^{14}C estimate of photosynthesis as the irradiance is reduced (as P tends to zero) means that the method cannot be used to estimate R by extrapolation to zero irradiance (cf. Harris, 1978). This is an important drawback of the method that should be more widely appreciated. Also it means that it is entirely possible to obtain positive ^{14}C uptake rates when the overall balance of ^{12}C is negative. Even at steady state the incorporation of a pulse of ^{14}C into cells undergoing balanced growth will exhibit some complex temporal behaviour depending on the pool sizes of photosynthetic intermediates and the rates of incorporation into the pathways of P and R (Smith and Horner, 1981). Rates of equilibrium between internal pools and between internal and external specific activities are crucial.

Many species of phytoplankton clearly have a bicarbonate pump in the cell wall (Raven and Beardall, 1981) which, together with the activity of carbonic anhydrase, serves to maintain a high internal concentration of CO_2. The cellular mechanisms of C uptake and transport interact with the speciation of DIC in the water and this interaction has a number of important physiological, methodological and ecological implications. It may, for instance, result in wide temporal disparities in the uptake of carbon and the release of oxygen (Gallegos *et al.*, 1983) and the bicarbonate pump is implicated in the operation of the active C4 pathway (Morris, 1981). Biochemical differences between species influence the ecological requirements of phytoplankton and may determine the nature of the phytoplankton assemblage in different waters (Talling, 1976; Jaworski *et al.*, 1981). Most of the carbon influx into cells results from the activity of photoautotrophic pathways although dark C uptake is important in some circumstances. The equations derived above state the necessary assumptions for the use of the ^{14}C method. The problems with the method lie in the incubation periods required (especially in oligotrophic waters), in bottle effects (Harris, 1978, 1980a,b) and in the assumption that all the terms of Fig. 6.2 are unchanged in time. It is also necessary to assume that ^{14}C uptake is linear with time and there is no rapid loss of recent photosynthate, either as excreted carbohydrate or as CO_2. This latter assumption appears to hold only at high relative growth rates (Harris and Piccinin, 1983). Rapidly growing cells appear to satisfy the requirements of the necessary assumptions (Dring and Jewson, 1982), but at low growth rates the release of recent photosynthate is very rapid (within a few minutes, Harris and Piccinin, 1983). Analysis of models and experimental data indicate that, if cells are growing rapidly, then ^{14}C is not made available to respiratory pathways within the cell until after a period of time in the dark (Bidwell,

1977; Harris and Piccinin, 1983), and *b* in Equations 6.3 and 6.4 is effectively zero. Thus ^{14}C uptake over periods of a few hours in the light is a measure of gross photosynthesis as long as the cells are growing at rates close to μ_{max}. At low relative growth rates ^{14}C rapidly appears in respiratory pathways, *b* is similar to *a* (Equations 6.3 and 6.4) and the method measures net photosynthesis. With a mix of species in natural assemblages growing at a variety of relative rates it is hardly surprising that the predominant result of *in situ* incubations is something between net and gross (Harris, 1978). In order to correctly interpret the result from a four-hour *in situ* incubation it is necessary to understand the physiology of the organisms rather well and to know how the labelled C is partitioned between the various intracellular pools.

6.1.2 ^{15}N UPTAKE

As the concentration of DIN in surface waters is frequently practically unmeasurable (especially in the oligotrophic ocean) the supply to the phytoplankton is frequently difficult to measure also (Goldman *et al.*, 1979; McCarthy and Goldman, 1979; McCarthy, 1981). The true rate of DIN influx into the cells is therefore rather difficult to quantify (McCarthy, 1981). The problem with the use of ^{15}N lies in the sensitivity of the assay. In order to produce a measurable effect either a large addition of ^{15}N or long incubations are required so that mass spectrometry or plasma emission techniques may be used. ^{15}N is not available as carrier-free isotope so some 'cold' N has to be added as well. McCarthy (1981) has discussed the methodological problems at length but basically they are similar to the ^{14}C method in that incubation times and bottle effects are important. Depletion of substrate during the incubation, linearity of uptake over time, overestimation of the background DIN concentration and the validity of the Monod uptake relationship are also topics of concern.

Numerous cases of uptake curves that do not fit the Monod relationship can be found in the literature (Murphy, 1980; McCarthy, 1981). There is good evidence (Collos and Slawyk, 1980) that the Monod parameters (V_{max} in particular) are dependent on the cell quota of N in a non-linear manner if, as is usual, an enhanced level of DIN is added to the experimental bottle. In freshwater, for example, additions of up to $100\ \mu g\ l^{-1}$ NH_4^+ and $500\ \mu g\ l^{-1}$ NO_3^- have been used (Murphy, 1980). A further problem with the DIN uptake experiments lies in the different properties of NH_4^+ and NO_3^-. NH_4^+ is the preferred substrate for many algae in that NO_3^- may not be taken up at all if more than $1\ \mu g\ l^{-1}$ NH_4^+ is present. The precise mechanism of the suppression of NO_3^- uptake is unclear (Syrett, 1981; McCarthy, 1981). The metabolism of NH_4^+ and NO_3^- is quite different in that NO_3^- requires reduction before metabolism. Cells taking up and metabolizing NO_3^- are required to exclude OH^- from the cells to maintain an internal pH balance, whereas cells metabolizing NH_4^+ produce one H^+ per NH_4^+ ion fixed. Both N species require C skeletons for fixation so an interaction with C-fixation is to be expected. All DIN species also have specific transport mechanisms built into the cell walls (Syrett, 1981) so that algae show a high affinity for substrate and rapid uptake. The transport mechanisms use ATP as an energy source and may take the form of proton linked cotransporters (Syrett, 1981). There is therefore a strong interaction between the energy metabolism of cells and the uptake and metabolism of C and N. Cells deficient in energy (light-limited) may not have sufficient energy to complete the reduction of NO_3^- to NH_4^+ within the cells and may thus accumulate NO_2^-. This is the explanation for the deep ocean NO_2^- maximum (French *et al.*, 1983) as the NO_2^- is released into the water. There is a spatial coincidence between the deep chlorophyll maximum and the NO_2^- maximum and the deep chlorophyll maximum occurs at or below the 1% light level. There is also a diel fluctuation in the NO_2^- concentration.

6.1.3 ^{32}P UPTAKE

P uptake experiments are made relatively easy by the availability of carrier-free ^{32}P. This isotope is highly radioactive and, while the half life is short, it is not inconveniently so. The high activity of the emitted radiation means that the samples may be counted by Cerenkov-counting and no fluor is necessary. As ^{32}P is available carrier-free, little or no perturbation of the external DIP concentration is necessary when uptake measurements are made. This is a useful feature as the external DIP concentration is extremely small in oligotrophic waters. DIP uptake is mediated by an ATP driven carrier in the cell wall and phytoplankton show both high affinity and rapid rates of uptake even at low DIP concentrations. All the evidence points to the fact that the phytoplankton are adapted to the patterns of DIP availability that normally occur and it is difficult to demonstrate P limitation of algal growth in natural waters (Peterson *et al.*, 1974; Serruya and Berman, 1975; Donaghay *et al.*, 1978; Nalewajko *et al.*, 1981; Smith and Kalff, 1981). There is a clear effect of cell size on ^{32}P uptake kinetics as small cells ($< 2\mu$) have the highest affinities for ^{32}P and the most rapid rates of uptake whereas the larger cells ($> 35\mu$) have lower affinities and slower rates of uptake (Lean and White, 1983). The relationships between TP and algal biomass so elegantly demonstrated by Vollenweider (Vollenweider, 1968, 1975, 1976; Janus and Vollenweider, 1982) must mean that there is some structuring of biomass and food chain dynamics by P availability.

The problems with the measurement of DIP uptake are similar to the problems with the measurement of other elemental fluxes. These problems include bottle effects on the mortality terms of Fig. 6.2, depletion of substrate and the non-linearity of uptake rates over time, estimation of the external DIP concentration and the influence of cell quota on the rate of uptake. Phosphorus storage is widespread in algae so that during periods of deficiency the 'bank balance' of stored polyphosphate is run down and, if the cells are presented with an external supply of DIP they may take up many times the amount of DIP which would be predicted from the C:P ratio of balanced growth (Lean and Pick, 1981). Subsequently, growth may continue for many generations without requirement for external DIP (Mackereth, 1953). Lean (personal communication) has noted that phytoplankton may take up sufficient DIP in one hour to last at least two weeks. As the process of DIP uptake requires ATP an interaction between the uptake of DIC, DIN and DIP is to be expected.

6.2 The integration of metabolic pathways

6.2.1 C AND N METABOLISM

Within the algal cell the major pathways of metabolism of the major nutrients are interconnected by virtue of the fact that macromolecular synthesis requires both C, N and P for structural components and a source of energy. Figure 6.4 gives some idea of the complexity of the pathways involved; even so, only the major pathways are outlined in diagrammatic form. Figure 6.4 includes only those interactions relevant to the measurement of productivity and growth by the most commonly employed methods. With the sole exception of CO_2, the uptake of all the major nutrient ions is controlled by specific uptake mechanisms. Cellular storage pools of C, N and P are maintained largely in the forms of HCO_3^-, NO_3^- and poly-PO_4. The metabolism of C and N is primarily linked by the pathways of amino acid synthesis in the chloroplast and the mitochondrion. In the light, most of the reducing power comes from electron transport in the chloroplast so that most of the amino acid synthesis takes place by that route. There is also a strong connection between the pathways of photorespiration and N metabolism as the pathway to glycolate involves the metabolism of serine and glycine (Harris, 1978).

An interaction between the pathways of C and N metabolism also occurs in the mitrochondrion. The activity of the tricarboxylic acid cycle (TCA cycle) in the mitochondrion serves to manufacture C skeletons for amino acid synthesis as well as being involved in the catabolism of carbohydrates and lipids and the generation of ATP. Thus it can be shown that if N is limiting, the activities of the TCA cycle are depressed, and if NH_4^+ is added to N-limited cells in the dark, a rapid fixation of DIC takes place through the TCA cycle intermediates (Yentsch et al., 1977; Morris et al., 1971a,b; Harris, 1978). A double isotope study of C and N uptake in the light and in

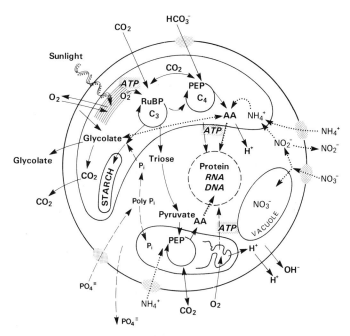

Fig. 6.4 A simplified version of some of the important pathways of C, N and P within a cell. Solid lines C; dots N; broken lines P. P_i, inorganic phosphate; AA, amino acids; PEP, phospho-enol pyruvate; RuBP, ribulose bis-phosphate. Active uptake sites in the wall dotted. Note the difficulty of determining the rate of growth from the uptake of C, N or P and the complexity of the overall balance of such major nutrients.

the dark can therefore be used to study the energy and nutrient requirements of natural populations.

Results such as this make it very difficult to interpret the results of *in situ* incubations with added nutrients. The addition of N may cause the rate of DIC uptake to decrease rather than increase. It was always assumed that the addition of nutrients to *in situ* incubation bottles would be a way to determine if the populations were nutrient limited – if the rate of DIC fixation increased after nutrient addition then the population clearly needed the added nutrient. This is a simplistic assumption (Harris, 1978; Lean and Pick, 1981) if the incubation time is of the order of a few hours, or of a time similar to the time required for nutrient uptake and metabolism. If the incubation is left long enough for growth effects to occur then the result would be quite different. This is but one example of the interaction of the time scales of methods, of the environment and of the physiology of the organisms. Non-steady state conditions in the environment, where DIN is supplied in the form of discrete pulses in time, can therefore lead to some problems of interpretation with DIC uptake data as the fluctuations in the internal storage pool of DIN will cause alterations in the metabolic pathways of C. Remember that such phenomena cannot be studied under steady state conditions as the uptake and storage terms of Fig. 6.3 disappear.

Even if the rate of DIC fixation goes down when DIN is added, it really does not give very much information about the limitation of growth by DIN. It merely indicates that the internal pool of DIN was run down at the time of the experiment and that the priority for uptake and metabolism was temporarily in favour of DIN rather than DIC. That merely indicates that the last pulse of DIN was received some time ago. If the population is capable of tracking the pulses of DIN accurately (and we shall see later that this is so) then using the 'bank account' analogy, no effect on growth is to be expected. Similarly, the lack of a DIN effect may indicate that the last pulse of DIN was received recently. It cannot be taken as evidence of a constant supply of DIN in sufficient amounts for growth. The whole picture is further complicated by the presence of a distribution of pulses of DIN over time (as discussed in Chapters 4 and 5) and a spectrum of different sizes of cells in the natural assemblage having different cell quotas, storage capacities and growth rates. It is hardly surprising that the literature documents differing results and opinions.

6.2.2 C AND P METABOLISM

The interactions between C and P metabolism are equally strong and complex. DIP uptake requires ATP so that if P limited cells are given an external supply of DIP then DIC uptake and fixation is temporarily suppressed while energy is used to take up the required DIP (Lean and Pick, 1981). Remember the 'bank account' analogy; if the account labelled poly-PO_4 is run down the cell has the option to preferentially restock that account by taking energy from the carbohydrate energy bank. Such effects will not be seen at steady state. The same restrictions on the interpretation of results apply. Harris and Piccinin (1983) found that populations grown under P limitation in continuous culture exhibited rapid loss of ^{14}C through dark respiratory pathways. This would indicate that cells growing slowly under P limitation rapidly transfer ^{14}C from the chloroplast to the cytoplasm and preferentially respire the recent products of photosynthesis. This can only be brought about by metabolic channelling of recent photosynthate to the mitochondria as the specific activity in the respired gas stream was high. This, of course, destroys most of the assumptions stated above regarding the relationships between the internal and external specific activities in ^{14}C uptake experiments. If the pool size of recent photosynthate is small then isotope equilibrium may be reached in minutes rather than hours and the ^{14}C method will give an almost instant measure of net photosynthesis so destroying the model assumptions of Dring and Jewson (1982) and Li and Goldman (1981). Significantly, rapidly growing, P sufficient, cells did not show the effect.

These considerations of metabolic pathways and the effects of N and P limitation raise some interesting general questions about the effects of nutrient limitation in organisms. The effects of fluctuating environments and fluctuating internal pools of stored nutrients are as often expressed in terms of deflected metabolism as they are in terms of a direct overall reduction in growth. Internal flexibility in pathways is as important as an adaptation to fluctuating nutrient supplies as is the presence of active uptake mechanisms and storage pools. Phytoplankton are clearly able to track fluctuations in the external environment and to buffer the effects of such fluctuations so that growth rates are little affected. One can argue that the steady state descriptions of K_s and V_{max} in different species and their use in theoretical models is only a partial description of reality, and that we may never be able to account for the diversity of species in natural communities by the use of such models. Harris' first law of ecology is that the organisms are much better adapted to their environment than we think. The corollary to this is that if an organism possesses an attribute that does not, at first, appear to be necessary, it is our understanding of the environment that is at fault not the organism.

6.3 Time scales and the effects of bottle containment

As Harris and Piccinin (1977) demonstrated, much of the C fixed by photosynthesis may be fixed in the first few minutes of the incubation period (particularly in the case of severe photoinhibition) so that the longer the incubation period the lower the apparent rate of photosynthesis. In the case of nutrient uptake experiments (particularly ^{15}N) the addition of 'cold' N along with the isotope may lead to rapid initial uptake with a consequent non linearity in the rate of uptake over time (Goldman *et al.*, 1981). In any case the longer the incubation the lower the measured rate, as the initial rate of fixation or uptake is a 'gross' rate or a measurement of the rate of transport of the element into the cell. Such uptake is controlled by transport processes (K_t and V_{max}) not growth processes (K_s and μ). Short term incubations also exaggerate the problem of isotopic equilibrium and consequently lead to high estimates of C, N and P uptake. Short term (minutes to hours) incubations measure gross photosynthesis and the replenishment of nutrient storage pools. Such measurements may be greatly divorced from the real population growth rate. The decline in the rate of C, N or P uptake over time during incubation means that kinetic data based on lengthy incubations will tend to underestimate such parameters as V_{max}. What is lengthy, evidently depends on the time scales and the growth rates of the organisms involved but 4 h incubations are evidently too long for most physiological investigations (Harris and Piccinin, 1977; Goldman *et al.*, 1981). Long term (hours to days) incubations should not be used to measure physiological processes (Harris, 1980a,b) as the time scale of the incubations is much greater than the time scale of the processes under study.

Short incubations should therefore be expected to produce higher rate estimates than long incubations (Vollenweider and Nauwerck, 1961). The decline in the measured rate over time depends on physiological processes (Harris, 1978; Goldman *et al.*, 1981); on the attainment of isotopic equilibrium (Buckingham *et al.*, 1975; Hobson *et al.*, 1976; Dring and Jewson, 1982; Harris and Piccinin, 1983; Li and Harrison, 1982); on respiratory and other losses (Eppley and Sharp, 1975; Marra *et al.*, 1981); on the death of cells, particularly delicate flagellates, in the bottles (Venrick *et al.*, 1977; Gieskes *et al.*, 1979); and on the distribution of the isotope or element into the food chain within the bottle (Jackson, 1983). Bottle size may have some effect on the outcome of the experiment as the use of smaller bottles may exclude larger grazing animals and may contribute to a greater than normal death rate (Gieskes *et al.*, 1979). There is no general agreement on the effect of bottle size however; in some experiments it had little effect on the outcome.

Comparisons of sets of short and long incubations are few but in the case of ^{14}C uptake (Rodhe, 1958; Vollenweider and Nauwerck, 1961; Hammer *et al.*, 1975; Gieskes *et al.*, 1979) the series of short term incubations gave results which varied from more or less the same as the long term incubations (Hammer *et al.*, 1973; Goldman and Dennett, 1983), to 5–15 times greater (Gieskes *et al.*, 1979). All available evidence points to a similar situation with ^{15}N uptake experiments (Goldman *et al.*, 1981). A pattern may be discerned however. The short and long term incubations appear to give similar estimates in the case of summer populations and blooms in eutrophic waters where the algae are robust and the rate of C uptake is linear over time. In the case of summer blooms in eutrophic waters, the physiological component of ^{14}C uptake (photoinhibition) is not important (Harris, 1978), the death rate in the bottles is not high, and the $P:R$ ratio exceeds 10. In oligotrophic waters, or in situations where photoinhibition is important (deep mixed populations in spring and autumn), the discrepancy is greater and short term incubations give much higher estimates.

Peterson (1980) reviewed the use of ^{14}C as a measure of photosynthetic activity and growth and showed that the interpretation of kinetic measurements was not easy. The relationship between short (2–4h) and long (24 h) incubations depends on the factors listed above but both Eppley and

Sharp (1975) and Peterson (1980) identified respiration as an important loss process. From the point of view of isotopic equilibrium the 2–4 h incubation must measure something near to gross photosynthesis and the 24 h incubation must be close to net photosynthesis (Dring and Jewson, 1982). The $P:R$ ratio of the populations is all important. If, in oligotrophic waters, the ratio of the 4 h:24 h ^{14}C uptake is high, the most parsimonious explanation is low-growth rates, low $P:R$ ratios and high proportional respiratory losses. Peterson (1978) showed that the 2 h:24 h ratio of ^{14}C uptake was about 5 at a growth rate of 0.1 d^{-1} in culture. The ratio fell towards unity as the growth rate increased and as the $P:R$ ratio rose (Fig. 6.5). Eppley and Sharp (1975) expressed the same result in different terms and also concluded that the growth rate of phytoplankton was low in oligotrophic waters. This is the standard equilibrium explanation.

Other loss processes will obviously contribute to losses of particulate ^{14}C in natural populations so the 2 h:24 h ratio may not be the same in cultured as in natural populations. Some recent observations in the Lund Tubes of Blelham Tarn follow the same trend in that the 4 h:24 h discrepancy is greatest at low biomass (Fig. 6.5). A comparison of the data from natural populations with that in culture indicated that the loss rates from natural populations exceeded the loss rates from cultures but the trend was the same. Reynolds *et al.* (1985) decided that it was possible to account for the losses of ^{14}C by invoking low $P:R$ ratios in the slow growing populations. As the losses were larger than those predicted solely from changing $P:R$ ratios in culture this result does not rule out the possibility of losses from nanoplankton which were not included in the microscope counts. Over a range of growth rates in the Lund Tubes it was clear that ^{14}C uptake heavily overestimated the actual growth rate of the slow growing populations (Fig. 6.6). This was explained by Reynolds *et al.* (1985) as the dark loss of ^{14}C by respiration, and hence by the labelling of a labile C pool in the light that was not reflecting the overall change of C in the cells due to growth. On one occasion the actual rate of increase of cell number in the larger species counted was only 3 % of the rate of ^{14}C uptake.

Waters of differing trophic state have different proportions of small and large species and the pattern of C flux in different waters is also different. Eutrophic waters with high biomass show a 4 h:24 h ratio of almost unity (Fig. 6.5) indicating high growth rates and high $P:R$ ratios as well as low losses due to grazing and mortality. In oligotrophic waters, rapid turnover of small phytoplankton may be observed (Fig. 6.7) along with a similar relationship between ^{14}C uptake and growth in cell number as that observed in the Lund Tubes. The measurement of growth in different size classes however (Fig. 6.7) indicates that the ^{14}C measurement is a gross average of very complex dynamics which are different in different size classes.

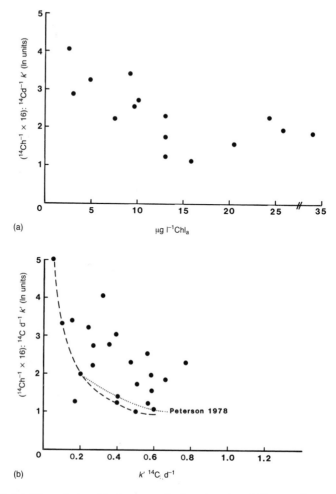

Fig. 6.5 (a) The relationship between the hourly rate of ^{14}C uptake multiplied by the daylength and the daily rate of uptake measured by 24 h incubations. Data from Blelham Tarn, English Lake District. At low biomass nightime losses of label ensure that short term incubations give higher rates than long term incubations.

(b) The relationship between the ratio of short term to long term incubations as a function of growth rate. The dotted lines are the model of Peterson (1978) who assumed that the discrepancy between the two incubations was due to respiratory losses dependent on growth rate. Clearly higher losses are incurred in field populations.

In oligotrophic waters the low biomass means that longer incubations are necessary. This conflicts with the requirement that the incubation be kept short in order to reduce mortality. The discrepancy between the short and

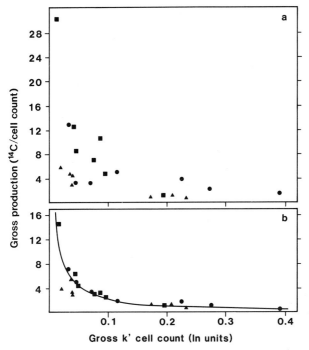

Fig. 6.6 The relationships between the ratio of gross production as determined by ^{14}C uptake and that determined by counting cells, and the actual rate of increase in cell number. Data from the Lund Tubes in Blelham tarn. (Reynolds *et al.*, 1985.)

long incubations is large and there is considerable uncertainty in the estimates of most of the terms in the relationships between ^{14}C uptake and growth (Eppley, 1980). There is a distinct possibility that in oligotrophic waters the predominance of nanoplankton leads to a pattern of C flux that looks like low growth rates and low *P:R* ratios when in fact the growth rates are high and the losses are really due to regeneration of ^{14}C from grazing and mortality. A simple comparison of 4 h and 24 h incubations will not distinguish between the two explanations because, if production and regeneration are tightly coupled in oligotrophic waters, the apparent patterns of ^{14}C flux will be indistinguishable (Peterson, 1980). It will all depend on how much of the labelled C is released from autotrophs and heterotrophs and over what time periods.

Discrepancies between short and long incubations can arise from cell death and mortality, high growth and grazing or low growth and low *P:R* ratios. Little work has been done to distinguish between these explanations. Cell death and mortality in the bottles is easy to check by looking at biomass (as chlorophyll, Gieskes *et al.*, 1979) or cell counts (Venrick *et al.*, 1977)

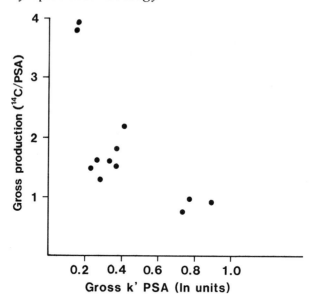

Fig. 6.7 The ratio of gross production as measured by ^{14}C uptake to that measured by particle size analysis (PSA) as a function of the actual rate of increase of particle volume. Data from experiments in the Tasman Sea.

during the incubation. Few people have done this. The last two explanations are much more difficult to separate; in order to decide which explanation is correct other evidence must be sought. The respiratory flux from the phytoplankton cannot be distinguished from the respiratory flux of the microheterotrophs which may have consumed labelled cells (Peterson, 1980). I suspect that the true answer may be a bit of both, in that different water types may show differences in emphasis depending on the populations concerned. In freshwater (Figs 6.5, 6.6, Reynolds *et al.*, 1985) it may well be that respiratory losses, low growth rates and low *P:R* ratios are the norm. In the oceans however, there is evidence for the rapid turnover of small species thus implicating grazing and mortality (Fig. 6.7). Again it is probably a case of not 'whether' but 'how often' each explanation applies. I will examine alternative evidence for rapid growth in oligotrophic waters in the next chapter.

6.4 From kinetics to growth rates?

Given the problems with the interpretation of kinetic measurements the comparison of specific growth rate measurements with actual, observed growth rates is fraught with problems. Few such comparisons have actually been made but those that have illustrate the problems well. Early on it was realized that significant regeneration of C, N and P occurred *in situ* and that estimates of growth from kinetic measurements exceeded the rate of biomass accumulation (Ketchum *et al.*, 1958a). For example Gibson *et al.* (1971) noted that the gross production of phytoplankton in Lough Neagh (as measured by the Winkler light and dark bottle method) exceeded the net production of biomass by a factor of three or four. The existence of high gross productivity estimates on occasions when the DIP concentration was low led these authors to conclude that significant regeneration of DIP occurred *in situ*. Estimates of gross productivity and of net productivity from the disappearance of C, P and Si all indicated similar discrepancies between the two measures and pointed to significant losses from gross productivity *in situ*.

Eppley (1972) discussed the conversion of isotope uptake data into growth rates (C or N specific growth rates) and noted that one of the major problems lies in the determination of algal biomass as C or N. The C or N specific growth rate k may be defined as:

$$k = 1/t \ln(C + \Delta C/C) \qquad 6.5$$

(and similarly in terms of N) where t is the incubation time, ΔC the C uptake during the incubation period and C the particulate C in the algal biomass. In such a relationship it is necessary to know C with some accuracy. The value may be estimated from microscope counts, from measurements of particulate organic carbon (POC) or from a carbon:chlorophyll (C:Chl) ratio. The commonest method is the use of a C:Chl ratio (Eppley, 1972). The equation then relates the assimilation number ($\Delta C \, Chl^{-1} \, t^{-1}$) directly to estimates of k. The commonest C:Chl ratios used lie in the range of 30–80 with many values quoted around 50. The use of a single number is dangerous as the ratio is both a function of environmental conditions and growth rate. The presence of detritus also interferes with the measured ratio in natural populations (cf. Chapter 7 below).

Li and Goldman (1981) and Harris and Piccinin (1983) compared the steady state growth rate of cells in continuous culture with the C specific growth rate measured under identical conditions. Even at steady state the two measures often differed by 100 % due, presumably, to handling and the physiological sensitivity of the populations. Small wonder therefore that Eppley (1981) concluded that, with all the associated errors of field measurements, there was an order of magnitude uncertainty in the measurement of the growth rate of natural populations using isotope kinetic methods.

In natural populations there is also a problem associated with the fact that only a small fraction of the total community may be physiologically active (Harris, 1978, 1984) and that much of the standing biomass may be doing very little. The overall C specific rate of increase will not be a strict function of C taken up divided by the total particulate C but will be a complex function of a number of exponential C specific growth rates each weighted by the contribution of each species to the total biomass (Fig. 6.7). The oceanic data discussed above showed that the total rate of C turnover was a complex function of C turning over at different rates in different size classes within the phytoplankton community. It should be remembered that it is not possible to simply average exponential growth coefficients to achieve an overall rate of increase for the whole community (Talling, 1983).

Knoechel and Kalff (1978) used autoradiographic techniques to measure the contribution of each species to the productivity of a lake and used the rate of ^{14}C incorporation during 2 h incubations to measure growth. These C specific growth rates were compared to the actual rate of increase in cell number. The model used also included estimates of death rates and sedimentation in order to achieve a mass balance for surface waters. In order to estimate the terms of the model it was necessary to make some assumptions about the cellular C contents of the species involved. In most cases the model underestimated the cellular C contents by as much as 50 % indicating some uncertainty in the magnitude of the terms of the mass balance. While the model was, in some respects, unsatisfactory the data revealed two significant features of the biology of the phytoplankton community. The autoradiographic data revealed that the individual species abruptly switched from low growth rates (0.1 d^{-1}) to sustained periods at higher growth rates (0.4–0.6 d^{-1}) and back again. Different species turned on and off at different times. Also the model results showed that loss rates were extremely important in determining the biomass in surface waters.

Jassby and Goldman (1974) and Tilzer (1983) have attempted to directly compare the results from 4 h *in situ* ^{14}C incubations with observed biomass changes. In each case, the data from 4 h ^{14}C incubations was integrated over 24 h and the C specific growth rate (d^{-1}) was compared to the observed changes in biomass. One thing is immediately obvious; the C specific growth rates far exceeded the observed changes in biomass and both Jassby and Goldman (1974) and Tilzer (1984) were forced to assume massive losses within the water column to balance the measured productivity against the appearance and disappearance of biomass. Jassby and Goldman assumed that, as grazing and sedimentation could not account for the overall mass balance, the cells were dying and lysing in the water column. Given the magnitude of the respiratory and other losses (Fig. 6.5) this conclusion is unnecessary. The loss rates calculated by Jassby and Goldman were highest in spring and autumn (0.8 d^{-1}) at low biomass and reached a minimum (0.2 d^{-1}) during the summer biomass peak. This same pattern was observed in Blelham Tarn. It is possible to achieve a rough mass balance for phytoplankton without invoking mass death and lysis *in situ*. The problem lies in the interpretation of the ^{14}C data. Pollingher and Berman (1977) published a similar set of observations from Lake Kinneret and concluded that the production and loss processes were tightly coupled.

Tilzer (1984) has attempted a similar comparison of ^{14}C data and growth rate estimates from Lake Constance. Tilzer used a form of the daily integral to estimate the daily production from short term incubations. He did not, of course, estimate the respiration rate by this method. Tilzer's paper nicely illustrated the close coupling between production and loss processes and showed that when production was high loss rates were also enhanced. Tilzer also showed that the photosynthetically fixed C was rapidly lost by respiration or by grazing and that changes in the biomass of phytoplankton in surface waters were brought about by relatively minor shifts in the balance between production and loss. Only rarely does the entire algal biomass increase at rates equivalent to the doubling times of the constituent species (rates > 0.8 d^{-1}). Such occasions are periods of strong 'new' production when soluble nutrient pools are being quantitatively turned into algal biomass. The fact that algal biomass changes slowly while production continues apace is evidence of the importance of the regeneration of materials *in situ*.

Tilzer's data shows only a weak trend towards high production to biomass ($P:B$) ratios when nanoplankton dominated the community but there was a noticeable slowing of the rate of C turnover when large 'K' strategists dominated the summer plankton. In most waters phytoplankton $< 10\mu m$ in diameter are responsible for the bulk of the productivity. Rai (1982) showed that the $< 10\mu m$ fraction of the phytoplankton was always present and always dominated the productivity, resulting in weak correlations between $P:B$ ratios and the apparent shifts in species composition. Tilzer (1984) attempted to write out the terms of the mass balance for phytoplankton by quantifying the terms in Fig. 6.2. The results were revealing. In the closed basin of Lake Constance, washout was negligible; this would not be so in oceanic waters with massive horizontal transport. Sedimentation rates were correlated with production rates but only accounted for about 15% of the total production during the summer period. If sedimentation rates account for only 15% of production on average, then it is clear that the bulk of the annual phytoplankton production must be processed in surface waters. This is consistent with much other data. The processing of materials in lakes shows similar trends to the oceans in that, on a seasonal basis, there is a shift from new production in the spring to regenerated production in summer. Sedimentation rates rose sharply in autumn to $> 50\%$ of production.

Grazing accounted for up to 82% of losses in spring and the activities of the grazers resulted in a clear water phase when the biomass of phytoplankton was severely reduced. Residual losses (respiration, excretion, mortality, lysis) were highest in late summer during and after the period of growth by the summer 'K' strategists. As Eppley and Sharp (1975) showed, respiratory losses of 5% h^{-1} lead to overall losses of 56–60% d^{-1} depending on the daylength. Loss processes for different species differ greatly but there is a discernible seasonal pattern. Grazing is more important early in the year and sedimentation is important in the autumn. These trends will appear again later and the seasonal distribution of loss processes will be seen to be an important force in determining the seasonal succession of species in surface waters. Sommer (1981) showed that for individual species, rates of increase and loss were highly correlated so that species that grew quickly also disappeared rapidly.

Given the complexity of the processes determining the rate of ^{14}C uptake *in situ*, the loss processes *in situ* and the interpretation of the data, it is misleading to plot 'growth' from integrated ^{14}C uptake against 'growth' from biomass accumulation. All that the data really reveals is that the processes of production and loss in surface waters are tightly coupled. It would be wrong to conclude that if the rate of change of biomass in surface waters is low then there is little recycling within the water column. The crucial information required is the balance between internal (recycling) and external (sedimentation) losses for different size fractions and species within the community at different times of the year. All in all, the data points to the importance of loss processes in determining the outcome of events in surface waters.

CHAPTER 7

The concept of limiting nutrients

The very rapid turnover of the soluble nutrient pools in surface waters has meant that for many years the term 'limiting nutrient' has been used too loosely in the literature on phytoplankton. There is often considerable confusion as to whether the nutrient limits the rate of growth or the phytoplankton biomass. It has often been stated, for instance, that phosphorus limits phytoplankton growth in lakes because the concentration of dissolved inorganic phosphorus (DIP) is very low. The same has been said to be true of dissolved inorganic nitrogen (DIN) in the oceans. This, wrongly, assumes a relationship between the *concentration* of the nutrient and the *rate* of algal growth. Such ideas grew out of culture work, particularly batch cultures, where the initial concentration of the nutrient represents the entire available supply. Large scale correlations between the spatial distribution of nutrients and production also impressed early oceanographers (Hentschel and Wattenberg, 1931) as did the temporal inverse correlation in the temperate spring. At smaller scales recycling mechanisms are in operation which restock the pools of DIP and DIN continuously and thus we are concerned with both the pool size and the rates of uptake and resupply.

A measurement of the concentration of a nutrient is not a sufficient test of whether or not it is limiting. What we really need to know is the pool size and the rate of turnover. We shall see later that there is a relationship between the stock of total nitrogen (TN) and phosphorus (TP) and the biomass of phytoplankton. This relationship is, however, a complex function of the distribution of TN and TP in the particulate pools of phytoplankton, zooplankton, fish (and even whales!) and is controlled by the planktonic food chain in question. The particulate pools of N and P in the food chain represent functional holons that turn over at quite different rates and there is a hierarchical sequence from small, fast (phytoplankton) pools to large, slow (fish) pools. We need to know both the rates of uptake from the soluble pool and regeneration of the nutrient from the particulate pools as, if they are not balanced, then depletion of the soluble pool may result. The partitioning of TN and TP between the particulate pools in the food chain is somehow linked to these rates of uptake and regeneration, but it is not, at present, clear how this is regulated. Evidently, all the *TP*, say, cannot be sequestered in the algal pool or else the zooplankton and the fish could not survive.

It is possible to study the recycling of nutrients in surface waters by an empirical ('large number') approach using properties of different lakes to seek general relationships. It is obviously very difficult to use kinetic data to make definite statements about the nutrient flux in surface waters so indirect approaches are necessary. At small spatial and temporal scales the size of the DIP pool may be out of balance with the requirements of algal uptake but such small scale perturbations average out and on the scale of annual average distributions the rapid recycling of N and P between the various particulate pools results in statistically significant relationships. These relationships will be investigated in this chapter.

As techniques of analysis have improved over the years the estimates of the pool sizes of soluble nutrients have decreased steadily. The latest estimates of the concentrations of DIN and DIP in oligotrophic surface waters of the oceans and of lakes indicate that the true concentrations are unmeasurable. In the deep oceans the concentration of DIN in surface waters is very low (McCarthy, 1980) and immediate analysis of the samples is necessary to reveal the true concentration of DIN. Storage, or worse, freezing of the samples results in release of DIN from the biota and a spuriously high value for the DIN in the water. DIP concentrations in the surface waters of oligotrophic lakes are also very low and have been overestimated for years. The normal method of DIP analysis overestimates DIP by virtue of the fact that organic-P is included in the result (Tarapchak *et al.*, 1982). The result obtained depends on the stoichiometry of the reagents used. Isotope uptake studies using ^{32}P (Rigler, 1966) revealed that in some cases in order to make sense of the Monod uptake kinetics, the true concentration of DIP had to be at least two orders of magnitude less than that measured by chemical methods.

7.1 Rate processes in oligotrophic waters

There is some controversy about the processes of nutrient recycling in lakes and the oceans and about rate processes in oligotrophic waters. This is not a new argument as there was a vigorous debate on this subject over thirty years ago (Riley, 1951, 1953; Steeman-Nielsen, 1952, 1954). Ryther and Yentsch (1957) used chlorophyll and light data to estimate primary production in the oceans and asserted that the best estimate of the assimilation number for oceanic phytoplankton was 3.7 mg C mg Chl^{-1} h^{-1}. If a realistic C:Chl ratio is employed, this translates into a growth rate of 1.1 d^{-1}.

The present debate still centres on events in the central oligotrophic oceans where, the equilibrium approach would argue, phytoplankton growth rates are low due to the lack of DIN in surface waters. Eppley *et al.* (1973) reported phytoplankton growth rates of 0.2–0.3 d^{-1} in the central gyre of the North Pacific Ocean. Such results were obtained by bottle experiments (cf. Chapter 6) and an examination of the uptake of C and various forms of DIN and DON. More recently McCarthy and Goldman (1979) and Goldman *et al.* (1979) suggested that the relative growth rates of the phytoplankton in such areas might be close to unity as it has frequently been observed that the elemental ratios of nutrients in the phytoplankton are identical to the Redfield ratio. This would mean absolute growth rates in excess of 1.0 d^{-1} at surface temperatures.

Goldman and others showed that phytoplankton only exhibited such elemental ratios at growth rates close to μ_{max} (Goldman, 1980) and therefore concluded that growth rates in the central oligotrophic oceans were similarly high (Goldman *et al.*, 1979; McCarthy and Goldman, 1979). For this to be possible it is necessary to invoke small scale nutrient recycling from small grazing animals at such a rate as to support rapid phytoplankton growth. The phytoplankton assemblage in the central oligotrophic oceans is dominated by microflagellates (Venrick, 1982) so all the organisms concerned are very small. The importance of such organisms to the overall productivity of the oceans has certainly been underestimated in the past as many of the standard techniques of measuring the uptake of C and N in the oceans tend to lead to the death of such organisms (Venrick *et al.*, 1977). While it is easy to understand the processes which lead to nutrient regeneration at a scale of days and weeks it is more difficult to grasp the importance of very small scale events. The arguments about N regeneration in the oceans concern events that are very difficult to observe directly and concern patchiness at scales which necessitate some assumptions about physical processes at those scales (Jackson, 1980). Jackson (1980) argued that in the open water the small pulses of regenerated nutrients would diffuse away before the phytoplankton could take them up. It is like the medieval arguments about the number of angels dancing on the head of a pin!

Goldman (1984) has suggested that the high productivity of the oligotrophic oceans may be maintained by rapid regeneration of nutrients in association with aggregates and particles in surface waters. Margalef (1963a) made a similar prescient prediction. This argument assumes that the microflagellates and small grazing organisms are physically associated with one another on the particles. In this way tight physical coupling is possible and the diffusion arguments of Jackson (1980) are avoided. This is a classic example of the use of non-equilibrium arguments as the particles are in an open system, so that migration of flagellates and grazers can occur between the 'islands' of biological activity across the intervening 'desert' of nutrient poor water. Rapid growth and grazing can occur on the particles and the cropping by the grazers can be high. In a statistical sense the overall system is at steady state but each aggregate has a different state as the phytoplankton grow unchecked on some particles until the grazers arrive to devour them. This is exactly analogous to Caswell's (1978) arguments for predator mediated coexistence; a non-equilibrium approach that allows large scale persistence and overall steady state. The problems of sampling such aggregates have yet to be tackled.

It is becoming increasingly clear that organisms in the < 10 μm size class are extremely important in production and nutrient recycling in surface waters. Recent observations in both the oceans and in freshwater indicate an abundance of small forms. Bacterial abundance frequently ranges from 10^6-10^7 ml^{-1} and it has recently been discovered that photosynthetic cyanobacteria are frequently present at densities of 10^5-10^6 ml^{-1} (Johnson and Sieburth, 1979; Waterbury *et al.*, 1979; Platt *et al.*, 1983). There are also a large number of both photosynthetic and heterotrophic nanoplankton in almost equal proportions with abundances in the range of 10^2-10^4 ml^{-1}. Intriguingly such abundances indicate an almost constant total biomass across all size fractions as, while the size of the organisms may decrease, the abundance increases. This almost certainly indicates trophic interactions between these size classes. Recent oceanic observations also indicate the presence of ciliates ($0-10^2$ ml^{-1}) and amoebae ($0-10$ or 10^2 ml^{-1}). Almost nothing is known of the patterns of occurrence of these organisms in space and time.

There is good evidence that there are systematic shifts in the cell size distributions in different waters and it is clear that nanoplankton dominate the phytoplankton in oligotrophic freshwaters (Watson and Kalff, 1981). The same appears to be true in the oceans (Semina, 1972). Not only do nanoplankton dominate the biomass in oligotrophic waters but they also dominate the productivity. Numerous studies have shown that the nanoplankton are responsible for at least 90 % of the total ^{14}C, ^{15}N or ^{32}P taken up (Yentsch and Ryther, 1959; Malone 1971a,b, 1977; Durbin *et al.*, 1975; McCarthy *et al.*, 1974; Friebele *et al.*, 1978; Kimmel, 1983; Lean and White, 1983) particularly at low concentrations. Indirect evidence points to rapid growth in such species even in the oligotrophic waters (Harris, 1980a; Goldman, 1980, 1984) and grazing must therefore be equally rapid (Fig. 6.7). Recent evidence points to the existence of the necessary microflagellates (Fenchel, 1982a,b; Azam *et al.*, 1983). As noted above, the importance of the cyanobacteria has only recently been realised. As was suggested earlier (Harris, 1980a) the events in oligotrophic waters may be likened to a small rapidly spinning wheel.

It is possible to obtain indirect evidence of high productivity in the oligotrophic oceans from a number of sources. By high productivity, I mean estimates that may be as much as an order of magnitude higher than that derived from measurements such as the uptake of ^{14}C or ^{15}N in bottle experiments. Evidence from physiological constraints, productivity models and food chain dynamics can be used to show that the recycling of essential elements in oligotrophic waters may be very rapid as long as sufficient energy is available. The small flagellates that dominate the oligotrophic oceans (Semina, 1972) show physiological features that are adapted to low nutrient concentrations and rapid regeneration in that they exhibit low half saturation constants for nutrient uptake (Malone, 1980a) and may also prefer NH_4^+ as a source of N (Malone, 1980b).

Direct evidence for N limitation of growth rates in the oceans is hard to find (Morris *et al.*, 1971a,b; Sheldon *et al.*, 1973; Yentsch *et al.*, 1977; Goldman *et al.*, 1979; McCarthy and Goldman, 1979; Morris, 1980b; Goldman, 1980). In the oceans, evidence which points to high productivity and a lack of N limitation has been obtained from a number of sources. Interestingly, none of these pieces of data (except Riley, 1951) comes from the usual bottle techniques used by biological oceanographers. Conservative (low) estimates of productivity (obtained by bottle techniques) in the central oligotrophic oceans vary but cluster around 20 g C m^{-2} y^{-1} (Steeman-Nielsen, 1954, 1963; Ryther, 1963; Eppley and Peterson, 1979; Eppley, 1980). Estimates of the productivity of surface waters obtained from sediment traps and the flux of N to deep water (Knauer *et al.*, 1979), from ocean box models of ^{14}C (Broecker, 1974), from oxygen measurements in surface waters (Shulenberger and Reid, 1981) and from the mixing of ^3H and ^3He in surface waters (Jenkins, 1980, 1982) all indicate productivity of the order of ten times the conservative estimates. Recent 'ultra-clean' ^{14}C experiments carried out by Marra off Hawaii indicate that with the best methods it is possible to obtain ^{14}C uptake rates about five times the previous estimates.

The higher productivity estimates can be converted to turnover times of the phytoplankton carbon (and hence specific growth rates) if the particulate C pool is measured and if the proportion of detrital C is known. From microscope counts Sharp *et al.* (1980) estimated that the detrital C accounts for about 50% of the particulate C in the North Pacific gyre. Using relatively low productivity estimates they argued that the relative growth rates of the phytoplankton in the gyre were of the order of 10% μ_{max}. The highest productivity estimates would therefore translate into at least 50% μ_{max} and the estimates from other (non bottle) methods would translate into rates apparently approaching μ_{max}. Eppley (1972) showed that temperature sets an upper limit to μ_{max} for phytoplankton in culture and that the absolute rates varied from about one doubling per day at 5°C to about four or five doublings per day at 25°C. It is of course possible that the phytoplankton in the North Pacific gyre have adopted low absolute rates of growth so that they maintain growth rates close to μ_{max} whilst growing at low absolute rates. There is little evidence for this however and the latest productivity measurements are consistent with the *P/B* versus *B* plots (Fig. 7.1) in indicating growth rates close to μ_{max} in oligotrophic ocean waters.

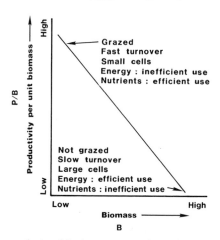

Fig. 7.1 The inverse relationship between productivity per unit biomass and biomass with some possible biological properties of the phytoplankton.

The whole debate about the productivity of the oligotrophic ocean depends on the relative magnitudes of 'new' versus 'regenerated' production. The balance between the two can be estimated by means of a so called 'f' number – where 'f' is the proportion of total production which is 'new' production, or the proportion of total nitrogen uptake which is nitrate uptake. In the previous chapter I touched on some of the methodological problems associated with the measurement of production by C or N uptake; neither method gives unequivocal results. The magic 'f' number works like this: if 'f' is small, say 0.05, then total production is 20 times the 'new' production. If, on the other hand, 'f' is larger, say 0.5, then total production is only twice the estimate of 'new' production. Note that this is a reciprocal relationship; small errors in the estimation of 'f', if it is small, will make a large difference to the estimate of total production. Also there is no need to assume that 'f' is constant in time or space so that limited sampling may produce a biased result.

The lowest (bottle) estimates of the production of the oligotrophic ocean *may* be about one tenth of the highest (indirect) estimates of same. Given that, in any debate, proponents of the two camps have a habit of slightly overstating their cases, it is more realistic to assume that the discrepancy between the two types of estimate is a factor of three or five. This is quite likely to be within the range of experimental and sampling uncertainty, particularly if 'f' changes with time. Pulses of entrained NO_3 entering the mixed layer from below as a result of wind events may be sufficient to ensure continued 'new' production even though the concentration of NO_3 in the mixed layer remains unmeasurable. This is the 'bank balance' analogy again. The time scales of such pulses remain unclear but a consideration of the results presented in Chapter 5 would tend to suggest that scales of a few days would be good scales to investigate first.

If sporadic nutrient entrainment events are shown to occur then the 'f' number may prove to lie, on average, in the 0.2–0.3 range. This would suffice to reconcile the major arguments of the opposing camps and to reduce the specific growth rates of the organisms to a more reasonable $1-2 \ d^{-1}$ at most. Such a nutrient regime would not obviate the need for aggregates and for the efficient scavenging of scarce resources, it would merely mean that continued but sporadic 'new' production was possible even in oligotrophic waters.

Whatever the outcome of the debate concerning the rates of growth of phytoplankton in oligotrophic waters, more work on the dynamics of oligotrophic waters is clearly needed. A number of conceptual models now exist which tend to favour rapid growth and recycling over slow growth in such waters. Whether this is the case in all waters remains to be seen. One way to test this is to examine the turnover of nutrients in a wide variety of water types and to study the fluxes of materials in surface waters. The link between elemental ratios and relative growth rates introduced by Goldman *et al.* (1979) and McCarthy and Goldman (1979) may be exploited if the elemental C:N:P ratios of natural populations and the effects of detrital contamination are known.

7.2 The cycling of N and P in lakes and the oceans

The sizes and rates of turnover of the DIN and DIP pools are very important variables which either determine, or are determined by, food chain processes. The phytoplankton and nutrient dynamics must therefore be seen in the light of events in the entire food chain in surface waters and the apparent complexity of the grazing and regeneration interactions (Lane and Levins, 1977) mean that an analytical approach is impossible at this time. The paper by Lane and Levins (1977) demonstrates very clearly that with grazing, nutrient regeneration and feed backs between trophic levels the effects of nutrient enrichment are almost impossible to predict and some very surprising results may be achieved.

A further complication arises from the fact that a number of the processes in surface waters operate over quite different time scales. The turnover of the DIP and DIN pools may be measured in minutes or seconds, the algal 'growth' processes operate over days and weeks, the size of the soluble nutrient pools and the loadings of DIN and DIP to surface waters may show seasonal components and there will be longer term components operating at scales of years coupled with basin residence times and sediment interactions. It is possible to identify a hierarchy of time scales in important processes. In this respect the methods of Vollenweider have represented a real conceptual breakthrough in that he deliberately chose to ignore the details of the temporal dynamics of each and every lake. When he began his empirical study of data from many lakes (Vollenweider, 1968, 1969, 1975, 1976) he deliberately sought patterns that cut right across the spectrum from oligotrophic to eutrophic waters. In doing this he was anticipating Rigler's (1982a,b) call for an empirical approach to ecology. Vollenweider's approach used annual average values for such variables as TP, TN and algal biomass for a number of lakes. This assumes that the small scale, rapid processes of DIP and DIN turnover, of grazing, growth and death average out over the longer term to produce average statistical properties that can be interpreted. In short Vollenweider's approach is to seek statistical properties of lakes as large number systems.

So if, for the time being, we overlook the detail and concentrate on a broad brush approach then a general conceptual framework is emerging. This framework makes use of empirical relationships between the particulate pool sizes of TN and TP and the regenerative flux rates of DIN and DIP. There appear to be great similarities between the situation in the oceans, where TN and the flux of DIN appear to be the crucial variables, and freshwaters where TP and the flux of DIP are important. It is possible to identify a number of properties of planktonic food chains which show systematic variations with trophic state. These may be summarized as follows:

(1) The pool sizes of DIN and DIP increase in size in eutrophic waters and show slow variability in space and time. i.e. they become more dependant on sources external to the surface mixed layer rather than on regeneration within surface waters.

(2) Rates of turnover of the DIN and DIP pools are faster in oligotrophic waters than in eutrophic waters as the pool sizes are smaller. Oligotrophic waters appear to be more dependent on internal recycling in surface waters.

(3) Oligotrophic lakes retain more P than eutrophic lakes (Dillon and Rigler, 1974a) mainly because the sediments act as a net sink in such systems.

(4) The phytoplankton of oligotrophic waters are frequently dominated by small flagellates, whereas the phytoplankton of eutrophic waters are frequently relatively large (Pavoni, 1963; Semina, 1972). There is also a relationship between trophic state and the balance between 'new' and 'regenerated' production (Fig. 6.1).

The dynamics of P in lakes can be described by a mass balance equation which relates external loads and outflow losses to sedimentation and regeneration within the basin. This is a development of the 'box model' approach outlined earlier. Unfortunately such an approach cannot, as yet, be attempted for TN in the oceans because of the uncertainty of the fluxes of N from upwelling and entrainment across the thermocline. Even the accurate measurement of the external P loading to lakes is not as easy as it might seem, but the fact that most of the P loading comes in the form of point sources (rivers and piped inputs) does help enormously. A mass balance equation may be defined as follows:

$$dP/dt = I - O - (S - R) \qquad\qquad 7.1$$

where dP/dt is the change in P over time, I is the external load, O is the outflow loss, S is the loss to sediments and R is the return from the sediments or the internal load. $(S - R)$ is not easy to measure directly but it may be obtained as a residual from the other variables. Such a model forms a one-box model (Imboden and Lerman, 1978, Chapter 4) and a two-box model may be defined if the surface and deep waters are treated as separate boxes (Imboden, 1974).

The steady state equation may be obtained by setting $dP/dt = 0.0$ but the discussion above makes it clear that restrictions must be placed on what is regarded as steady state. Lakes are rarely, if ever, at steady state as loadings fluctuate over time and components of the dynamics of P show a wide range of time constants. Vollenweider (1969, 1975, 1976) chose to assume that annual average values for I, O, S and R could be used and that steady state at such time scales could be assumed. This effectively collapses all the short term dynamics into annual average patterns of behaviour. The annual time frame is an important assumption that must be borne in mind in all that follows. It is assumed that at scales of a year the lakes may be treated as completely mixed reactors (cf. Chapter 4).

Vollenweider's (1969, 1975, 1976) and Vollenweider and Dillon's (1974) equations have been modified and developed by a number of workers (Dillon and Rigler, 1974a,b; Larsen and Mercier, 1976; Canfield and Bachmann, 1981) who have shown that it is possible to predict the surface concentration of total phosphorus (TP) from the loading and a small number of easily measured variables. The majority of the variability between lakes can be accounted for by morphometric and physical variables (mean depth, water residence time).

Whatever the precise form of the equations used it is evidently possible to determine the average in lake *TP* concentration from the loadings using a mass balance approach. It is important to remember that these relationships make some assumptions about the lakes in question. The assumptions explicitly require that the annual rainfall exceeds the annual evaporation and that the annual flow in the input equals the annual output. It is also necessary to assume that the average annual flow is a valid quantity. In lakes with great seasonal fluctuations in hydraulic load these relationships may break down. The assumptions also require a net sedimentation velocity of about 10–20 m y^{-1} (Larsen and Mercier, 1976). This assumes some stratification in the lake i.e. temperate latitudes. In short, these relationships may work well in temperate lakes but need to be tested in tropical situations.

Vollenweider and Kerekes (1980, 1981) and Janus and Vollenweider (1982) have undertaken an analysis of nutrient data from a set of lakes sampled in an international OECD programme. The results are consistent with the summary of major conclusions listed above and some interesting trends have been revealed. An analysis of the *TN* (total nitrogen) and *TP* (total phosphorus) data from 55 lakes revealed that the *TN:TP* ratio varied systematically from over 200 in oligotrophic lakes to less than 15 in the eutrophic ones. The ratio of 200 in the oligotrophic lakes indicates that these lakes are heavily dependent on the rapid recycling of the *TP* in order to provide the growth requirements of the algae and gives indirect confirmation of the ^{32}P data which shows rapid turnover of the DIP pool in such lakes. Further confirmation of this trend comes from an analysis of data on the relationships between the *TP* and DIP pools.

If the DIP data is plotted against the *TP* data the trend is not a constant ratio between the two (Fig. 7.2). In oligotrophic lakes the DIP is a small fraction of the *TP* (< 10%) as indicated by the data of Lean (1973a,b), whereas at high *TP* the DIP pool becomes almost 100% of the total. This indicates a change in the way in which the *TP* is metabolized within the system with rapid turnover of the DIP at low *TP*, and slow turnover of a much larger pool at high *TP*. Such conclusions are consistent with the ^{32}P kinetics described by White *et al.* (1982) (Chapter 4). These changes must be coupled with alterations in the food chains in these lakes; the results are consistent with a change over from a food chain dominated by grazing in oligotrophic waters to one dominated by detrital metabolism in eutrophic waters.

A very similar situation exists in the oceans but the metabolism of *TN* appears to be the critical determinant of events. (Data for the turnover times of the DIP pool in the oceans is scant (Eppley *et al.*, 1973; Perry and Eppley, 1981; Nalewajko and Lean, 1980), so perhaps we should be wary of too

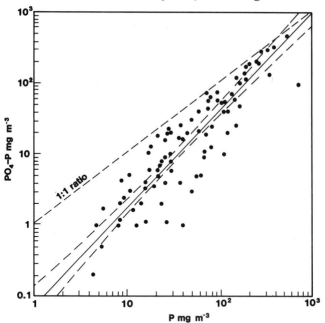

Fig. 7.2 The relationship between total phosphorus and PO₄–P in a set of lakes sampled by an OECD study. Data courtesy of Richard Vollenweider. As the total P content of surface waters rises the proportion of PO₄–P rises also.

broad a generalization!) TN data for the oceans is hard to find but the same qualitative patterns exist for TN as exist for TP in lakes. NO_3^- concentrations become a larger proportion of the TN in eutrophic, upwelling areas as the NO_3^- concentration is below measurable levels in oligotrophic waters (McCarthy, 1980). All the available evidence points to an inverse relationship between 'specific growth' rates and biomass in ocean waters. Data by Sutcliffe *et al.* (1970) and Koblentz-Mishke *et al.* (1976) indicate that the turnover of the algal biomass is more rapid in oligotrophic waters (Fig. 7.1). These data indicate that turnover may be as high as $6 - 10$ d^{-1} in oligotrophic waters, falling to 0.1 d^{-1} in eutrophic waters. Rates of $6 - 10$ d^{-1} may be rather high but the trend is nevertheless present. This trend is consistent with the balance of 'new production' and 'regenerated production' described in Eppley and Peterson (1979) (Fig. 6.1) and indicates a dependence on rapid turnover and regeneration of DIN in oligotrophic waters. The same type of cell size shifts appear to occur in both the oceans and in lakes (Pavoni, 1963; Semina, 1972) and these appear to be similarly correlated with alterations in the structure of marine food chains.

Vollenweider (1975) took the concept of input/output 'box' models (Chapter 4) a step further to include the turnover times of TN and TP in lakes. The results of these studies illustrate the importance of biological

processes in determining the cycling of nutrients. Vollenweider developed
two indices of the way in which *TN* and *TP* are being processed in lake
basins. By defining indices that were measures of the way in which the lake
was acting on the load by mixing, dilution, chemical and biological reactions and export, Vollenweider obtained two indices of turnover velocity
which did not depend solely on hydrological factors. An examination of
data from a number of European lakes led Vollenweider (1975) to conclude
that there was a systematic change in the ratio of the turnover velocities of
N and P in passing from oligotrophic to eutrophic lakes. The ratio declined
from 5–10 in oligotrophic waters to <1 in eutrophic waters, indicating
that the rate of cycling of N speeds up (or the rate of cycling of P slows
down) in eutrophic waters. This is entirely consistent with observations
which show that the proportion of DIP in the *TP* is higher in eutrophic
waters (Fig. 7.2). Cycling of DIP is more rapid in oligotrophic waters and,
because of the high *TN/TP* ratios in such waters the behaviour of *TN* is
more nearly conservative. Vollenweider's analysis demonstrates the importance of flux rates and turnover times rather than concentrations. An
analysis of data from the US EPA National Eutrophication Survey shows
that the ratio of the turnover velocities of N and P is positively correlated
with the *TN:TP* ratio in the water (Fig. 7.3). The ratio of the turnover

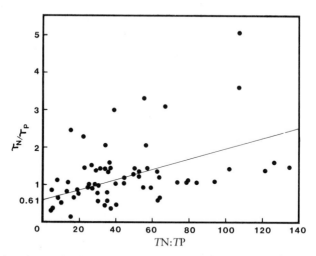

Fig. 7.3 The relationship between the rates of turnover of N and P τ_N and τ_P and the
ratio of the concentrations of *TN* and *TP*. Data from US EPA lakes reports. The ratio
of the rates of turnover is about 1 when the ratio of concentrations is about 30.

velocities is unity when the *TN:TP* concentration ratio is approximately 30
indicating that the turnover of *TP* in lakes is indeed faster than the turnover
of *TN* as the uptake ratio for phytoplankton is about 7 (in mg).

By measuring the disappearance of C, N and P in the surface waters of lakes during the stratified period, Vollenweider demonstrated the presence of allometric trends in turnover times in lakes of varying trophic state. The changes in C and P (Fig. 7.4) indicate that C:P uptake ratios (in mg) in lakes

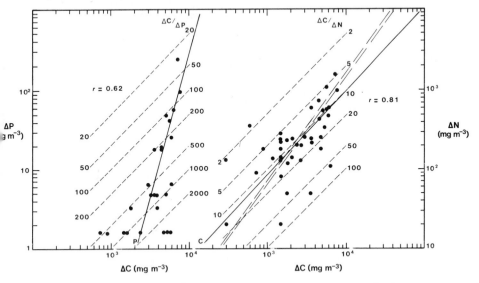

Fig. 7.4 Ratios of the rates of depletion of dissolved inorganic C, N and P from a set of lakes. Rates of depletion measured as the decrease in concentration in surface waters during the summer stratified period.

varied from over 1000 to about 50. Thus in eutrophic lakes the P is being taken up at about the same rate as C; a result in accord with the observation that the majority of the TP is DIP (Fig. 7.2) and that turnover times are slow. In oligotrophic lakes the C:P uptake ratios lie in the range of 1000 or more, indicating that the P is cycling at least twenty times during the stratified period. The N:P uptake ratios follow the same pattern (Fig. 7.5). They ranged from about 200 in oligotrophic lakes to almost 5 in eutrophic lakes. By weight the Redfield ratio for N:P uptake should be about 7. Again the N:P uptake ratios indicate that in eutrophic lakes N and P are taken up at the same rate but that in oligotrophic lakes the P cycles at least twenty times faster than N. As would be expected the C:N uptake ratios remained constant between 5 and 10 and showed little change with trophic state. (Fig. 7.4)

These data give clear indications of the patterns of nutrient turnover in lakes of varying trophic state and indicate clearly the role of biological processes in determining the turnover of the major nutrients in these systems. All together, the rates of removal of C, N and P may be plotted in

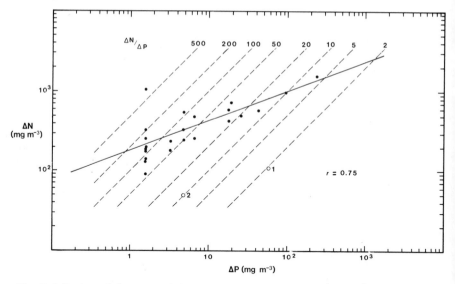

Fig. 7.5 Ratios of the rates of depletion of dissolved inorganic N and P from the surface waters of lakes. Rates of depletion measured as the decrease in concentration in surface waters during the summer stratified period.

the form of an allometric 'depletion line' which summarizes the results from a large number of lakes (Fig. 7.6 *et seq*. Vollenweider's analyses highlight one very important parallel between lakes and the oceans as they indicate the importance of the relationship between uptake in surface waters and the release of nutrients in bottom waters by decomposition. The time scales of regeneration in lakes and the oceans are different as are the basin scales and water residence times. Fundamentally, however, the regenerative processes are the same. It is a pity that there is little *TN* data for the oceans so that these types of analyses cannot be carried out for such systems. There are, however, sufficient parallels between oceanic and freshwater systems that a clear picture is emerging.

7.3 Nutrient cycling, elemental ratios and the 'Redfield ratio'

In a recent paper, Vollenweider and Harris (1985) have reviewed the available data on the elemental composition of particulate matter in lakes and the oceans. A clear allometric trend was observed which closely followed the uptake ratios. If the particulate elemental ratios are plotted in C:P and N:P space then the data from a number of natural marine (Fig. 7.6) and freshwater (Fig. 7.7) samples lie clustered around the 'depletion' line. When plotted in this way, P limitation appears as a spreading of the points diagonally upwards and to the right, N limitation drives the points at right angles; upwards and to the left. Culture data showed that the effects of N and P limitation are independent and do indeed move the elemental composition of the cells at right angles (Fig. 7.8). This is good evidence of the fact that nutrient limitation is not multiplicative but is dependent on separate effects and separate minimum cell quotas (Talling, 1979). The fact that the particulate elemental compositions cluster around the depletion line immediately raises some interesting questions about the relationships between uptake ratios and particulate ratios in such populations. Banse (1974) pointed out that the relationships between particulate ratios and uptake ratios may not be a simple one. He argued that the observed rates of disappearance of nutrients are not the sole result of uptake but represent a balance between uptake, release and regeneration processes.

While there are undoubtedly many physiological processes which lead to much short term dynamic behaviour in the uptake of nutrients by phytoplankton populations, when the uptake ratios are averaged over a whole stratified period then the stoichiometry is remarkably similar to the particulate composition ratios. Averaging short term dynamic behaviour by averaging nutrient uptake at seasonal scales and averaging physiological processes by examining particulate elemental ratios in the biomass reveals statistical properties not evident in the behaviour of lower level processes.

The nutrient depletion data and the elemental composition data from natural populations reveal that the stoichiometric ratios of C:N:P change with the trophic state of the water body. Curiously they change in the same way. Not only does the particulate elemental composition data follow the depletion line closely if single samples are taken from diverse water types, but the temporal sequence of composition ratios follows the line over time in single basins. The temporal trend in composition ratios is noticeable in marine data from Sakshaug *et al.* (1983), (Fig. 7.9), from freshwater data from Blelham Tarn (Jones, 1976; Fig. 7.9) and in the Great Lakes data (Vollenweider and Harris, 1984). Sakshaug *et al.* (1983) noted that a number of different species in their study exhibited distinctly different minimum elemental ratios and showed that the minimum cell quota of P varied significantly between species. Sakshaug *et al.* (1983) showed that the mean C:P ratio in *Skeletonema* was 59, that in *Staurastrum* was about 90 while that in *Amphidinium* was 167. As the C content of phytoplankton cells is remarkably constant at about 45 % of ash free dry weight the species specific C:P ratios indicate interspecific differences in minimum P cell quotas. Minimum cell quotas are also highly correlated with cell size (Shuter, 1978). Thus the changing elemental ratios in the biomass can be brought about by the substitution of species as readily as by the alteration of the ratios with an unchanging species complement.

Eutrophic waters lie in the bottom left portion of the graphs whereas

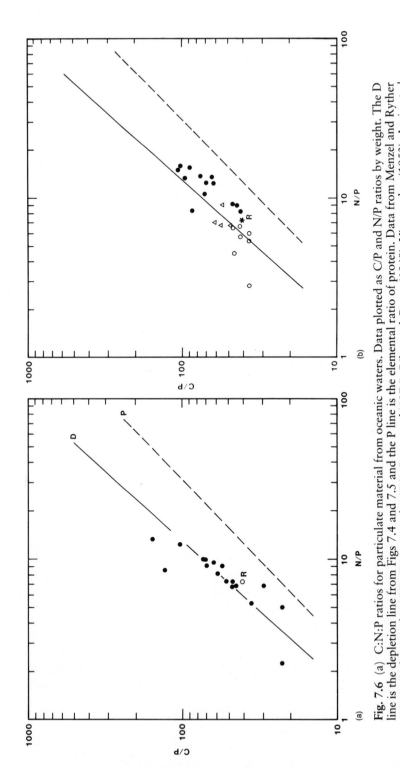

Fig. 7.6 (a) C:N:P ratios for particulate material from oceanic waters. Data plotted as C/P and N/P ratios by weight. The D line is the depletion line from Figs 7.4 and 7.5 and the P line is the elemental ratio of protein. Data from Menzel and Ryther (1960), Bishop *et al* (1977), Knauer *et al* (1979), Perry *et al* (1976), Riley and Gorgy (1948), Vinogradov (1953), Antia *et al* (1963) and Herbland and Le Bouteiller (1981). R is the Redfield ratio.
(b) Data from oceanic particulate material near Hawaii. Unpublished data courtesy of Richard Eppley; plotted as Fig. 7.6(a). R is the Redfield ratio.

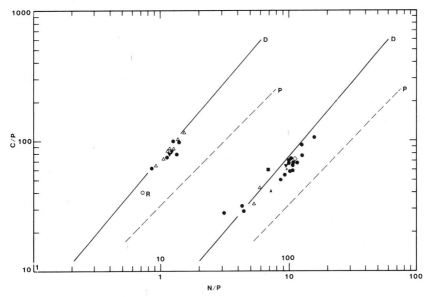

Fig. 7.7 Elemental ratios in particulate material from the Great Lakes. Plotted as Fig. 7.6a. Left: Lakes Huron and Superior. Right: Lake Erie.

oligotrophic waters occupy the top right portion of the scatter plots. The average cell sizes of populations in eutrophic waters are known to be larger than those in oligotrophic waters (for a review cf. Harris, 1984) so the trophic state, the composition ratios and the cell size distributions are consistent. The greater spread in the P dimension may be indicative of the fact that minimum P cell quotas are more variable than minimum N cell quotas (Shuter, 1978) so there is more possibility of intra- and interspecific variability in that dimension. In recent papers, Harris *et al.* (1983) and Harris (1984) have summarized the overall strategies of phytoplankton from oligotrophic and eutrophic waters and have shown that species from oligotrophic waters should be very efficient users of scarce nutrients and should be small, highly grazed and capable of rapid growth. In eutrophic waters the cells should be larger, slower growing, relatively free from grazing and should be inefficient users of nutrients. The evidence for these assumptions will be reviewed later (Chapter 8) but the broad strategy is consistent with the nutrient flux data and the elemental compositions.

If it is true that the changing nutrient regeneration rates lead to the selection of species that can grow well under the prevailing conditions (and a lot of information now points to this conclusion), then there must be some important feed back relationships between the phytoplankton and the zooplankton as the rapid regeneration of nutrients requires that the rates of

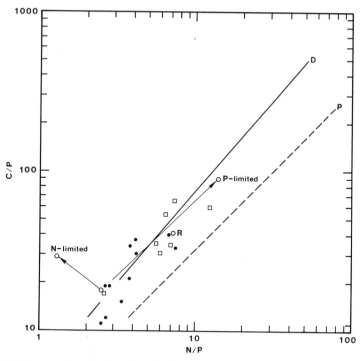

Fig. 7.8 Elemental ratios in cultured phytoplankton. Plotted as Fig. 7.6. Data from Parsons *et al* (1961) and Strickland (1960).

uptake and release be tightly coupled. If we assume that the phytoplankton community in surface waters is composed of a large number of opportunist species which grow rapidly when conditions are favourable and merely persist when conditions are unfavourable, then the rate of nutrient recycling at various trophic states should be broadly correlated with the growth strategies of different phytoplankton types. We shall see that this is true. There is an important, testable, consequence of this assumption of opportunism in the phytoplankton community. If the species which are growing, are growing rapidly at rates close to their maximum relative rates, then it should be possible to demonstrate this. Such evidence may have to be indirect, however, as the species which are growing may only comprise a small proportion of the total biomass.

The substitution of opportunist species should occur in all water types and hence the relative growth rates of phytoplankton should be close to the maximal values in all water types, even the most oligotrophic. This assumption has been made by Goldman *et al.* (1979) and has been confirmed by them using elemental composition ratios. Goldman *et al.* (1979) noted that

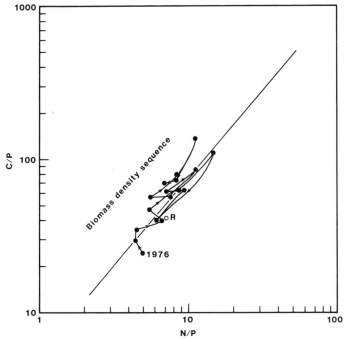

Fig. 7.9 Elemental ratios from marine phytoplankton showing the sequence of ratios associated with the development of an algal bloom. Data from Sakshaug *et al* (1983).

only those populations growing in culture at rates close to the maximum exhibited Redfield ratios. The common occurrence of this ratio in natural populations was taken as evidence of rapid growth, even in oligotrophic waters. As it happens the composition data of Goldman *et al.* (1979) can be overlaid on the C:P and N:P plots as contour lines of relative growth rates. The close fit between the depletion line and the composition ratios of the cultured populations is remarkable (Fig. 7.10). Reference to all the previous figures which show the composition ratios of natural populations immediately reveals that the natural populations are, almost without exception, growing at rates close to the maximum. Nutrient limitation of growth cannot be a frequent occurrence in the surface waters of lakes and the oceans. As critical nutrient elements become scarce in surface waters during the period of summer stratification the whole planktonic community is driven to rapid nutrient regeneration and rapid recycling rates and there is a change in the species composition of the phytoplankton. Maximum relative growth rates are maintained.

The particulate composition data reveals little evidence of nutrient

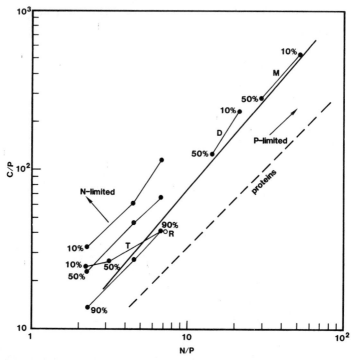

Fig. 7.10 The elemental ratios of cultured phytoplankton as recorded by Goldman *et al* (1979) together with the relative growth rates of the cultured populations. Relative growth rates as % μmax.

limitation of growth in natural waters but it does reveal that both N and P limitation occur rarely in both marine and freshwaters. The received version of phytoplankton ecology is not borne out by this data. P is as frequently limiting as N in oligotrophic oceans and N limitation also occurs in freshwater. Clearly it is dangerous to assume that, just because the concentration of DIN is frequently unmeasurable in the oceans and DIP is frequently unmeasurable in freshwater, these nutrients are the only candidates for rate limiting effects in these waters.

The response to nutrient depletion in surface waters is therefore a community response and the only time that severe nutrient depletion and a reduction in growth should be observed is when a single species bloom occurs (Harris, 1980a; Sakshaug *et al.*, 1983). Under these conditions the rates of uptake and regeneration become temporally unbalanced and the population runs out of nutrients: or to put it in ecological terms, the minimum cell quota of the bloom population is too high to allow the population to grow rapidly on the depleted stocks of nutrients left after the bloom. About the only case of elemental ratios in the literature that does not fit the high growth rate relationship is that from dinoflagellate blooms in Lake Kinneret (Serruya and Berman, 1975). These dinoflagellate blooms appear to be close to competitive equilibrium (Harris, 1983) so that ecological theory and limnological observation are in accord. It is just that such conditions are rare.

Tilman *et al.* (1982) reviewed the relationships between limiting nutrients and the community structure of phytoplankton. Theirs was essentially an equilibrium approach. Their approach stressed the role of nutrient concentration ratios and the Monod growth parameters $\mu(d^{-1})$ and $K_s(\mu M)$. The theory of Tilman (1982) is a steady state theory which takes nutrient concentrations ratios into account rather than the flux ratios and uses external concentrations rather than internal cell quotas. The use of fluxes and internal quotas allows for a more realistic approach to the dynamics of the real world. At steady state the two approaches are equivalent. Tilman's (1982) theory requires that there is a simple relationship between concentrations and fluxes and that, as the nutrient concentration increases the turnover rate decreases. The data appears to be consistent with this assumption. The results given in Tilman *et al.* (1982) are also consistent with those shown above in that blue green algae occur at low *TN*:*TP* ratios. As Tilman *et al.* (1982) used concentration ratios to predict the occurrence of blue-green algae there must be a simple monotonic relationship between turnover rates and concentrations.

Vollenweider and Harris (1985) attempted to assess the role of detrital material in the particulate elemental compositions observed in natural waters. The partitioning of the particulate material into living and dead material is almost impossible but estimates of the elemental ratios in the detrital fraction, obtained in a number of ways, all indicated that the detritus contained C:N:P ratios indistinguishable from the living fraction. This indicates that microbiological activity, which scavenges nutrients from dead cells, and the regeneration of nutrients by grazing, proceed very rapidly in the water column and that all the particulate fractions soon come to have the same elemental composition. This conclusion is borne out by an examination of the elemental ratios in particulate material from sedimentation traps in lakes and the oceans. Recently sedimented material has elemental ratios like the plankton of surface waters, but material from deep water and from the sediment interface is depleted in nitrogen and phosphorus.

All the data serves to reinforce the conclusion that lakes and the oceans are basically similar in their nutrient regeneration processes. The scales of the processes differ as the water column in lakes is much shallower than the oceans and the residence time of sedimenting material in the water column is consequently less. Material from the sediment interface of eutrophic waters rapidly becomes depleted in N as a result of denitrification. Oceanic data from Bishop *et al.* (1977, 1980) and Knauer *et al.* (1979) indicates that the C:N and C:P ratios of particulate material rise rapidly as particles fall through the water column and that at least 90 % of all particulate C, N and P is returned to dissolved forms within the top 400 m. C:P ratios rise more rapidly with depth than do C:N ratios indicating that the P is recycled to dissolved forms faster than N. In the data reported by Bishop *et al.* (1980) the C:P and N:P ratios rose rapidly with depth until the depth of the oxygen minimum and then changed more slowly.

Vollenweider and Harris (1985) also reviewed the elemental composition ratios of zooplankton and other planktonic material. The composition ratios of zooplankton and of the larger phytoplankton fractions (> 75μ) in the ocean show lower C:P and C:N ratios than the phytoplankton and consequently the allometric C:P and N:P line passes through the Redfield ratio (Fig. 7.11). It should be noted that the Redfield ratio is significantly

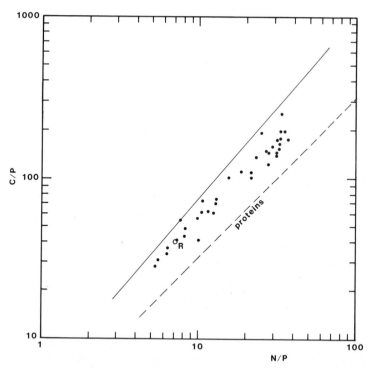

Fig. 7.11 Elemental ratios of plankton larger than 75μ. Data from Suess and Ungerer (1981). Note that the trend of this data does pass through the Redfield ratio.

lower in C than almost any of the normal planktonic data and it is significant that the Redfield ratio was derived from the ratios of regenerated nutrients in deep water. In this respect the Redfield ratio is not representative of most phytoplankton composition ratios. This raises the possibility that the original Redfield ratio data was biased in favour of the larger phytoplankton and zooplankton fraction of the plankton. The discrepancy between the Redfield ratio (106 C: 16 N: 1 P by atoms) and the more representative phytoplankton ratio (166 C: 20 N: 1 P) shown here (Vollenweider and Harris, 1985) is significant but not sufficient to alter the interpretation of Goldman's data on composition ratios and relative growth rates. At low relative growth rates the ratios may be distorted to the extent that C:P ratios of 1000 and N:P ratios of 50 are observed.

The mean C:P and N:P ratios are a function of trophic state in lakes and the mean ratio quoted above only applies to mesotrophic waters. Vollenweider and Harris (1985) averaged all the data to produce regression equations for the correlations between particulate C, N and P. The least squares regressions by weight were:

$$C = 88 \ P^{0.882} \qquad (r = 0.996) \qquad\qquad 7.2$$

$$N = 12.1 \ P^{0.882} \qquad (r = 0.996) \qquad\qquad 7.3$$

with the exponents being significantly different from 1.0 at the $p < 0.01$ level. Accordingly the C:P and N:P ratios become a function of P:

$$C:P = 88/P^{0.118} \qquad\qquad 7.4$$

$$N:P = 12.1/P^{0.118} \qquad\qquad 7.5$$

with the C:N ratio constant at 7.3 by weight. Thus oligotrophic waters have C:P ratios above 75–80 and eutrophic waters C:P ratios below 55–60.

A comparison of the variability of C:N:P ratios in nutrient depletion studies in lakes, with the spread of those ratios in particulate matter, indicates that there is much greater variability in depletion rates than in particulate ratios. Fig. 7.12 shows the two regression lines (one for depletion ratios, one for particulate composition ratios) plotted on the same axes. The slopes of the two lines are in fact statistically indistinguishable although they appear to intersect at a particulate composition ratio which corresponds to a total P content of 100 μg l^{-1}. The stretching of the depletion line has to be understood in terms of the recycling of P under natural conditions. Below 150 μg l^{-1} *TP*, P cycling becomes more and more rapid as the *TP* is reduced. Above a value of 150 μg l^{-1} *TP* recycling of DIP is probably not limiting and a case can be made for a limitation of growth by

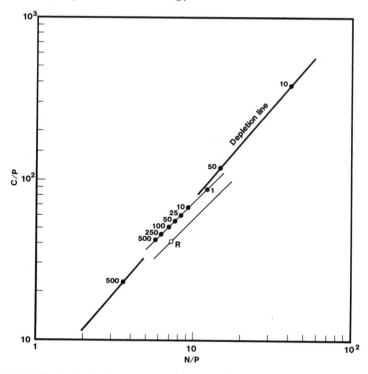

Fig. 7.12 The depletion line compared to the lake mean ratios at a variety of total phosphorus concentrations. The depletion ratios (Figs. 7.4 and 7.5) are much more variable than the mean elemental ratios in the particulate material. Plotted values are concentrations of total phosphorus in mg m^{-3}.

C, N or other elements. This is consistent with the data presented above concerning the proportion of *TP* in the form of DIP and the probable regeneration rates of DIP in lakes of differing trophic state. In relation to C, it appears (Fig. 7.12) that at a *TP* of 1–2 μg l^{-1}, P turns over 15–25 times faster, at intermediate levels (10–20 μg l^{-1} *TP*) P turns over 4–6 times faster and that at high *TP* levels (100–200 μg l^{-1}) removal rates for the C and P will be comparable.

There is, unfortunately, insufficient data to carry out this kind of analysis on oceanic nutrient recycling. The trends in the freshwater and oceanic data are the same, and, with the exception of a few data points, the spread of oceanic data is greater in the direction of the P limitation axis than in the N limitation axis. This may well indicate that P plays as important a role in the oceans as N. Significantly, the whole concept of nutrient limitation in the oceans is based on reports such as that of Hentschel and Wattenberg (1931) who correlated the spatial distribution of total plankton with the spatial distribution of phosphate in the South Atlantic. There are some basic physiological differences in the uptake, storage and metabolism of N and P in phytoplankton and it is these processes which determine the correlations between the depletion ratios and the particulate elemental compositions. Cellular P quotas vary more widely than cellular N quotas and this probably accounts for the difference in the spread in the directions of N and P limitation.

By examining the elemental composition of particulate material from surface waters, and knowing that the detrital material is not seriously biasing the results, the observer is, in effect, averaging much of the spatial and temporal dynamics of the system. The same is true of the use of seasonally averaged depletion rate data. In both cases the data obtained corresponds to a higher, slower holon than the data associated with nutrient uptake by the phytoplankton and with short term (hours and days) fluctuations in the rates of regeneration. Vollenweider and Harris (1985) noted that in those cases where there was physiological data available for particular sets of particulate composition data, the uptake ratios bore no resemblance to the composition ratios. Physiological processes show rapid temporal fluctuations and the particulate elemental composition represents a time weighted average of the uptake ratios in the previous hours or days. The rate of growth of the phytoplankton is a high level property of the cell and the cell 'filters' the uptake signal.

It is true, as Banse (1974) pointed out, that in the short term, depletion rates may bear little resemblance to the particulate composition. Harris (1983) reported some weekly nutrient uptake data for Hamilton Harbour and these show much more 'noise' than the seasonally averaged data. By averaging over time, however, repeatable patterns emerge as biological processes dominate the processes of nutrient recycling in surface waters. As both the growth rate and the particulate composition ratios are high level averaged parameters (higher level holons rather than physiological processes), they contain statistically significant information. There is more information to be gained by correlations between holons at the same level than between holons at different levels. Thus physiological information is a poor predictor of growth rate. Allen and Starr (1982) concluded that there was much information to be gained by looking at the same process at different levels. That assertion is borne out by this data.

7.4 Patterns in temporal fluctuations

If the temporal variability in the particulate C, N and P is compared to the algal biomass as chlorophyll then the lakes of varying trophic state appear to show quite different patterns of behaviour. Data from the Great Lakes indicates that oligotrophic waters show little temporal variation in all parameters (Fig. 7.13). Even in Lake Erie, which is mesotrophic, the fluctuations are small but there is some indication that the data points skew across the lines of constant C:Chl, N:Chl and P:Chl. This trend becomes more evident as the trophic level of the water body increases until, in hypertrophic prairie potholes, the data shows a very strong tendency to skew across the lines of constant ratios (Fig. 7.14; Vollenweider and Harris, 1985).

This trend towards repeatable patterns is not at all obvious in the actual week to week data from these lakes. Indeed the weekly data on particulate C, N and P and chlorophyll is highly variable in eutrophic and hypertrophic waters and almost no patterns are decipherable. From week to week and from year to year biological processes in these lakes are very dynamic indeed. Once again the pattern of small scale variability which is constrained by higher level processes is repeated. The C:Chl, N:Chl and P:Chl ratios indicate that at low algal biomass (as Chl) the ratios are high (about 250, 50 and 5 respectively) whereas at high biomass the ratios decrease to about 40, 8 and 0.3. When these ratios are compared to the ratios in oligotrophic waters it is evident that the dynamic trends in eutrophic waters extend upwards from the oligotrophic state (ratios in oligotrophic waters are about 150, 20, 1.5 respectively) in a curvilinear fashion. According to Goldman (1980) the C:Chl ratio may also be interpreted in terms of a relative growth rate. For the species Goldman studied the C:Chl ratio decreased as the relative growth rate increased, so that ratios of 150 or greater were associated with low relative growth rates (< 0.2 μ_{max}) and ratios of 40 were associated with high relative growth rates (> 0.8 μ_{max}). This would appear to suggest that the relative growth rates of phytoplankton in oligotrophic waters are low whereas the relative growth rates in the most eutrophic waters approach unity.

This result appears to contradict the results stated above concerning the elemental composition of plankton and the relationship with relative growth rates. According to that data the only natural populations to exhibit nutrient deficiency and a reduction in relative growth rates are bloom populations in the late stages of the bloom (for a review cf. Harris, 1980a; Sakshaug *et al.*, 1983). The explanation of the apparent contradiction lies in the detrital component of natural plankton assemblages. In oligotrophic waters, or in eutrophic waters at low algal biomass, the detrital component of the total particulate C, N and P is high; even though the overall elemental ratio of that detritus is similar to that of the phytoplankton at high relative growth rates. Thus the C:Chl, N:Chl and P:Chl ratios are all inflated relative to the living algal biomass. At high algal biomass the detrital component becomes relatively less important, until at a biomass of 200–300 mg m^{-3} Chl it becomes almost insignificant. Thus the skewed

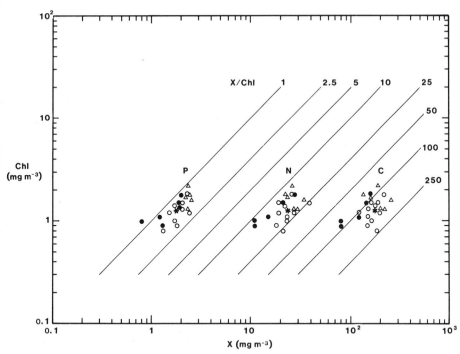

Fig. 7.13 A plot of chlorophyll concentration versus particulate C, N and P for the oligotrophic waters of the Great Lakes. Different symbols represent data from Lakes Superior, Huron and Georgian Bay. Mean ratios represented by stars. X represents P, N or C.

trend of the C:Chl, N:Chl and P:Chl ratios is brought about by the presence of a variable detrital component. In a study of the oligotrophic waters of the North Pacific gyre, Sharp *et al.* (1980) found that as much as two thirds of the C in surface waters was detrital and hence the C:Chl ratios of the phytoplankton in such waters are entirely consistent with high relative growth rates.

In passing, therefore, from oligotrophic to eutrophic waters the elemental compositions of the phytoplankton are consistent with high relative growth rates in all water types. As the DIP and DIN pools decrease in size in oligotrophic waters the rates of turnover of those pools must increase. The data from natural populations suggests that the maintenance of rapid nutrient recycling and high relative growth rates is, at least partially, accounted for by substitution of species of phytoplankton of varying minimum cell quotas of the critical nutrient. The oceanic data indicates that P limitation is a strong selective force in such waters and that the variation in the P quota of cells is as important in the oceans as it is in freshwater. For a

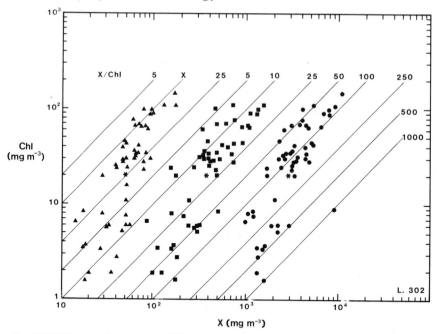

Fig. 7.14 Elemental ratios and chlorophyll contents of particulate material from the surface waters of prairie pothole lakes. X represents P, N or C. L302 is the identification number for this particular set of prairie pothole lakes.

number of reasons small cells should do better in oligotrophic waters. This trend in cell size will be discussed more fully in a later chapter. By indicating maximum relative growth rates in all water types this chemical composition data has a significant contribution to make to the discussion about the nature of the phytoplankton community. Constant, high relative growth rates would be indicative of opportunistic species growing at maximal rates whenever possible. The bewildering complexity of the week to week variation in the particulate C, N and P, particularly in eutrophic waters, can also be explained by the rapid growth of numerous species at different times. The phytoplankton community must be seen in terms of the growth of opportunistic species constrained by an outer envelope of physical, chemical and biological factors.

7.5 Whole basin averages

Vollenweider and Harris (1985) showed that if annual average ratios of C:Chl, N:Chl and P:Chl were examined then there was no deviation of the exponent from unity in the case of C:Chl and N:Chl. This means that the relationship between the variables is linear with mean C:Chl and N:Chl ratios of 114.3 and 15.3 respectively. These ratios did not change depending on trophic state. The P:Chl ratio did, however, show a significant deviation of the exponent from unity; indicating that the ratio is dependent on both TP and biomass (Fig. 7.15). Such a non-linear relationship explains the

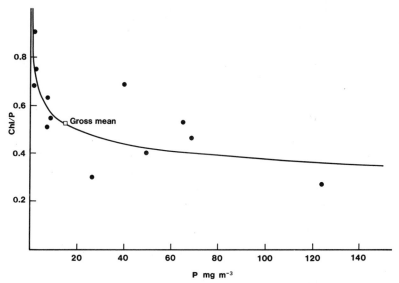

Fig. 7.15 Lake mean values for Chl/P ratios for a set of lakes of varying total phosphorus content. The Chl/P ratio decreases with increasing eutrophy.

reason for the non-linearity in the *TP* versus chlorophyll plots which Vollenweider has used for the management of algal biomass in lakes. The use of these plots will be discussed in a later chapter but the underlying processes which produce these plots are now clear. The trophic state of the lake determines the fraction of *TP* which is in the DIP pool (Fig. 7.2) and, as the lake becomes more eutrophic the fraction of DIP increases. The recent plots of Janus and Vollenweider (1982) show that as the trophic state of the lake changes so does the slope of the *TP* versus Chl plot with the biomass yield per unit *TP* decreasing as the trophic state becomes more eutrophic. As already noted there are systematic changes in the cell size distributions in lakes of differing trophic state and the phytoplankton community changes in response to trophic state.

CHAPTER 8

Physiological scales: non-steady state conditions in the field

Harris (1980a) reviewed the evidence for non-steady state conditions at a number of scales and showed how the processes of horizontal and vertical diffusion linked the scales of space and time in the planktonic environment. Goldman (1984) has invoked the dynamics of growth and grazing on small aggregates in the ocean to explain the close coupling between autotrophs and heterotrophs and the regeneration of nutrients at very small scales. That argument is a classic case of small scale, non-steady state events averaging out over larger scales to produce statistically discernible trends.

There are a number of physiological responses at time scales of the order of hours or less (Harris and Piccinin, 1977; Harris, 1980a,b,c; Falkowski, 1980; Marra, 1980; Marra and Heineman, 1982). The most completely understood are responses to fluctuations in the light regime which appear to be associated with alterations in photosynthetic characteristics. The fluorescence which arises from the photosynthetic pigments (Harris, 1978) shows two time constants – a fast response of a few seconds and a slow response over time periods of a few minutes to half an hour. Harris (1980c) described the response of a fluorescence ratio (F + DCMU/IVF) derived from the ratio of the fluorescence produced by DCMU poisoning divided by the *in vivo* fluorescence. This ratio displays the response of the cells to the underwater light field over periods of half an hour or so as they are mixed vertically in surface waters. As the water column becomes more stable the F + DCMU of the surface populations becomes depressed; a process which takes about 40 minutes to go to completion. Because the cycle times of cells in surface mixed layers are mostly less than this when vertical mixing is active (Table 3.1; Denman and Gargett, 1983) the effect is not seen under mixed conditions. Thus the fluorescence ratio may be used as an index of mixing in surface waters. Figure 8.1 shows that the fluorescence ratio of natural populations may be related to the mixed layer Richardson number (Equation 3.16) and indicates that the phytoplankton populations are very sensitive to the turbulence in the mixed layer. The time constants of the algal

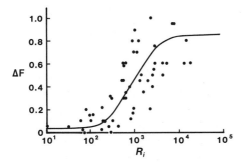

Fig. 8.1 The relationship between the change in the fluorescence ratio with depth (ΔF) and the mixed layer Richardson number (R_i). As the mixed layer becomes stratified so vertical differentiation in the fluorescence becomes evident.

response are closely matched to the characteristic time scales of physical processes in the mixed layer (Table 3.1). Also, the biological response indicates very clearly the difference in the physical regime which occurs as the Richardson number rises from less than 500 (when vertical mixing predominates and the thermocline will be destroyed in 1–10 days) to 10 000 or so (when a strong thermal gradient inhibits vertical mixing, and internal wave motions result; cf. Chapter 3, Fig. 3.4). The phytoplankton may clearly be used as monitors of the physical regime (Harris, 1984).

The vertical motions brought on by eddies in the mixed layer are responsible for the difference in response between contained populations and populations floating free in the water. Static bottle chains, which are commonly used to measure *in situ* productivity, underestimate the actual free water productivity if the water column is freely circulating. The reason for this is the fact that while the populations in the bottles become photo-inhibited, the free floating populations circulate in and out of the high irradiance at the surface sufficiently frequently that the damaging effects of that high irradiance are avoided (Harris and Lott, 1973a; Harris, 1973a,b, 1978, 1980a,b; Harris and Piccinin, 1977; Marra, 1978a,b). Estimates of integral productivity are often underestimates for this reason. Marra (1980) has shown that the time constants of the photosynthetic response to changes in irradiance are of the same order of magnitude as the fluorescence response; indeed there is good evidence that the two measurements reveal different facets of the same process. Falkowski (1980) has recently shown that phytoplankton also adapt to fluctuations in the irradiance regimes over slightly longer time scales (mixing times in deep mixed layers, Table 3.1) by the synthesis and degradation of chlorophyll. In this way the adaptation to 'sun' and 'shade' conditions through changes in the size and number of the photosynthetic units interacts strongly with the turbulence in the mixed layer. There are also diurnal changes in such parameters (Harris, 1978), and these changes interact with the time scales of mixing.

The vital question remains however: how do I know if the static containment effect in *in situ* incubations is significant on any given day and how often is this likely to be a problem? This question is essentially the same as the one raised earlier by long term nutrient uptake experiments. If the length of the incubation is much longer than the time constant of the algal response then the result will always be a function of the incubation technique and there will be no way to tell if the result is an artifact. In the case of *in situ* productivity measurements there is, I believe, a way to tell if the photo-inhibition experienced by the cells in surface bottles is an artifact of the method. The fluorescence ratio of the cells in the water column will show if the water column is stable or actively mixed. If the water column is stable then the cells in the water column will be photoinhibited in the same way as the cells in the bottle. If the water column is vertically mixed then the bottle result may differ from the true result and the fluorescence ratio of the cells in the water column should show this. There are changes in the mixing regime in surface waters from day to day (Fig. 5.4) so that there is a spectrum of motions in surface waters (Figs 5.3 and 5.5). It may therefore be possible, on any given occasion, to determine whether or not the bottle incubations are giving a biased result.

The situation is further complicated if the data from such *in situ* incubations are used to analyse the photosynthesis versus irradiance relationships of the algae. Much effort has been made of late to derive good curve fitting systems to obtain the necessary photosynthetic parameters for integral productivity models from such data (Platt and Gallegos, 1981). The results to be fitted are, however, the results of long term incubations which suffer from all of the above problems. Marra and Heinemann (1982) have recently shown that the time constants for changes in the apparent quantum efficiency of photosynthesis are roughly the same as those for the fluorescence changes but that the time constants differ depending on whether the irradiance is increasing or decreasing. Marra (1980) and Marra and Heineman (1982) have also shown that the photosynthetic parameters of the phytoplankton are variable in time and the only true analysis of such phenomena is that made by time series analysis so that both the magnitude and the time history of the irradiance regime is required. The problem lies in the fact that the temporal resolution of the ^{14}C and O_2 methods is coarse in relation to the true scales of variation experienced by free living populations during the course of the day (Harris, 1980a).

At a scale of hours to a day or so, the ecology of phytoplankton in surface waters begins to be dominated by growth and loss processes. The P:R ratio is a measure of the relative gain and loss functions of populations suspended in a mixed water column. The important external parameters are the surface irradiance (I_o), the depth of the photic zone (Z_{eu}) and the depth of the mixed layer (Z_m). Bannister (1974a,b, 1979) developed equations for gross daily production in a mixed layer which were extended to the analysis of balanced growth in cells in continuous culture (Laws and Bannister, 1980; Bannister and Laws, 1980). These models make a number of assumptions which will be detailed as necessary but the most important restrictions lie in the assumption of steady state and the use of parameters which are averaged over time scales of a day. Bannister's models arise from a class of critical depth models for production in a mixed layer. Such models date back to the ideas of Riley (1942) and Sverdrup (1953) in relating integral column photosynthesis to column respiration on a daily basis. The critical depth of Sverdrup (1953) was defined as the depth at which column respiration equalled total column photosynthesis. If the depth of mixing exceeds the critical depth for any extended period then the growth and survival of the population in question is impossible. By definition, the critical ratio:

$$q_c = \frac{\text{Column photosynthesis per day}}{\text{Column respiration per day}} \qquad 8.1$$

must be equal to, or greater than unity. There are many forms of the equation to calculate column photosynthesis on a daily basis (Talling, 1957a,b, 1970, 1971; Vollenweider, 1965, 1969). All models assume a homogeneous distribution of biomass in the vertical, a constant respiration rate in light and the dark and a particular form of the curve which relates photosynthesis to irradiance. Bannister's (1974a,b) model adapted and rewrote the equations in a form which explicitly included the factors which most directly control production in a column. Bannister's aim was to develop equations which were 'important both in the analysis of field data and for the eventual development of a general theory of population dynamics' (Bannister, 1974a). The chosen factors of importance were:

(1) The incident irradiance I_o.
(2) The amount of light absorption by the water; the extinction coefficient due to the water k_w.
(3) The concentration of phytoplankton pigments, conveniently measured as chlorophyll concentration Chl.
(4) The extinction coefficient of the phytoplankton k_c.
(5) The efficiency of the conversion of the absorbed quanta to photosynthetic products, the quantum yield ϕ_{max}.

The photosynthetic efficiency is contained in the parameters of the photosynthesis/light curve of P versus I. A number of forms of the P versus I curve have been proposed in order to obtain the integral column photosynthesis.

The integral daily rate calculation requires two parts: a depth integral of the P versus I curve and a time integral of the fluctuations in surface irradiance, I_o. Vollenweider (1970) formulated a number of such daily rate integrals and Bannister adopted a subset of these.

If π is the gross daily production (gC m^{-2} d^{-1}) this quantity can be found from:

$$\pi = \psi k_c \gamma / (k_c \gamma + k_w Z_m) \qquad 8.2$$

In this expression $k_c \gamma / (k_c \gamma + k_w Z_m)$ is the fractional absorption by the phytoplankton, where γ is the areal chlorophyll concentration (mg m^{-2}) and Z_m is the depth of the mixed layer. ψ is the upper limit to production: the daily production that would occur if all the under water irradiance were absorbed by the phytoplankton. Bannister showed that ψ may be found from:

$$\psi = 12\phi_{max} I_o(0.7) \lambda \int_{-1/2}^{+1/2} \sinh^{-1}(I_o/I_o(0.7)) \, dt' \qquad 8.3$$

with units of gC m^{-2} d^{-1}. In this relationship λ is the day length in days, t' is the time of day such that $t = -1/2$ at sunrise and $t = +1/2$ at sunset. I_o, the irradiance just under the surface is a function of t'. The factor $12\phi_{max}$ converts the maximum quantum efficiency from units of moles to grams. $I_o(0.7)$ is the irradiance at which the photosynthesis rate reaches 70% of its maximum rate. This is analogous to, but not identical to, Talling's (1957a,b) I_k.

In a homogeneous, totally absorbing column, daily production may be estimated from:

$$\pi(\text{gC m}^{-2} \text{ d}^{-1}) = P_{max} \text{ Chl } z' \qquad 8.4$$

where P_{max} is the maximum specific rate of photosynthesis per unit chlorophyll at light saturation and z' is the depth of maximum production which usually occurs at about 10 % I_o so that $z' = 2.3(k_c \text{Chl} + k_w)$. Bannister also showed that the upper limit to production, $\psi = 2.3 P_{max}/k_c$ occurred when all the incident irradiance was absorbed by the phytoplankton.

For growth in a mixed layer of depth z_m, containing γ mg Chl m^{-2} we may write:

$$d\gamma/dt \text{ (mg Chl m}^{-2} \text{ d}^{-1}) = \pi/\theta - (R + S + G)\gamma \qquad 8.5$$

where θ is the C:Chl ratio in the phytoplankton, and R,S and G are carbon specific loss rates (d^{-1}) due to respiration, sedimentation and grazing respectively. The loss due to sedimentation may be estimated from v/z_m, where v is the sedimentation velocity in m d^{-1} (Fig. 5.2). The instantaneous specific growth rate μ may be found from:

$$\mu = \pi/\theta\gamma - R \quad (d^{-1}) \qquad\qquad 8.6$$

and at steady state, assuming no *in situ* death:

$$\pi = \theta\gamma(R + G + S) \quad (gC\ m^{-2}\ d^{-1}) \qquad\qquad 8.7$$

and also therefore:

$$\mu = G + S \quad (d^{-1}) \qquad\qquad 8.8$$

At steady state these equations give a solution for the areal standing crop and daily production in terms of a series of fundamental environmental and algal parameters. The driving environment variables are: I_o, λ, k_w, z_m, v and G and the algal parameters are: ϕ_{max}, $I_o(0.7)$, θ, R, S and k_c. Bannister regards ϕ_{max} and k_c as constants so the algal response is restricted to changes in $I_o(0.7)$ or I_k, the saturating irradiance (Talling 1957a,b), θ and R. These algal response variables may be altered in an adaptive fashion in response to changes in the environment. As the foregoing model is restricted to nutrient saturated environments the adaptive changes by the phytoplankton will be adaptations to reduced surface irradiance (either by reduction in I_o or by reductions in day length, or both), to changes in the depth of the mixed layer or to changes in the attenuation of the light by substances other than chlorophyll (Wofsy, 1983). The effects of reduced underwater irradiance are to increase the cellular chlorophyll concentration (decrease θ, the C:Chl ratio) and to reduce the saturating irradiance (reduce I_k), (Harris, 1978). Harris (1978) also demonstrated that the $P:R$ ratios of natural phytoplankton populations are adapted to the ambient light penetration and mixing depth in any particular mixed layer.

The major problem with the theory of growth in the mixed layer lies in the assessment of losses due to grazing G. There is no real rationale for the assumption of steady state grazing pressure on the phytoplankton crop; indeed the field data on nutrient regeneration supports a non-steady state argument as there are seasonal components to the depletion of dissolved nutrients and the proportion regenerated within the water column. It appears that, as there are empirical relationships between such parameters as TP and algal biomass (Chapter 11), this is another case in which small scale disequilibrium may be averaged over large temporal and spatial scales to produce statistical patterns.

It is clear that in nutrient saturated waters there is a distinct tendency for the phytoplankton to grow to 'fill' surface waters until light limitation sets in. Seliger *et al.* (1975) noted that in Chesapeake Bay there was a high correlation between the depth of the photic zone (Z_{eu}) and the depth of the mixed layer (Z_m). Wofsy (1983) has modelled the relationship and shown that there is a general tendency for phytoplankton populations to grow to a steady state where the optical depth of the mixed layer is constant. This assumes that there are sufficient nutrients for the required biomass to be reached and that conditions are sufficiently stable for the phytoplankton to be able to track the changes in Z_m and to grow accordingly. The relationship between P:R and the Z_{eu}/Z_m ratio shown in Harris (1978) is another way to present the same trend. The phytoplankton populations track the changing mixing conditions in surface waters if possible, adapt physiologically and grow until the biomass rises to the point where the light limited 'envelope' in surface waters is filled. This is another consequence of the rapid, opportunistic growth of the individual populations in surface waters. The tracking ability of the populations will be discussed below but it should be noted here that immediate responses are impossible and that severe perturbations in Z_m at a scale of days, interfere with this general relationship. Wofsy (1983) concluded that it took up to fifteen days to 'fill' the mixed layer with chlorophyll.

Bannister and Laws (1980) extended the analysis of nutrient saturated growth by examining the relationship between the parameters of the *P* versus *I* curve and growth rates in light-limited situations. Data from the literature indicates that the ratio of $I(0.7)/\theta$ is approximately constant (Myers and Graham, 1971; Steele, 1962) so that cells with a high cellular chlorophyll content saturate at low irradiance. Also Laws and Bannister (1980) presented data which showed that the dark respiration rate *R*, was a direct function of the growth rate so that in continuous cultures both θ and *R* are a function of the dilution rate (Fig. 8.2). According to the model, P_{max} is independent of nutrient limitation over a wide range of dilution rates but is very sensitive to the light limitation of growth in culture. This conclusion permits the interpretation of field observations of P_{max} and essentially predicts that in well illuminated water columns P_{max} should be relatively high and constant. The fact that this is so also seems to argue for a general lack of nutrient limitation in natural waters.

The effects of light limitation are quite different from the effects of nutrient limitation if θ, the C:Chl ratio is used as a master indicator of the algal response. At low light, and hence reduced growth rates, the cells build

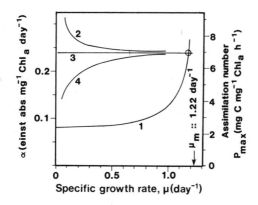

Fig. 8.2 The modelled relationship between growth rate and photosynthetic parameters for: 1 nutrient saturated growth and 2, 3, 4 nutrient limited growth P_{max} may vary little under nutrient limited growth. α is the light saturation parameter of photosynthesis. From Bannister and Laws, 1980.

in more pigments to compensate for the reduced supply of incident quanta. Under conditions of nutrient limited growth the C:Chl ratio rises as the growth rate is reduced. This effect seems to be the result of the accumulation of carbon in the cells.

Laws and Bannister (1980) extended the work on light limited growth by deriving a general model of nutrient limited growth in continuous culture. They were able to show that the Droop model (Equation 5.4) is a reasonable model of the behaviour of algal cells if, and only if, the assimilation rate of the limiting nutrient is a linear function of Q, the cell quota. Such a model can be used to make correlations between the growth rate of the cells and the elemental ratios of the major nutrients within the cells. This approach may only be used to assess relative growth rates for each species, but the relationship between relative growth rates and the elemental ratios of the phytoplankton has become an important tool in phytoplankton ecology and figures largely in the present debate about growth rates in oligotrophic waters.

Goldman (1980) has reviewed the relationships between relative growth rates and the ratios of cellular constituents in cells grown under nutrient limitation in continuous culture. Under nitrogen limitation the cellular chlorophyll contents, C:N ratios and C:Chl ratios all change in a predictable fashion. Elemental ratios become increasingly different from the

Redfield ratio as the relative growth rate declines. Figure 8.3 shows some representative results. Under phosphorus limitation a similar relationship exists (Fig. 8.3). Such correlations do not, unfortunately, give any information about the absolute growth rates of phytoplankton in nature. They do, however, give information about the physiological health and relative growth rates of such populations. Cellular ratios of major nutrients that are observed to be close to the Redfield ratio do seem to indicate high relative growth rates (Goldman *et al.*, 1979; McCarthy and Goldman, 1979) and do seem to be a frequently observed phenomenon. Such elemental ratios indicate that phytoplankton are, in reality, protein synthesizing organisms and that cellular ratios of protein, carbohydrate and lipid are approximately 50–60% protein, 30–40% carbohydrate and 10–20% lipid when the cells are growing rapidly. This was noted by Myers (1962) who also commented on the ironic situation that an organism such as *Chlorella*, which was selected by the early researchers on photosynthesis as an organism for the study of carbohydrate metabolism, is in fact an organism which preferentially synthesizes protein when healthy.

The short term (hours) response of the phytoplankton to patchy supplies of nutrients and to fluctuations in the irradiance regime is nested into a longer term response at a scale of days because the phytoplankton are also tracking the day to day changes in the turbulence structure and the depth of the mixed layer. Here we meet another problem; limnologists tend to make *in situ* nutrient uptake and productivity measurements once a week and the time between oceanographic measurements may be even longer. The true scales of motion in each water body need to be resolved before it is possible to say how often such measurements should be made.

High frequency (daily) productivity data is hard to come by but what exists shows up a high degree of day to day variability. Daily productivity measurements have been published by Rodhe *et al.*, 1958; Spodniewska, 1969; Coté and Platt, 1983; Sephton, 1980; Sephton and Harris, 1984, and all these papers show a surprising degree of day to day variability. Surprising, that is, if weekly data be taken as the norm. I will use examples from my own work and that of Sephton and Harris (1984) to illustrate what happens in a physically perturbed water body (Fig. 8.4). The day to day fluctuation in such parameters as P_{max} and I_k is greater than would be expected from weekly observations. The algae are responding to changes in the Z_{eu}/Z_m ratio at a scale of days and in most temperate lakes the situation is completely non-steady state as the period between storm events is roughly equivalent to the response time of the algae. In our time series work in Hamilton Harbour (Sephton and Harris, 1984; Harris, 1982, 1983) the atmospheric forcing function had a characteristic period of about 200 hours and the algal response ranged over the 100–125 h time scales (cf. Figs 5.3 and 5.4). The paper by Harris *et al.* (1980a) showed that the photosynthetic

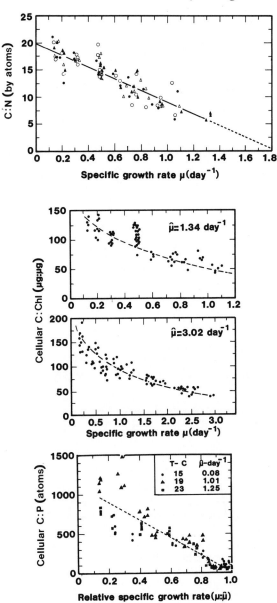

Fig. 8.3 The relationships between elemental ratios and relative growth rates of cultured phytoplankton. Data redrawn from Goldman (1980).

characteristics of the phytoplankton were best related to the environment one week before, but that result was based on weekly data. An analysis of the daily data by Sephton and Harris (1984) indicates that the true lag is

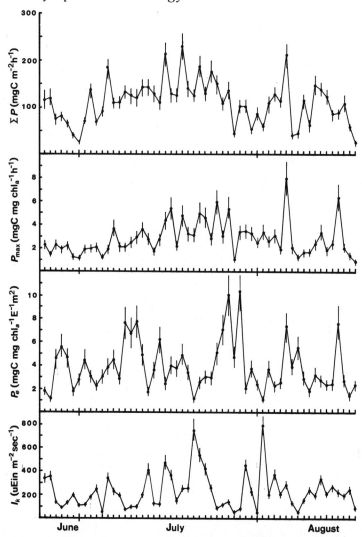

Fig. 8.4 Day to day fluctuations in photosynthetic parameters of a natural phytoplankton assemblage in Hamilton Harbour. From Sephton and Harris (1984). Integral productivity (ΣP); assimilation number (P_{max}), photosynthetic efficiency (P_e) and saturation intensity (I_K).

nearer five days or 125 h. This is entirely consistent with the recent work of Jones (1978). It looks as though a lag of one or two generation times is the time constant for responses which require growth responses on the part of the cells (Harris, 1982a). Standard sampling and incubation techniques resolve the hour to hour and day to day fluctuations in physical and biological parameters rather poorly.

8.1 The effects of environmental variability on growth rates

Harris (1983) recently presented data on the apparent growth rates of phytoplankton in Hamilton Harbour. Apparent doubling times were about one day, under conditions of continuous mixing but ranged from two to eight days during the intermittently mixed conditions of summer. The results clearly indicated that the populations in Hamilton Harbour were saturated by nutrients and light and the uptake ratios indicated that the populations were growing at, or close to, their maximum relative growth rates. Why should the apparent maximum growth rates of the summer populations have decreased when conditions in surface waters were apparently favourable? To explain this it is necessary to distinguish between the mean level of a resource and the accompanying variance. The somewhat slower rates of growth in the perturbed physical conditions of summer meant that the disturbances in the environment had reached the point where it had become difficult for the organisms to track the variability and that the populations were being limited by the variance in resources.

It takes at least one generation time for phytoplankton populations to adapt to changed conditions (Harris, 1978) and some recent work by Brand (personal communication) indicates that it may take five to twenty generations for the new growth rate to be established. The match between the significant scales of perturbation and the growth rates of the organisms explains the slower growth rates observed in summer. It is known that temporal variance causes instability in ecological systems (Levins, 1969; May, 1974) and it would appear that the algae, by averaging over time periods equivalent to the characteristic frequencies of perturbation, have evolved a strategy to minimize the effects of such variance (Slobodkin and Rapoport, 1974; Harris, 1980a). The biological cost of tracking a more variable environment was a reduced growth rate. Such was predicted on theoretical grounds by Levins (1968) and Heckel and Roughgarden (1980). Once again the properties of filters and holons are demonstrated.

One simple way to envisage the effects of environmental variability on growth rates is to use the model proposed by Levins (1979). In that paper, Levins showed that species could coexist in a variable environment by using both the mean level of a resource and the variance of that resource as separate components. The critical relationship concerned the growth response to changes in resource levels. If the organism showed a linear relationship between growth and resource availability then no amount of variability would affect the growth rate. Non-linear responses led to enhanced growth in the presence of resource variability if the response was exponential and reduced growth if the response was saturating (Fig. 8.5). As most phytoplankton growth curves are of the Monod type which show a saturating growth rate at high nutrient concentrations, then an environment

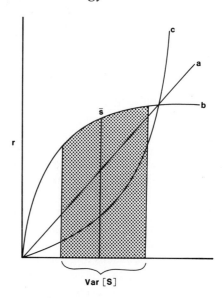

Fig. 8.5 The effect of environmental variation depends on the shape of the functional response. Saturating relationships between growth (r) and resource level (S) lead to a damped, integrative response (b). Linear responses leave growth unaffected (a) whereas exponential responses (c) may lead to increased rates of growth in fluctuating environments. Most phytoplankton seem to be typically characterized by response (b).

with high environmental variability should result in a reduced growth rate. Monod type responses are typical of the integrating and filtering behaviour of cellular holons. The key relationship between growth and the mean and variance of the resource levels depends on the fact that, for non-linear processes, the average resource level over time is not the same as the expected value – the integrated value over time.

8.1.1 DIVERSIFICATION BETWEEN SPECIES

It begins to look as though the first thing that must be done in any water body is to characterize the motions in the mixed layer and to design a sampling and analysis scheme accordingly. This is really only good science; a failure to appreciate the true sources of variance in the system can only lead to horrendous problems of interpretation. Aliasing problems can be caused by sampling at too infrequent a time interval (Fig. 8.6) and such problems appear to be common in limnology and oceanography. We have, in the past, underestimated the true degree of high frequency variance in planktonic environments and we have not appreciated the importance of the short time constants of the algal response. Theoretical arguments may be made to show that coexistence in a non-equilibrium phytoplankton commu-

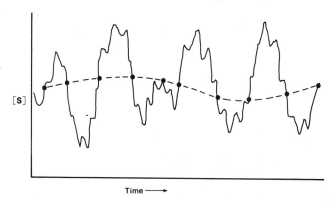

[S]

Time ⟶

Fig. 8.6 A simple representation of the effects of aliasing. Sampling a resource (S) at too infrequent intervals (dots) leads to a severe underestimate of the true degree of variability.

nity may be most easily facilitated by variations in the time constants between species (Levins, 1979; Schaffer, 1981; Harris, 1983). Data presented by Harris (1983) indicates that each of the species in the water column has a specific set of responses to both the mean and variance components of the environmental fluctuations. So what physiological and biochemical strategies may be diversified to produce this result?

(a) *Nutrient uptake characteristics*

Two parameters are of importance; the mean nutrient flux (as determined by the trophic state of the water body, external inputs and regeneration, Chapter 4) and the 'patchiness' of that flux in time. The diurnal periodicity shown by many features of phytoplankton physiology (Harris, 1978) is one mechanism by which phytoplankton may coexist. Stross and Pemrick (1974) showed that phytoplankton species had different diurnal periodicities of nutrient uptake over the day and could therefore utilize the same nutrient flux in quite different ways. When coupled to a diurnal periodicity in the regenerated nutrient flux (from the vertical migration of grazing animals, for example) coexistence is theoretically possible. Recent papers by Turpin and Harrison (1979, 1980) have shown that it is possible to manipulate the species composition of perturbed continuous cultures of phytoplankton by varying the periodicity of nutrient pulses. The dominant species in their continuous cultures could be altered by changing the periodicity of the nutrient pulsing. They compared true continuous cultures at steady state with cultures in which everything was held constant and the same amount of nutrient was added but it was added in the form of periodic, discrete pulses. Frequent additions of small pulses of nutrient favoured small phytoplankton and as the pulses became larger and less frequent the regime favoured larger and larger cells (Harrison and Turpin, 1982). These relationships (Fig. 8.7) indicate that differences in the uptake kinetics and cell quotas of limiting nutrients in different species can influence the outcome of competition.

Quarmby *et al.* (1982) showed that a number of features of the nutrient uptake kinetics of phytoplankton were changed by a patchy nutrient regime. They showed that not only were there basic physiological differences between the species studied but that the physiology of the species altered in response to the pulsed nutrient supply. Their work shows that there is a close coupling between the periodicity of the nutrient pulses and the physiology of the phytoplankton. The organisms tracked the pulses of nutrients with surprising precision as the concentration of NH_4^+ in the cultures became unmeasurable just before the addition of the next pulse. In short, the phytoplankton adjusted their uptake and storage capabilities and growth rates to suit the environmental periodicity. The addition of a single daily pulse of NH_4^+ to continuous cultures of *Chaetoceros* and *Skeletonema*, as compared to a continuous supply, resulted in increases in the velocity of nutrient uptake, the appearance of biphasic uptake kinetics, decreases in the value of K_s and the appearance of a diurnal periodicity in the photosynthetic activity in the cells. The effect of the addition of a single daily pulse of ammonia led to biphasic kinetics as the cells exhibited an initial very rapid uptake of nutrient when the pulse was added (V'_{max}), followed by a longer uptake phase controlled internally (V_i). The initial surge of nutrient uptake presumably represents the rapid replenishment of internal storage pools. As the initial phase of uptake is determined more by transport through the wall than by cellular growth processes the apparent half saturation constant is lower, as it is the true K_t not K_s. When coupled with internal physiological events the uptake, storage and metabolism of the nutrient produced a strong diurnal periodicity in other physiological pathways, notably photosynthesis. The physiological differences between the cells were maintained and increased in the patchy environment.

Turpin *et al.* (1981) recently modelled the process of phytoplankton growth in a patchy nutrient environment and showed that growth rates are greatly influenced by the patchiness in the nutrient supply. Species which dominated the populations in the cultures under steady state conditions were eliminated under patchy conditions. Turpin *et al.* (1981) used a Droop model in which the growth rate was determined by nutrient uptake rate, cell quota replenishment and the magnitude and timing of the nutrient pulses. Figure 8.8 shows how the model represents the growth rate of two hypothetical species as a function of S_{av}, the average nutrient concentration and P, the patch periodicity. Patch duration was kept constant. By assuming that the effective K_s of the population was reduced under patchy conditions (i.e. the substrate affinity was increased over time as the period between the

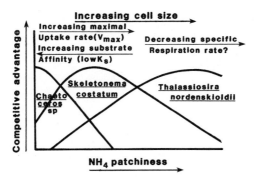

Fig. 8.7 The influence of environmental patchiness on the selection of phytoplankton of differing size and physiological properties. (Harrison and Turpin 1982.)

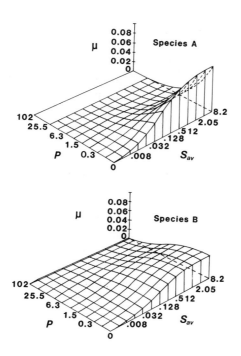

Fig. 8.8 The growth response of two hypothetical phytoplankton to changing culture conditions. The variables controlling the growth rate (μ) were the average nutrient concentration (S_{av}) and the patch periodicity (P). Model by Turpin *et al* (1981).

pulses increased), Turpin *et al.* (1981) were able to produce a growth response for a patch adapted population which showed increased growth rates in the patchy environment and a reduced average K_s. Thus physiological adaptation to the patchy environment results in altered outcomes for the competition experiments in culture (cf. Chapter 5). The steady state analogy cannot be applied as the physiology of the species measured at steady state does not represent the situation in the field. All this raises the question of the extent of steady state nutrient fluxes in the field. One guess might be that in oligotrophic water, where the background concentration is low and regeneration is crucial, patchiness and fluctuations are the rule.

The ecological consequences of this result are considerable as community structure will be a result of the physiological properties of the species under fluctuating nutrient regimes. The resulting community structures will be a complex function of the environmental spectrum of variability (Figs 5.3–5.7) and the spectrum of cell sizes and cell quotas. The integrative nature of cellular physiology is important in that the various pathways of metabolism are buffered from the external environmental fluctuations by the storage pools and the ability to deflect metabolism from one pathway to another. The tracking ability of cells is not perfect however, and there is a time-lag between the filling of a depleted storage pool and the expression of the nutrient uptake as an altered growth rate. It has been shown that time-lags are required in order to correctly account for the growth and metabolism of phytoplankton in patchy environments (Caperon, 1969; Cunningham and Maas, 1978; Burmaster, 1979a,b). Grenney *et al.* (1973) showed that coexistence between species in chemostats could be produced by fluctuations in the dilution rates and inflow concentrations. Cell size appears to be an important determinant of function in such systems.

A number of physiological properties are a function of cell size. Larger cells appear to be favoured in environments in which the nutrient pulses are large and few and far between as, despite the fact that they exhibit larger K_s values (Malone, 1980a), they exhibit higher uptake rates (V_{max}) and have larger storage capacities than smaller cells. Small cells appear to be favoured by constant environments or environments in which the nutrient pulses are small and frequently available. Small cells have low K_s values but small storage capacities. The impact of these conclusions on methodology is also considerable as the use of increased nutrient concentrations and extended incubation times in patchy environments will lead to some complex artifacts. The problems will differ in oligotrophic and eutrophic waters as the systematic changes in cell size, background nutrient concentration and periodicity of pulsing in the different water types will affect the results in different ways. It appears that results from eutrophic waters would be more reliable than those from oligotrophic waters for a number of reasons.

Eutrophic waters have a higher background nutrient concentration and appear to be less dependent on small scale regeneration processes.

(b) *Photosynthetic characteristics*

Vollenweider has recently derived an equation to describe the response of phytoplankton in the mixed layer subject to a given set of physical conditions. The equation is only strictly applicable at steady state but the relationships between the variables conveniently summarize the important interactions. Once again it will be necessary to examine both the effects of the mean level of the underwater irradiance and the effects of temporal variation in that mean. The maximum biomass in the photic zone may be expressed as:

$$\gamma_{max} = 1/k_c [1/r \; 1/Z_m \; \ln(2I_oP_e)/P_{max} - k_b]$$
8.9

where k_c is the attenuation coefficient due to the phytoplankton, and k_b is that due to background absorbtion in the water. The respiration rate, r, is expressed as a fraction of the maximum photosynthetic rate per unit biomass (P_{max}), Z_m is the depth of the mixed layer, I_o the surface irradiance and P_e the slope of the initial linear portion of the P versus I curve (mg C mg Chl^{-1} E^{-1} m^2). Phytoplankton growing in the surface mixed layer have the possibility of responding to changes in I_o, k_b and Z_m by physiological and structural adaptations. Once again I_o and Z_m are shown to be the master driving functions. Z_{eu} (through the interaction of k_c and γ_{max}) is the other master variable though it is a complex function of background attenuation and self-shading by the phytoplankton (Wofsy, 1983). Bannister (1974a,b, 1979) regards k_c as constant so the algal response is limited to changes in r, P_{max} and P_e. P_e is a function of ϕ, the quantum efficiency, and cellular chlorophyll contents as it relates the amount of C fixed to the incident irradiance.

Published data clearly show that phytoplankton adapt to the mixing regime in which they grow (Harris, 1978) but the precise degree of adaptation will depend on the ability of the cells to track the changes in I_o and Z_m. The first line of defence of cells when confronted by changes in the irradiance regime lies at the level of electron transport (Harris, 1980b,c). Subsequent changes in physiology effect changes in P_e, P_{max} and I_k through changes in the number and size of the photosynthetic units (Falkowski, 1980). Longer and larger perturbations go deeper into the hierarchy of cellular control mechanisms. The steady state response of phytoplankton to changed irradiance levels has been well modelled by Bannister (1974a,b, 1979; Laws and Bannister, 1980; Bannister and Laws, 1980) and was reviewed in Harris (1978).

The limits of adaptation of the various species in the mixed layer are less than might originally be supposed so that the effects of strong fluctuations in I_o and Z_m at a scale of days are to effect changes in the species composition of the phytoplankton assemblage rather than physiological adaptations by the same set of species. Data presented in Harris *et al.* (1980) showed that perturbations in Z_m drove lagged changes in all of the above physiological parameters and we now know that these physiological changes were, in fact, due to changes in the species composition in surface waters (Harris *et al.*, 1983). The variability in the daily data (Fig. 8.4) indicated the extent to which the phytoplankton were being forced to track the environmental changes in Hamilton Harbour. Time series analysis (Harris *et al.*, 1983) showed that the major species in the Harbour were switching growth on and off in response to the environmental fluctuations (Fig. 8.9) and that the community response was a function of these individual specific responses.

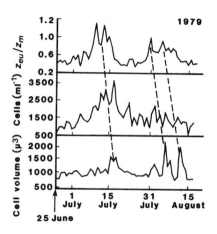

Fig. 8.9 The daily fluctuations in the ratio of Z_{eu}/Z_m, population density and cell volume of *Oocystis borgei* in Hamilton Harbour (Harris *et al* 1983). The time lag between peaks in the Z_{eu}/Z_m ratio and peaks in cell volume (representing a decrease in growth rate) was equal to one generation time.

Harris (1978) presented evidence that certain groups of algae showed systematic differences in $P:R$ ratios. Diatoms had the highest $P:R$ ratios and dinoflagellates the lowest; green algae were intermediate between the two. The adaptation of natural populations to the mixing conditions in surface waters appears to result from the selection of species that grow most rapidly under the prevailing conditions. The sequence of reducing $P:R$ ratios from diatoms, through green algae to dinoflagellates mirrors the seasonal progression of reduced mixing depth and points to a physiological explanation for the seasonal progression of species.

Changes in P_e in the Harbour were found to be due to changes in both the mean cell size of the entire phytoplankton assemblage and to changes in the mean cell size of the sum of the individual species present (Harris *et al.*, 1983). These changes in cell size are sufficient in themselves to negate the assumptions of steady state as the uptake and storage terms of Fig. 6.3 are themselves functions of cell size, as are the sedimentation velocities (Fig. 5.2). From week to week the cell volume of the dominant green alga (*Oocystis*) fluctuates as the growth rate fluctuates (Fig. 8.9) and the time-lag between the environmental change and the alteration in volume is, indeed, equal to the doubling time of the cells (Harris *et al.*, 1982). The mean cell volume of phytoplankton appears to reflect the growth rate; rapidly dividing cells tend to be smaller. Small flagellates in Hamilton Harbour have lower quantum efficiencies than larger cells (Fig. 8.10) and the systematic

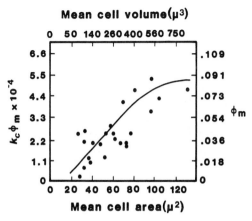

Fig. 8.10 The relationship between mean cell volume and the product of specific chlorophyll absorbance and quantum efficiency. Data from natural populations in Hamilton Harbour (Harris, *et al.*, 1983). The maximum quantum efficiency, ϕ_m, was obtained by assuming that the value of k_c was $0.008 \text{m}^2 \text{ mg}^{-1}$ (Bannister, 1974).

changes in P_e from week to week indicate that groups of species of varying size are selected for under different environmental conditions.

Laws (1975) showed that there was a size dependent relationship between P_{max} and r, in which large cells suffered smaller proportional respiratory losses. Laws (1975) used the relationship between P_{max} and r to predict the size spectrum of phytoplankton in various parts of the ocean. Harris (1978) denied the existence of such a relationship because 'systematic changes in the net growth efficiency . . . must not occur in cells of different sizes. If such systematic changes did occur, then marked seasonal changes in the mean cell volume of natural populations would occur' (Harris, 1978, p. 118). The reason that I and Banse (1976) denied the existence of such a relationship was that the calculation of net growth efficiencies from growth rates is fraught with danger (Malone, 1980a,b) and little evidence of seasonal cycles of cell size was then available. I have recently presented evidence of such cycles and have had cause to point out the error of my previous assertion (Harris *et al.*, 1983). What this cautionary tale does indicate is that the measurement of efficiencies is as important as the measurement of growth rates and that both types of information are necessary. If anything, the data presented here indicate that efficiencies give more valuable ecological information than absolute or relative growth rates.

The size spectrum of phytoplankton in the Harbour is not smooth but has 'bumps' which correspond to certain groups of species (Fig. 8.11). Significantly the papers of Sheldon and Parsons (1967) and Sheldon *et al.* (1972)

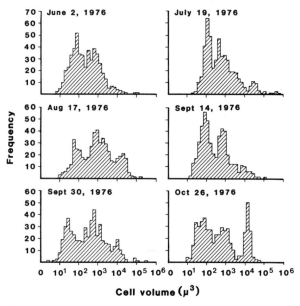

Fig. 8.11 Representative size spectra of natural populations in Hamilton Harbour. (From Harris *et al.*, 1983.)

show that the size spectra of phytoplankton in the oceans are not smooth and that peaks occur which can be identified with certain groups of organisms. The abundances of the 'small' and 'medium' groups in the Harbour appear to be mutually exclusive and the mechanism of the switch appears to be fluctuations in the underwater irradiance regime brought on by changes in the stability of the surface water, and the interaction of water column stability with photosynthetic efficiencies. The appearance of the 'large' group of phytoplankton in the Harbour is clearly the result of the interaction between cell size, vertical mixing and sedimentation velocities. The 'large' cells are a large centric diatom (*Stephanodiscus*) which sinks rapidly and only occurs when the Harbour water is strongly mixed vertically. There are therefore a series of complex interactions between the physics of the water column, the observed photosynthetic parameters, the interpretation of productivity data and the phytoplankton assemblage present.

The fluctuation in the size of individual species and the fluctuations in the photosynthetic efficiencies of phytoplankton of differing size raise some important questions about the interpretation of productivity data. The use of chlorophyll as a measure of algal biomass has led to the widespread use of such indices as the photosynthetic capacity (or assimilation number) where the photosynthetic rate is expressed per unit biomass. The use of such indices glosses over a great deal of significant variation in cell size and the characteristics of the phytoplankton assemblage, even when the species composition does not change. When the species composition is also changing, does this not lead to the comparison of apples and pears?

The use of chlorophyll to scale the productivity data in situations where the phytoplankton assemblage is changing leads to confusion as to whether differences in the estimate of P_{max} in any given situation are due to physiological changes or changes in the species present, or both. It has recently been shown (Trimbee and Harris, 1983) that the motion of phytoplankton populations with lake basins occurs over time scales equivalent to the time scales for adaptive changes in photosynthetic parameters. Samples taken in a small reservoir on a daily time series showed that advection of species within the basin was a prevalent feature of the lake and that events at a single station were a complex function of events at that station plus advective effects from populations washed in from elsewhere. The advective effects occur over time periods of two to three days and further complicate the interpretation of data from weekly sampling schemes. If, as appears to be true, many of the species present in the water column at any one time are, in fact, doing very little (Watt, 1971; Maguire and Neill, 1971; Stull *et al.*, 1973; Harris *et al.*, 1983) then the productivity expressed per unit biomass may change dramatically for reasons due more to shifts in which species are active than to changes in physiology. Stull *et al.* (1973) concluded that some minor species contribute more heavily to productivity than to biomass and that the seasonal pattern of primary productivity in Castle Lake was closely correlated with the changes in biomass of several of the less abundant species (Amezaga *et al.*, 1973). Stull *et al.* (1973) also showed using autoradiographic data that there was a correlation between cell size and C specific growth rates. It is now widely accepted that small cells turn over much faster than large ones (Malone, 1980a,b). The diversity data analysed in Harris *et al.* (1983) showed that the diversity of phytoplankton in the mixed layer is high due to the presence of many persistent populations. Rapid growth and persistence during unfavourable conditions will lead to high diversity and many inactive populations. This will be particularly true in highly variable environments. Given these problems perhaps it is time to stop dividing the rate of productivity by the biomass to obtain such indices as P_{max} unless the species composition of the phytoplankton assemblage and the contribution of individual species to the total productivity is well known.

I have already discussed the varying abilities of different species to utilize the various forms of carbon in surface waters. The ability of species such as blue–green algae and dinoflagellates to utilize bicarbonate at high pH allows such species to grow at high densities in late summer (Talling, 1973, 1976). Conversely the susceptibility of diatoms to carbon limitation at high pH (Jaworski *et al.*, 1981) limits the growth of such species to periods of active vertical mixing early and late in the year. It is becoming clear that broad groups of phytoplankton possess suites of physiological attributes which permit growth in certain waters at certain times of the year.

(c)   *Temperature effects*

The effects of temperature on the physiology of phytoplankton are as complex as the effects of nutrients and light. The time scales of variation in temperature are less rapid than either of the other parameters as, while variance in light and nutrients are influenced by temperature (and hence density) differences, the mean temperature of surface water changes rather slowly in a seasonal pattern or not at all. The temperature 'signal' is one of slow seasonal change in temperate waters but, even so, there are discernible effects of both the mean and the variance components. There is a clear overall temperature effect on the growth rate of phytoplankton (Eppley, 1972; Goldman and Carpenter, 1974) and the maximum rate of growth doubles, roughly, every 10°C increase in temperature (a Q_{10} of approximately 2.0). Temperature, therefore, sets an upper limit to the rate of growth.

Phytoplankton show marked physiological adaptation to the ambient temperature and data from natural populations indicates that when respiration rates are expressed per unit chlorophyll there is a tendency to maintain a rate of approximately 1 mg O_2 mg Chl^{-1} h^{-1} (Ryther and Guillard, 1962; Harris and Piccinin, 1977; Jewson, 1976; Jones, 1977). Changes in the respiration rate at different temperatures appear to be effected by changes in the cellular protein contents. The examination of the effects of temperature on photosynthesis has been somewhat confused by the effects of culture conditions (Morris and Glover, 1974; Harris, 1978) but there are clear temperature effects on the cellular composition of cells. The effect of temperature on photosynthesis appears to be invariable (Harris and Piccinin, 1977) so the only adaptive mechanism available to the cells is to change the ratios of the major cellular constituents. Temperature has a direct effect on the ability of cells to track other environmental fluctuations. Temperature presumably affects physiological processes which affect the synthesis and degradation of macromolecules.

The cell quota of a number of cellular constituents shows a minimum value at the optimum temperature of growth (Goldman, 1977a, 1979; Goldman and Mann, 1980) and a U-shaped function best describes the response of cell quotas to temperature (Fig. 8.12). As a general rule it

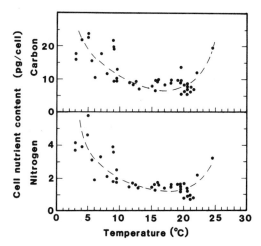

Fig. 8.12 The relationship between temperature and the cell quota of nutrients in cultured populations of *Phaeodactylum tricornutum* (Goldman and Mann, 1980).

appears that the minimum cell size coincides with the maximum growth rate. Goldman and Mann showed that the effects of temperature on cell quotas could only be resolved with accuracy under nutrient saturated conditions and that the complexities in the interactions between nutrient uptake, cell quota, growth rate, cell size and temperature made the interpretation of Droop type models very difficult. The complex interactions between the nutrient uptake and growth parameters at different temperatures can lead to complex competitive interactions between species (Goldman and Carpenter, 1974) and can certainly lead to changes in the mean cell size of phytoplankton assemblages. Goldman and Ryther (1976), Goldman (1977b), and Goldman and Mann (1980), showed that seasonal temperature cycles led to distinct periods of dominance by different species of phytoplankton at different seasons in the Woods Hole mass culture experiments. The outcome of the competition between the species (*Skeletonema, Phaeodactylum* and *Thalassiosira* and others) depended on cell size, nutrient uptake and growth rates but the results in the mass cultures were not the same as the results in laboratory cultures, indicating that not all the factors were completely understood. The data from laboratory cultures indicated a U-shaped temperature effect on growth rates and cell quotas and, as in previous examples, the progression to competitive exclusion took a number of days (Fig. 8.13). Field experiments indicated further complexities as the species shifts, which were associated with the seasonal temperature changes, occurred at different temperatures depending on whether the temperature was rising or falling.

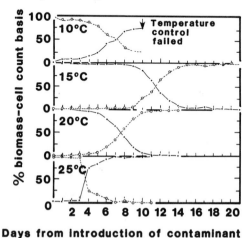

Fig. 8.13 The time course of competition between *Thalassiosira pseudonana* (closed circles) and *Phaeodactylum tricornutum* (open circles) at various temperatures in mass cultures of algae (Goldman and Ryther, 1976).

There is some indication that certain species of phytoplankton exhibit temperature thresholds. For example, *Ceratium hirundinella* does not appear to grow at temperatures below 12 °C (Harris, 1983; Heaney *et al.*, 1983). If no other factor were limiting, the slow seasonal temperature cycle would be sufficient to produce significant changes in the species composition of natural phytoplankton communities as the temporal scale of the seasonal cycle is much longer than the time required for competitive exclusion. The overall biomass in the mass cultures was relatively independent of temperature but the mean cell volumes were greater at colder temperatures as the species composition was changed.

8.2 Energy, nutrients and cell size: a synthesis

All the data presented point to a classical case of non-steady state conditions in the phytoplankton assemblage. Physical processes in the atmosphere and in the water column lead to environmental fluctuations, which the cells attempt to track by means of adaptations and deflections in physiological pathways and/or by changes in growth rates and species composition. The time scales of the environmental fluctuations which drive physiological changes interact strongly with the time scales of the algal response. Processes which cause changes in the species composition occur at longer time scales but interact strongly with growth processes. This would predict that there would be cells in the phytoplankton assemblage in surface waters that would be in a state of physiological stress or reduced growth; these would be species for whom the environment was temporarily unfavourable or species which had been washed in from elsewhere in the basin. This aspect of phytoplankton ecology was identified by Hutchinson (1941, 1953, 1961, 1967) many years ago. All we have done is to identify the time scales and to formalize the arguments.

Under non-steady state conditions the storage components of the physiological pathways become very important and the phytoplankton cell can be looked on as an integrator which is attempting to buffer the internal machinery from external fluctuations whilst growing at rates as close to μ_{max} as possible. Under non-steady state conditions short term measurements of the flux of nutrients into cells may overestimate the actual rate of growth if what is being measured is the temporary replenishment of storage pools (Morris, 1980b; Eppley, 1981). With rapid uptake and storage the phytoplankton may only require nutrients for a brief period during the cell cycle. Populations may continue to grow rapidly as long as they are supplied with N or P once a week (Eppley, 1981). While growing on depleting internal reserves natural populations may simultaneously show physiological evidence of nutrient depletion and near maximal growth rates. The actual situation will depend on the precise combination of growth rate, storage pool size, uptake characteristics, lag-times and the mean and variance of the nutrient flux and irradiance (Fig. 7.1). It is unlikely that all species will show the same characteristics and it is dangerous to assume that all the species present are doing the same thing at the same time. As many of the above characterics are correlated with cell size, I believe that such diversification is the major reason for the spectrum of cell sizes usually observed.

The recent revolution in thought has led to some new insights into the way these organisms interact with their environment (Fig. 7.1). The problems of measurement and the interpretation of data have forced us to reconsider many features of the ecology of phytoplankton and have demonstrated that unless we fully understand the dominant time and space scales of different environments we cannot properly interpret the data obtained by *in situ* methods. We now see that these organisms are sensitive to a wide range of environmental perturbations and that we have in the past tended to underestimate the role of small scale, rapid events. The apparent dynamic, non-equilibrium status of the planktonic ecosystem requires that we match the time scales of physical, chemical and biological processes before we can measure anything with confidence. The improved understanding of phytoplankton ecology goes hand in hand with improving knowledge of nutrient regeneration in different water types (Chapter 7) and is entirely consistent with such ideas. The integrative nature of the cellular physiology of phytoplankton is a clear indication that the organisms have the ability to track the changing external environment and to make adaptive changes which enhance their chances of survival. The adaptive changes that are possible range from alterations in the flow of materials in intracellular pathways, through changes in cellular composition, to changes in cell size and the species composition of the phytoplankton community in surface waters. The importance of such strategies for those measuring productivity and growth lies in the fact that the time scales of *in situ* experiments overlap the time scales of many of these responses and the mere fact of containing the cells during the incubation may induce adaptive responses; responses which are not typical of the unconfined populations. The relative disregard of the mortality terms has led to severe problems with the interpretation of data from bottle experiments in oligotrophic waters and will certainly lead to new methods of measuring productivity which will not require the use of bottles (cf. Chapter 6). A variety of such methods has begun to appear and the search for new methods will doubtless continue.

CHAPTER 9

Seasonal patterns of distribution and abundance

At scales much longer than the generation times of the organisms (weeks to months) it is possible to observe large scale patterns in the distribution of individual species and of biomass in space and time. Physical processes operating at these scales include seasonal thermoclines, persistent fronts, rings and eddies. Biomass accumulations and persistent physical features lead to large scale changes in the supply and demand of macronutrients (particularly N,P,Si) and strong seasonal changes in the soluble pool sizes of such elements.

9.1 Spatial distribution of biomass

The accumulation of phytoplankton biomass is brought about by the maximization of factors favouring growth and the minimization of loss factors. Furthermore, appreciable biomass accumulation will only take place in areas where sustained new production is possible. In fronts and thermoclines the necessary conditions are met as the environment is stable for times in the order of months, turbulent diffusion is reduced by the sharp density gradient and nutrients may be available from the entrainment of deeper, nutrient rich water. This combination of factors is responsible for the deep chlorophyll maximum in the oceans (French *et al.*, 1983) and in the Great lakes (Mortonson and Brooks, 1980). The deep chlorophyll maximum can arise by organisms sedimenting out of surface waters and accumulating in the thermocline. A small decrease in the sedimentation velocity will be sufficient to cause an accumulation of organisms in the stable layer (Hutchinson, 1957a). Growth in the layer is also possible if the thermocline or front is in the photic zone. Many species of buoyancy regulating phytoplankton take advantage of such situations to achieve considerable biomass accumulations. Such species include blue-green algae and dinoflagellates which cause noxious scums and red-tides. Buoyancy regulation and vertical migration serve to minimize the loss of cells by sedimentation and to maximize the growth rate by virtue of the fact that the organisms may move vertically to obtain the optimal levels of light and nutrients (Harris *et al.*, 1979). Such organisms display strong covariance relationships with indices of water column stability (Harris, 1983).

The interaction of phytoplankton growth with mesoscale physics produces strong horizontal patchiness in biomass distributions at scales of kilometres. Such patterns are clearly visible in satellite and space craft images of large lakes and the oceans (Figs 3.5 and 3.6). In Fig. 3.5, the thermal bar of Lake Michigan is clearly visible in the thermal image. The phytoplankton biomass is visible as lighter nearshore bands which display higher reflectance in the green. These nearshore patterns are most evident in the region of the front. Phytoplankton also accumulate in tidal fronts in areas such as the English Channel (Fig. 3.6) (Holligan and Harbour, 1977) and accumulation of dinoflagellates occurs in these fronts below the surface when a combination of stability and nutrient availability favours growth.

The satellite image of the east coast of the United States obtained by the Coastal Zone Colour Scanner (CZCS) was a striking example of the interaction of mesoscale physics and phytoplankton biomass (Matthews, 1981). The oligotrophic waters of the Gulf Stream and the warm core rings were clearly visible, as was the shelf-break front which runs parallel to the edge of the continental shelf. Phytoplankton biomass was highest in the shallow nutrient rich waters on the shelf. The image was obtained in Spring when the Spring burst of new production was at its height in shallow water.

9.2 The seasonal cycle of phytoplankton growth

The seasonal cycle of phytoplankton biomass is determined by seasonal fluctuations in high level variables such as solar irradiance and total nutrient pools. Given sufficient nutrients, the phytoplankton biomass rises until the mixed layer is 'filled' (Wofsy, 1983; Chapter 8) as long as the mixed layer is not much deeper than the depth of the photic zone. The relationship between phytoplankton growth rate and energy availability leads to a strong seasonal effect at high latitudes in the oceans and an effect of both latitude and of altitude in freshwaters. Deep mixing and/or low insolation reduces the growth rate of phytoplankton in high latitude and high altitude waters in winter. The effects of deep mixing will be eliminated in shallow water so the seasonal cycle of biomass is different in inshore and offshore waters. If the pool size of N or P limits the biomass (a common occurrence in stratified water) then algal biomass rises until limited by TN or TP. The seasonal cycle of biomass is a smoothed sum of the individual abundances of all the species present so the biomass will fluctuate less than the abundances of the individual species. The biomass will fluctuate as the species overshoot the ceiling or if the level of the ceiling itself varies as a result of the removal of nutrient stocks by sedimentation or by changes in the mixed layer depth due to storms. Such variations are clearly visible in Hamilton Harbour (Harris *et al.*, 1983) where, due to shelf-shading and dissolved organic carbon in the water, energy is limited and large fluctuations in mixed layer depth occur. Because of the many species which make up the overall algal biomass simple empirical relationships between annual averages of *TP* and biomass may be found. As the ceiling on the algal biomass is a smooth function of the seasonal availability of light or nutrients, the seasonal biomass cycle rarely exhibits more than three or four peaks in the year. By far the most common pattern is that of a unimodal or bimodal cycle.

Different sampling strategies may give quite different views of the seasonal cycle and may introduce spurious fluctuations if horizontal and vertical patchiness is not fully taken into account. Seasonal cycles reconstructed from data from single stations or from single depths may not be representative. Deep chlorophyll maxima, which often occur in summer, will be missed entirely if surface samples are used to reconstruct the seasonal cycle of biomass. Large scale advection may also modify the seasonal cycle as observed at a single point if the seasonal biomass cycle differs in the different water masses. As the seasonal biomass cycle is smoothed over time weekly sampling is usually adequate to define the overall pattern, although care must be taken to ensure good spatial coverage each time a sample is taken.

One of the most complete sets of oceanic data which gives good spatial and temporal coverage and demonstrates the effects of physics on the seasonal cycle of phytoplankton is that taken from the Continuous Plankton Recorder (CPR) surveys. The CPR was designed by Sir Alistair Hardy before the War (Hardy, 1935b) and has been towed at 10 m depth on standard routes around the British Isles since 1948. The CPR is basically an automatic plankton net. Plankton is filtered off onto a continuously moving band of silk with a mesh size of 270 μm and the plankton is preserved as the silk is wound into a tank of formalin. After the journey the CPR is returned to the laboratory and the silk is cut into sections each representing ten miles of tow. Data from the ten mile sections of tow are blocked up into standard areas (Fig. 9.1). for analysis. The two tracks and analysis methods have remained constant for nearly 40 years so a long block of comparable data is available.

The CPR was originally designed to sample zooplankton and the mesh size is far too large for phytoplankton. Many algal cells do get trapped in the mesh, however, as the silk is spun from finer threads and many algal cells get trapped on larger particles. Despite the problems, it is clear that counts of the major phytoplankton species trapped in the sampler do reflect the gross

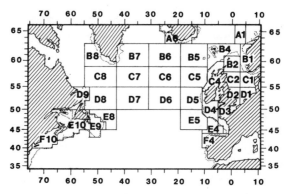

Fig. 9.1 The standard areas used in the analysis of the Continuous Plankton Recorder data. Data from individual tows are blocked up into these standard areas for subsequent analysis. Marginal notation: degrees latitude and longitude.

changes in abundance associated with seasonal cycles and the larger year to year fluctuations in abundance (Robinson, 1970). Phytoplankton data is usually expressed in the form of counts of diatoms or dinoflagellates. A visual assessment of phytoplankton 'colour' is also made, as chlorophyll derivatives remain on the silk from trapped cells and broken chloroplast fragments. This 'colour' index is also an assessment of total algal biomass.

The plots of the seasonal fluctuations in total diatoms, dinoflagellates and 'colour' bear little or no resemblance to one another. The spring diatom outburst usually comes earlier than the 'colour' peak and the dinoflagellate peak comes later. Reid (1978) noted this discrepancy and suggested that an unidentified group of species was responsible for the 'colour'. From the timing of the 'colour' peak it seems clear that the spring outburst of '*r*' strategist species is responsible for the majority of the 'colour' so that we are, in effect, looking at a diatom, flagellate, dinoflagellate sequence. This argument can be bolstered by a more detailed examination of the causes of year to year variations (Chapter 12).

In a series of papers Colebrook and Robinson (Colebrook, 1979, 1982; Colebrook and Robinson, 1961, 1965; Robinson, 1970) have used the CPR data to show the links between hydrography and the seasonal abundance of phytoplankton and zooplankton. Their data reveal the factors which control the seasonal growth patterns of phytoplankton and allow some inferences to be drawn about the nature of planktonic food chains. When the CPR data is averaged over standard ocean areas month by month regular patterns are revealed. In the temperate waters of the North Atlantic and the North Sea, phytoplankton growth begins in February in the shallow waters of the North Sea and reaches a peak in March and April. The peak abundance of phytoplankton occurs later in offshore waters so that a wave of 'colour' appears to move away from the continental shelf (Fig. 9.2). When almost the whole of the North Atlantic is studied (Robinson, 1970) there is a clear trend for the spring outburst of phytoplankton to occur earlier in shallow water and for the biomass peak to be larger. There is also a clear tendency for the length of the growing season to be longer in inshore waters (Fig. 9.3). Robinson showed that the timing of the spring outbreak was highly correlated with an index of temperature difference between March and the average of May, June and July. Thus the seasonal cycles of phytoplankton growth are very much a function of latitude and hydrography (Fig. 9.4; Colebrook and Robinson, 1965; Colebrook, 1982). Colebrook's (1982) analysis showed how a wave of phytoplankton growth moved north in spring with the peak phytoplankton abundance occurring one month later at 55°N than at 35°N. The close correspondence of phytoplankton growth and the cycle of stratification can be seen in Fig. 9.4. These data are an elegant confirmation of the early 'critical depth' ideas of Atkins (1924), Gran and Braarud (1935), Sverdrup (1953) and Riley (1957) concerning the spring phytoplankton bloom in temperate waters. The calculation of 'critical depth' was discussed in Chapter 8 and the interaction of photosynthetic and physical parameters was presented there. Transient stability in the water column allows phytoplankton growth to begin before full stratification develops (Colebrook, 1979).

Early work by Atkins (1930), Cooper (1933a,b, 1938) at Plymouth clearly identified the nature of the spring outburst of production in temperate waters. The onset of stratification leads to physical conditions which make it possible to rapidly convert soluble nutrient pools into biomass. The soluble nutrient pool size is therefore a result of the balance between supply and demand. Early in the year the supply is large, as deep, nutrient-rich water is brought to the surface by vertical mixing. At the same time the demand is low because the phytoplankton are energy limited. As stratification sets in, the supply is reduced and the demand is increased. There is a period of quantitative transfer of soluble N and P into biomass and this was, in fact, used by Atkins and Cooper as an index of production. As the organisms of the spring outburst are either grazed or sedimented out of surface waters algal biomass decreases and the soluble and particulate nutrients are transferred to deep water and regenerated there. Hence the simple stoichiometric relationships between C:N:P uptake and particulate ratios displayed in Chapter 7. Redfield (1958), Redfield *et al.* (1963) and Richards (1965) showed similar simple relationships between the composition ratios of sea water and showed that the regeneration of nutrients in deep water followed the same stoichiometry. More recently Harris (1983) and Lam *et al.* (1983) have examined the relationships between algal biomass (as chlorophyll) and phosphorus uptake and have shown that it is

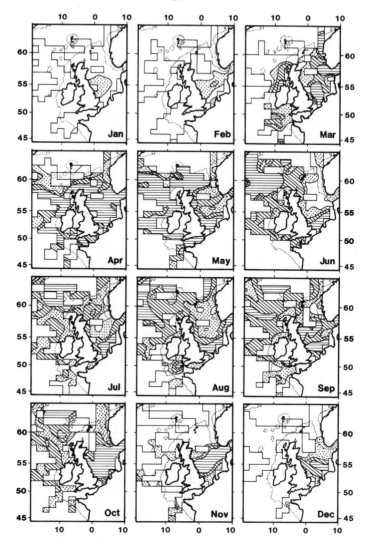

Fig. 9.2 The spatial and temporal distribution of phytoplankton 'colour' as determined by the CPR network in the North Sea and the North Atlantic. Contoured at three levels (Contours in arbitrary units: no shading, <2; stipple, 2–6; cross hatch, 6–13; horizontal shading, >13). From Colebrook and Robinson (1965).

possible to model the process of algal growth by measuring the disappearance of DIP in surface waters. This is, of course, only possible in situations where the recycling of DIP is not an important part of the overall flux (Chapters 4 and 7).

As the nutrient status of the surface waters of the oceans and large lakes

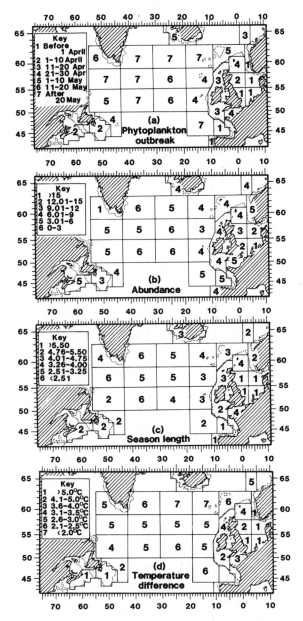

Fig. 9.3 Measures of the timing, abundance, season length and temperature difference associated with the spring bloom of phytoplankton in the North Atlantic (Robinson, 1970). Temperature difference determined by the difference in temperature between March and the mean of May, June and July.

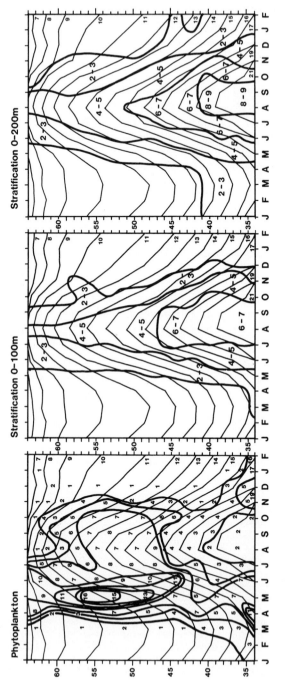

Fig. 9.4 The temporal and latitudinal distribution of phytoplankton 'colour' (arbitrary units) and stratification in the top 100 m and 200 m. Stratification measured as the temperature difference between the stated depths (Colebrook, 1982).

declines the production processes switch over from predominantly new production to regenerated production. This change in the nutrient pathways is associated with a decline in algal biomass and a change in the cell size distributions of the organisms. In shallow and inshore waters, and in lakes with significant nutrient loadings, there may be a much more continuous distribution of new production over the summer and a consequently high summer algal biomass. Nutrients may be supplied to surface waters in summer from transitory upwellings, from rivers and from municipal and other effluents (Chapter 4). At autumn overturn, the supply of nutrients from entrained deep waters may produce a second, smaller pulse of biomass production.

Colebrook and Robinson (1965) and Colebrook (1979) also used the CPR data to examine the seasonal fluctuations in the abundance of grazing animals. Once again the resolution of the data is low because of temporal and spatial averaging, but some broad trends were visible. In northern areas of the North Atlantic, the copepods showed low numbers, late growth and short seasons whereas in the North Sea and in shelf waters, the copepods began to increase earlier, reached higher abundances and persisted longer. Colebrook and Robinson (1965) identified the key parameter to be the timing of the spring phytoplankton increase. The timing of the copepod increase was determined by the timing of the phytoplankton increase indicating a dependence of the animals on food supply and some degree of 'bottom up' control in the food chain. Cushing (1959) and Heinrich (1962) discussed the seasonal patterns of nutrient depletion, phytoplankton growth

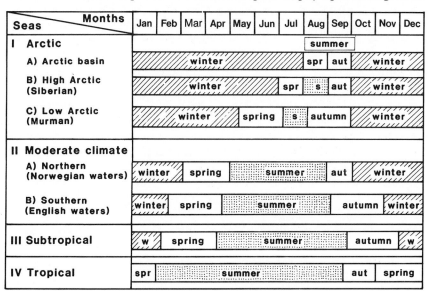

Fig. 9.5 The duration of biological seasons as summarized by Bogorov (1958).

and grazing by zooplankton in the oceans and showed how the seasonality of climate at different latitudes affected the outcome (Fig. 9.5). Cushing (1959) described the pattern in polar and temperate waters as unbalanced, as the disparity between the growth rates of the phytoplankton and zooplankton coupled with the delay in egg production leads to unrestricted growth by the phytoplankton in spring. The spring biomass may be 10–100 times larger than the zooplankton biomass. If the spring phytoplankton bloom is not grazed then sedimentation of the algal biomass is a likely result when the stabilization of the water column occurs. Colebrook (1982) showed that while the phytoplankton peak was one month later in the northern North Atlantic (55–60°N) as compared to more southerly areas (35–40°N), the peak zooplankton abundance occurred two months later. One month of the lag was due to a lack of food supply and one month to the effects of temperature on rates of growth. The abundance, timing and duration of the copepod biomass was well correlated with the timing of the spring phytoplankton outburst and the length of the phytoplankton bloom.

Colebrook's (1979) analysis suggested that the copepods were incapable of responding to the spring phytoplankton increase in areas where overwintering stocks were low and generation times were long. This would account for the poor correlation between phytoplankton abundance and zooplankton abundance early in the year. The great disparity in the generation times of the two groups of organisms will ensure that the phytoplankton can easily outstrip the grazing pressure once conditions become favourable for growth. Colebrook (1982) suggested that during both the period of spring increase and of autumnal decrease the zooplankton were primarily food limited. Colebrook's (1979) analysis did indicate some causative relationships between the abundances of zooplankton and phytoplankton late in the summer but others (Checkley, 1980; Durbin *et al.*, 1983) have suggested that zooplankton production was food limited even during this period.

Colebrook (1982) carried out an extensive analysis of the seasonal cycles of phytoplankton and zooplankton abundance in the North Atlantic and concluded that the spring zooplankton increase was stimulated by an increase in the food supply. Overwintering stocks of zooplankton appeared to be food limited. Colebrook identified three broad zones in the North Atlantic in which there were differences in the seasonal dynamics. South of 47°N the summer biomass of both phytoplankton and zooplankton was low and presumably limited by the TN and TP pool sizes. Such biomass limitation was particularly noticeable south of 40°N; in waters showing permanent stratification in the top 200 m. North of 60°N there were large stocks of *Calanus* and Euphausiacea but low water column stability. Nutrients were not limiting but the phytoplankton may well be energy limited by virtue of the low solar irradiance and deep mixing. Between 48°N and 60°N, which is a large fraction of the North Atlantic, there appears to be a wobbly steady state between phytoplankton growth and grazing. This implies that 'the system over this very considerable area is almost entirely determined by the specific rates and fluxes of processes as opposed to the sizes of stocks, whether of organisms or of nutrients' (Colebrook, 1982). It must be borne in mind that the CPR data is heavily averaged over space and time and that such averaging will produce a result that looks to be closer to steady state than it really is. It is interesting to note, however, that a sampling method designed to measure biomass produced a result that stressed dynamics. The CPR data has revealed that the most important variable in the seasonal cycle is the timing of the spring phytoplankton bloom and has shown that organisms with greatly differing growth rates and generation times will rarely, if ever, reach equilibrium in a seasonally changing environment.

The factors which control the seasonal cycle of biomass in freshwaters are similar in essence to those which operate in the oceans. In deep lakes vertical mixing and 'critical depth' considerations are important in winter and early spring. As in the oceans, there is a period of new production in early summer after stratification followed by a decrease in the soluble nutrient polls. Lakes, like the oceans, may be affected by horizontal transport in that washout after rainfall may impose significant seasonal variability. Lund (1949) documented the effects of heavy rainfall on the seasonal cycles of phytoplankton biomass in the English Lakes. In the extreme, in rivers, the whole biological system may be entirely dependent on the rate of water flow. Talling (1969) showed how vertical mixing and washout controlled the seasonal cycle of phytoplankton biomass in the Nile Reservoirs. In tropical lakes the seasonal cycle of rainfall may be the major determinant of the annual hydrological cycle. In oligotrophic freshwaters there is a characteristic seasonal pattern with peaks in spring and autumn. Summer biomass is limited by the low levels of TN and TP. As the nutrient loading is increased, these biomass peaks become larger until a summer peak emerges, when the nutrient loading exceeds the capacity of uptake by the spring bloom. At this point stocks of TN and TP are appreciable in surface waters in summer. This change in seasonal cycle will be discussed more fully in Chapter 12 and the pattern is illustrated by Fig. 12.7. As the stocks of TN and TP are increased the algal biomass rises until it becomes light limited.

The seasonal light fluctuations are a function of solar declination and latitude, and the fluctuations in the nutrient pools are a function of seasonal thermoclines, sedimentation and resuspension. There is therefore a strong correlation between latitude and the variability in seasonal biomass patterns (Fig. 9.5). This was well documented by Melack (1979). Whether or not the algal biomass actually reaches the ceiling set by the resources will depend on the number of opportunist species which grow in sequence to fill out the envelope below the ceiling and on the relative rates of change in resource levels compared to the growth rates of the organisms. That the envelope is normally filled, is demonstrated by the observations that the seasonal biomass cycle is much smoother than the seasonal cycles of the individual species (Kalff and Knoechel, 1978; Talling, personal communication); by the fact that under light limitation biomass rises to fill the mixed layer (Seliger *et al.*, 1975; Wofsy, 1983) and by the fact that in the absence of light limitation, Vollenweider's models of chlorophyll and TP give highly predictable results. Wofsy (1983) showed that the growth response of the phytoplankton dictated that a period of up to two weeks was necessary to fill the mixed layer. There are therefore occasions when the biomass is lower than expected. These occasions occur when the resources change faster than the growth response of the species; at spring and autumn overturn when the mixing depth changes rapidly or at other times of the year as a result of storm activity.

There is evidence in freshwaters also that the zooplankton react to the availability of food and that they track the spring phytoplankton outburst. There is also increasing evidence that outside the spring period populations are food-limited; as in the oceans (Lampert and Schober, 1980). By inducing artificial spring blooms in the phytoplankton (by artificially mixing and restratifying large enclosures) George (personal communication) was able to produce multiple zooplankton outbursts during the summer. In each case the time-lag between the phytoplankton and zooplankton peaks was a function of the relative generation times. As with the phytoplankton, the total zooplankton biomass is made up of a number of species of varying size and reproductive rates. As might be expected there are also simple relationships between such variables as TP and zooplankton biomass (Chapter 11). Once again these relationships arise from the compounded effects of the many species tracking food availability.

9.3 Manipulation of the seasonal cycle of biomass

Data from the operation of large enclosures (the Lund Tubes) in Blelham Tarn demonstrated the relatonships between the P loadings and mean summer algal biomass as well as the effects on the seasonal cycles of that biomass (Lund and Reynolds, 1982). These authors reported a mean Chl:P ratio of 0.5 which is consistent with the data shown in Fig. 7.15 for eutrophic waters. The effects of P-loadings on the seasonal cycles of biomass are visible in the work of Lund and Reynolds (1982) and in the long-term data from Blelham Tarn (Fig. 12.7). The seasonal cycle of biomass depends to a large extent on the seasonal cycle of TP and DIP. Maximum DIP values occur in late winter in temperate waters and the sedimentation of the spring diatom bloom may result in an almost total removal of N, P and Si from surface waters after stratification. Summer populations are largely dependent on recycled N and P. The addition of P to surface waters will have a great effect on the seasonal biomass cycle if the rate of P loading throughout the stratified period is sufficient to support the growth of large phytoplankton populations. The effects of P loadings which are confined to winter and spring are therefore different from continuous loadings (Lund and Reynolds, 1982).

Additions of P in winter led to a larger spring peak but had little effect on the summer biomass as long as the amount added was less than that which could be removed by 'new' production and sedimentation in spring. Limitation by light, N or Si finally limited the spring biomass peak leaving a surplus of P in surface waters to stimulate the accumulation of summer biomass. Due to seasonal rainfall patterns the majority of the natural P loadings enter water bodies in winter or spring so that the effects of changed land use are initially a rise in the spring algal pulse followed by a rise in the autumn pulse. This can be seen in Fig. 12.7. The autumn pulse increases in size because of the reintroduction of hypolimnetic P to surface waters at overturn. When the P loading exceeds the capacity of the spring pulse or significant P is added in summer (from sewers) a large summer pulse of biomass appears. Thus alteration of both the total P load and the seasonal distribution of that load can be used to manipulate the seasonal cycle of biomass. If summer loads were reduced in proportion to total loads this would clearly have beneficial effects on nuisance blooms in summer and, by reducing the sedimentation of algal material in summer, hypolimnial anoxia would be reduced accordingly. Lund and Reynolds (1982) showed the difference in seasonal patterns which resulted from alterations in the distribution of P loads over the summer.

The evidence from the CPR data indicated that physical perturbations also influence the seasonal cycle of biomass. In particular, the timing of the spring stratification, through an interaction with critical depth effects, has a large effect on the timing and magnitude of the spring bloom. Physical mixing processes in summer can also stimulate the growth of larger biomass pulses by the entrainment of nutrients from deeper, nutrient rich water. As biomass responds easily to nutrient manipulation then no-one appears to have tried to manipulate biomass directly by physical means.

9.4 The influence of mixed layer physics on biomass distributions

Harris (1983) recently published an analysis of the effects of physical processes on the spatial and temporal distribution of algal biomass in Hamilton Harbour. This could be said to be a system limited by excess variance. The mean summer chlorophyll in any year was found to be correlated with the variability in the mixing depth, Z_m, but not with the mean value (Fig. 9.6). When sufficiently perturbed, the algal biomass did not reach the 'ceiling' set by light and nutrients. The average summer value of N^2 was correlated with both the mean and the standard deviation of the summer surface algal biomass. Greater stability in surface waters caused an increase in the mean surface chlorophyll but a decrease in the variance about the mean. The analysis of data from a 77-week time series indicated that the biomass data was composed of three separate lognormal distributions with winter, spring and autumn, and summer components (Fig. 9.7). Each of these biomass distributions was correlated with a particular physical regime and with a particular phytoplankton assemblage. Details of the phytoplankton assemblages in Hamilton Harbour can be found in the paper by Harris and Piccinin (1980).

Vollenweider and Kerekes (1980) also noted that algal biomass data from lakes was lognormally distributed. The reason for the lognormal distribution is now clear. The non-equilibrium conditions lead to the growth of many opportunist species which, when compounded together, give lognormal distributions (cf. Equation 11.2). The algal biomass distributions in the Harbour were statistically well-behaved and resulted from the low level non-equilibrium dynamics. The TP levels in the Harbour gave an accurate prediction of the mean biomass values to be expected in the water (Vollenweider, 1968, 1975, 1976; Vollenweider and Kerekes, 1980; cf. Chapter 11 below) so, the observed TP, when combined with a measure of vertical stability, could be used as a predictor of both the mean and the maximum biomass. The algal biomass was thus *yield* limited by the TP, a feature of lake biology that has been exploited for management purposes (Harris, 1980a). The predicted maximum biomass is perhaps more use for management purposes than just the mean expected biomass as, while the mean value may be tolerated, the maximum value may not be (Harris, 1983). At constant TP there is a trade-off between high biomass and low variance in physically stable lakes and the lower mean, but higher variance, of biomass in physically perturbed waters.

The correlations between mixed layer physics and the biomass in surface waters appear to be brought about by two mechanisms – one, a simple physical effect of restricted vertical diffusion, and the other, a more complex interaction with algal growth rates and diversity that was only revealed by time series techniques. The phytoplankton in the surface waters of the Harbour were clearly not *growth* limited by nutrients. It appears that nutrient recycling is rarely insufficient to satisfy the demands of algal growth because the scales and holons are closely matched. Because of a lack of information the effects of fluctuations in nutrient fluxes in nature are hard to quantify (cf. Chapter 5 and the discussion of perturbation experiments in Chapter 8). The effects of fluctuations in the underwater

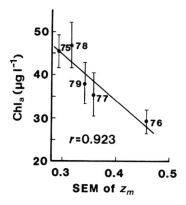

Fig. 9.6 The effect of fluctuations in the mixing depth, z_m, on the mean summer chlorophyll in Hamilton Harbour (Harris, 1983). Data from the summers of 1975–1979.

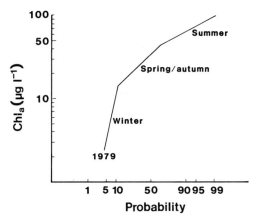

Fig. 9.7 The cumulative probability plot of the concentration of chlorophyll a for different seasons in Hamilton Harbour in 1979. The logarithmic scale means that the distributions are lognormal.

light climate are much better documented and must serve as a model system to illustrate the general principles.

The seasonal changes in the underwater light climate and the effects on the ecology of phytoplankton were discussed by Talling (1971). There are, however, some very important effects of fluctuations in the Z_{eu}/Z_m ratio which are not concerned with light limitation in the classical 'critical depth' sense. When light is not limiting the overall biomass, fluctuations in light availability from day to day can still influence the species composition and hence the biomass and the seasonal succession of species. Species of differing niche requirements grow whenever the environment is suitable and coexistence is possible by virtue of the interaction between the spectrum of environmental perturbations and the growth responses of the algae (Harris, 1983; Harris *et al.*, 1983). The normal seasonal cycle is therefore one of a combination of internally generated (autogenic, competitive) shifts in the species composition alternating with externally forced (allogenic, catastrophic) perburbations.

9.5 Ecological succession

'Mixed populations or biotic communities in a steady state are the exception in nature. Commonly there is a continuous shift in their structure, which is continuously readjusted to a changing environment' Margalef (1963).

The seasonal cycle of phytoplankton species in the surface waters of lakes and the oceans is, in many respects, analogous to the successional sequences of other plant species in other habitats. The seasonal succession of phytoplankton species progresses at an absolute rate which is much faster than the successional sequences of higher plants but, when the relative sizes and generation times of the organisms are taken into account, there appears to be little relative difference. The nature of the ecological factors which determine the structure of phytoplankton communities has been the subject of debate for a number of years. It is a sobering thought to realize that many of the basic observations were made in the latter half of the nineteenth century and that the main hypotheses regarding the nature of the seasonal succession were advanced around the turn of the century (Hutchinson, 1967). Most of the basic questions were formulated in modern terms thirty or forty years ago (Hutchinson, 1941, 1953; Vollenweider, 1948, 1950, 1953). Margalef is another who has contributed greatly to the understanding of phytoplankton ecology and theoretical ecology in general. In a series of books and papers Margalef (1961, 1963a,b, 1964, 1967, 1968, 1975, 1978b,c) has integrated the study of phytoplankton successions into the broader framework of plant successions and has shown that the processes which underlie the successions in lakes and the oceans are not significantly different from each other or from those in terrestrial systems.

The hierarchical models of community structure (Allen and Starr, 1982) relate the responses of the organisms to environmental fluctuations and to interactions at various levels. Whether or not a successional sequence is seen to be deterministic depends on the degree of nesting in the hierarchy. Fully nested hierarchies will be quite deterministic in their behaviour. Hierarchies which are not fully nested and those which show external perturbations at high level will display a sequence of species which is apparently non-equilibrium. Such approaches do not deny the existence of competition, but they temper the effect of competition by assuming that exclusion takes time and that the environment may change the rules of the game before the required time has expired. By incorporating more realism into the theory, hierarchical ideas make more sense to the field ecologist who must wrestle daily with the vagaries of the real world. Real world systems are rarely closed systems. At the extreme, non-equilibrium ideas represent one end of the spectrum and to deny that competitive exclusion never occurs would be a rash statement. Competition is tempered by environmental change but not perhaps eliminated. Such ideas are very much in line with the work of Margalef (1963a,b, 1968, 1978b) who has written about the successional sequences and community structure of phytoplankton. Margalef sees the community organization as resulting from both the influence of external perturbations and from the interactions between species. The presence of many species in the phytoplankton community produces 'information' which is carried forward in time.

9.6 The seasonal succession and community structure of phytoplankton

The successional sequences of both marine and freshwater species have been reviewed by Margalef (1963a, 1978b), Smayda (1980) and Reynolds (1980a, 1984a,b) and these authors have identified many of the major trends and features. It is now possible, in the light of the foregoing discussions of physical, chemical and biological mechanisms, to do more than simply describe the main trends. A discussion of the underlying processes is now in order.

The most recent complete discussion of the phytoplankton community and the patterns of seasonal fluctuation found in lakes is that by Hutchinson (1967) and the most widely quoted statement of the problem of the apparently anomalous diversity of phytoplankton communities is that by Hutchinson (1961) in his paper on the 'paradox of the plankton'. Hutchinson recognized that the patterns of relative abundance of phytoplankton populations did not match the patterns to be expected from equilibrium interactions and discussed a number of possible ways in which high diversity may be maintained. The anomaly in the diversity lies in the coexistence of many species in an apparently isotropic and homogeneous environment. Coexistence in such an environment can only be brought about by resource partitioning (Stewart and Levin, 1973; Peterson, 1975), by limit cycles (Hutchinson, 1965), or by commensalism (Hutchinson, 1967). Alternatively there must be significant variability in the environment which results in the presence of a number of niches or the variability must be sufficient to prevent the approach to competitive equilibrium.

The ubiquitous presence of non-equilibrium processes at the population level might be expected to produce patterns of abundance and species occurrences which, in the extreme, would approach chaos. On the other hand, the ubiquitous operation of equilibrium processes would not account for the observed diversity. As discussed in Chapter 2, the real world may be thought of as a wobbly compromise between equilibrium and chaos. Thus, while competitive exclusion and equilibrium are not the rule, neither is chaos. There are distinct patterns in the phytoplankton communities of lakes and the oceans (Hutchinson, 1967; Smayda, 1980) and both equilibrium and non-equilibrium situations may be found (Harris, 1983). The old debate about the mechanism of successions centred around the nature of the processes which drove the changes over time. Historically, the analysis of seasonal cycles has reflected the prevalent equilibrium viewpoint. The views of Clements had a great influence on ecological thinking, particularly in North America, and the view that successional sequences were dominated by autogenic factors came to be the prevalent view. In the last decade this view has been somewhat modified (Drury and Nisbet, 1973) as reductionist approaches came to dominate thinking and in this respect the modern view of phytoplankton successions (Reynolds, 1980) is in accord with the work on higher plant and forest successions.

Non-equilibrium communities and successional sequences might be expected to show a number of features which distinguish them from equilibrium situations. Such communities should be characterized by the presence of opportunistic species and should show some non-deterministic features. Round (1971) noted that many seasonal sequences of species show chance effects of immigration and colonization. The 'correct' species did not always appear in any given lake due, in part, to the chance effects of migration between the bodies of water. Lakes are ecological islands in a terrestrial sea and the movement of species between lakes is much affected by the vagaries of climate and animal vectors (Talling, 1951; Harris, 1983). This feature of migration between open universes is a feature of non-equilibrium explanations (Chapter 2) and can lead to apparent ecological voids if the resources are present but the species which use those resources are not (Trimbee and Harris, 1984). Such voids might be expected to occur in the seasonal sequence if species grow slowly in comparison to the rate of environmental change. Thus there are periods in the year when the phytoplankton assemblage changes dramatically and diversity is reduced; for example when the environment of the mixed layer undergoes major physical restructuring in spring and fall (the 'cardinal points' of Round, 1971).

Opportunistic species show rapid growth until an upper limit is reached or until the environment changes and becomes unfavourable. Such communities are characterized by sharp fluctuations in the numbers of individual species but there will be a more or less constant total biomass if the upper limit is set by an external factor and there are sufficient species to fill in the seasonal cycle. Talling, in a recent lecture, noted that whilst the abundances of the individual species fluctuated markedly the overall biomass showed much less variability. The population strategies of the individual species are essentially opportunistic but constrained by 'envelope dynamics' within boundaries. This is precisely the result to be expected if the overall biomass was set by seasonal conditions in the environment. In that case the maximum biomass would be a seasonal function of such parameters as temperature, mixed layer depth, total nutrient and light availability (Chapter 7). In a disequilibrium state there is no relationship between abundance and growth rate and the limiting factor (or factors) will set a limit to biomass not growth rates. Periods of growth limitation by nutrients may be rare but important. Ecological 'crunches' may be infrequent but may nonetheless determine the sequence of events in the seasonal succession of species.

The partitioning of the biomass between the various particulate pools and the partitioning of the nutrients between the various soluble and particulate pools will be expected to hunt around some long term mean determined by a mixture of thermodynamics and chance. Heuristically however, the system will have predictable long-term statistical properties that arise from the trial and error processes within the food chain: long term (annual) averages will be required. The food chain might be expected to show seasonal components as the environment fluctuated and the nutrient stock changed with time, but all the TP cannot be permanently tied up in the phytoplankton, nor can it be permanently tied up in the zooplankton or the fish. Margalef (1975) showed how the cycling of essential materials through pools of various turnover rates (holons) had a stabilizing effect on ecosystem dynamics. Opportunist species fill in the seasonal cycles and grow until a ceiling is reached. That ceiling will be set by an external factor such as light or mixing depth or by internal factors such as the slower cycling of N or P through a higher, slower holon. The reduction of the total nutrient stocks during the period of the summer stratification (or in waters with a permanent thermocline) results in the selection of small, energy efficient cells. Heuristic community processes also account for the presence of diffuse competition. If the mixed layer maintains a large standing stock of phytoplankton in the summer, even if only a few of them are active at any one time, the mere presence of a large number of cells in the water column will influence the underwater light climate. Because the majority of the chlorophyll is contained in the largest cells the larger cells have a disproportionate effect on the underwater light climate.

The succession of phytoplankton species in surface waters may be regarded as a non-equilibrium sequence at the population level. At the community level, however, there is, in Margalef's terminology, communication of structure through time. The rare species, the community 'safety net', persist in time and by their very presence feed information to successive communities. Thus the response to environmental change will not be random as the population responses will be constrained by the nature of the species present. As noted by von Bertalanffy (1952) the ecosystem is a loose, hierarchical system that carries information through time. The presence of species which persist through time gives the phytoplankton community both a 'feedback' capability to react to environmental change and a 'feedforward' ability to anticipate environmental change if the community contains species with perennation mechanisms which operate at seasonal scales.

Shugart and West (1981) examined the long-term dynamics of forest successions and showed that the succession of communities could be classified as equilibrium or non-equilibrium, depending on the scale of the community in relation to the scale of the disturbance. This is essentially the same as the arguments of Allen and Starr (1982), as the outcome depends on the level of perturbation in the hierarchy. Small-scale perturbations leave the community largely undisturbed and the process of competitive exclusion can continue to completion. Large-scale perturbations disrupt the approach to competitive equilibrium and produce non-equilibrium dynamics. Phytoplankton communities are perturbed by turbulence and vertical mixing and the impact of such perturbations depends on the scale of the perturbation. We have already seen that in most waters the scales of perturbation are a function of the interaction between atmospheric effects and basin size. We might therefore reasonably expect to see a range of successional types in different water bodies depending on the interaction of physical, chemical and biological scales.

9.7 Sampling and counting problems: time and space scales

In the past the role of rare species has always presented something of a problem (cf. the discussion in Williamson, 1981). In opportunist communities, the rare species have a vital role to play in that they are the 'safety net' for the community. When conditions change the species composition will change in response, and unless there is a pool of rare species that can grow to fill the voids, major readjustments in community structure will not occur. The rare species are those which grow for brief periods or for which growth is only possible on rare occasions. The presence of rare species has always provided counting problems for phytoplankton ecologists as the abundances of the species present in surface waters may range over several (6–8) orders of magnitude depending on season. Common species are easy to count but the rare species may be overlooked. Even at abundances of one cell per litre there is a significant innoculum of cells in the smallest lake. Counting methods that will detect abundant species will not detect rare species. It is rare to find reports of changes in phytoplankton abundance which range over more than 3 or 4 orders of magnitude. In this respect the recent study of *Ceratium* by Heaney *et al.* (1983) is worthy of note in that the seasonal fluctuations in the abundance of the organism were followed over 8 orders of magnitude. This heroic feat involved searching for individual cells in 50 l volumes in winter. Increased counting effort will always increase the accuracy of the enumeration of the common species (Lund *et al.*, 1958) and will also turn up more rare species. Thus different data sets are not necessarily comparable; especially if they are used to calculate diversity indices which are sensitive to the number of species and their relative abundances.

A further problem arises if the taxonomy of different workers differs. Kalff and Knoechel (1978) pointed out that taxonomists have always found more species than ecologists and that this results in wide disparities between workers and between data sets. In Lake Erken, for example, they noted that an ecologist found 151 species whereas a taxonomist recorded 440 species. Comparisons of diversity indices between data sets and between workers are therefore practically useless. Kalff and Knoechel suggested that if such indices must be used they should be calculated on the basis of biomass or volume rather than cell numbers so that the smallest flagellate is not given the same weight as the largest dinoflagellate. While it is not possible to compare diversity indices from different data sets, I will show below that by using consistent counting techniques such indices may be used to seek temporal trends within one basin.

The small size of some phytoplankton has led to problems with their enumeration in the past, as very small cells pass through the filters normally used for biomass determinations and are very difficult to count under the light microscope. This, together with the fact that many of the normal preservation methods lead to the disruption of naked, flagellated cells, has led to the importance of such cells being underestimated in both lakes and the oceans. Recent counts made by epifluorescence methods reveal that the phytoplankton smaller than about 3 μm (the so called picoplankton) are ubiquitous and very common (Platt *et al.*, 1983). In terms of abundance they are about two orders of magnitude more common than the larger phytoplankton and about an order of magnitude less common than the bacteria. This size fraction accounts for about 80–90 % of the total primary productivity in some waters. In many waters the smallest phytoplankton are cyanobacteria; little or no data exists on the seasonal successions of such organisms. Much is known about the phytoplankton which are easier to count, but in many cases the dominant organisms in terms of biomass (volume) are not the dominant organisms in terms of metabolism and productivity. Plots of *P/B* versus *B* reveal that the smallest cells have the highest metabolic rates per unit biomass.

In his review of the growth and succession of algal populations in freshwaters Round (1971) included a diagram which summarized the time-scales of the growth periods of algae. Such periods range from a few days to post glacial time scales, but in considering seasonal successions we are most interested in the time scales of the seasonal thermoclines and similar processes. Time scales of up to 100 days are therefore of great significance. I have already shown that such time scales encompass great environmental heterogeneity and large changes in phytoplankton communities are therefore to be expected. With organisms that double in a matter of a few days, significant changes in relative abundance can be achieved in periods of a week or less (Kalff and Knoechel, 1978), so weekly sampling is the minimum necessary to adequately resolve such events and daily sampling may be desirable in some circumstances.

As would be expected from the smooth spectra of fluorescence (Chapter 5) spatial patchiness occurs at a number of scales from centimetres to entire basins. Patchiness at scales of centimetres and metres (Cassie, 1959; Harris and Smith, 1977) can heavily influence numerical estimates of abundance and can introduce noise to the signal of interest. Horizontal advection may be an even more severe problem and it may be sufficiently severe to completely mask the changes of abundance under study (Lund, 1949). Trimbee and Harris (1983) showed that daily sampling was necessary to resolve the advection of phytoplankton within the basin of a small reservoir. The horizontal motion of the cells between sampling stations would have made the interpretation of weekly sampling data very difficult. Heaney (1976) and George and Heaney (1978) discussed the influence of wind on the redistribution of *Ceratium* within Esthwaite Water and showed that the motion of the cells caused large errors in the estimates of abundance if the cells happened to be at the other end of the lake when the sample was taken. Thus the sampling scheme must be designed in such a way as to ensure that statistical reliability is maintained and, as noted above for such parameters as *P/B* ratios, the interpretation of data collected infrequently from a single station in a basin may be well nigh impossible.

The time series plot of the 1982 El Nino (Fig. 3.7) graphically illustrates another point concerning the relationships between sampling strategies and the sequence of phytoplankton communities. The 1982 El Nino arrived as an internal wave so the temperature of the surface waters (and hence the physical and chemical conditions) changed dramatically overnight. Daily sampling is required to resolve such a phenomenon and it is clear that much of the previous phytoplankton sampling in the Peruvian system has been carried out at too infrequent an interval. Furthermore, as the wave travels south along the Peruvian coast the associated coastal currents cause large scale horizontal motions in the coastal phytoplankton assemblages. Thus in such an area the sequence of species recorded over time is the result of both a succession in the true sense and a translation of water masses (Margalef, 1963; Smayda, 1980). In oceanic areas, where horizontal advection is the dominant physical process, a considerable knowledge of hydrographic conditions is necessary in order to reconstruct the seasonal succession of phytoplankton species.

Physical and chemical scales interact to produce characteristic scales of change in different water types. In tropical and subtropical waters where there is a more or less permanent thermocline the concentration of nutrients in surface waters may be continuously low and a strong seasonal temperature signal will be lacking. There is, consequently, a weak seasonal succession of species in such waters (Smayda, 1980). In tropical lakes there may be periodic overturns which cause the species composition to oscillate in an irregular fashion throughout the year. In mesotrophic and oligotrophic temperate waters there is both a strong seasonal temperature signal and a strong seasonal change in total nutrient stocks. These signals combine to produce large-scale changes in the species composition of phytoplankton in surface waters at different times of the year. In temperate eutrophic waters the signals differ as the nutrient input may be high and continuous; thus removing the seasonal nutrient signal entirely. In polar waters the seasonal cycles may again be weak (cf. Fig. 9.5).

9.8 The seasonal successions of species: the ideal sequences

9.8.1 ARCTIC AND ANTARCTIC WATERS

In cold polar waters, and in lakes in winter, deep mixing leads to greatly reduced energy inputs to the phytoplankton in the water column. The lack of density gradients in the water leads to rapid circulation and brief residence times in surface waters. The polar environment may be totally dark for six months and this will restrict the period of phytoplankton growth to the sunlit period. Heterotrophic nutrition will assist survival in low-energy environments. According to Smayda (1980) there is little information about the seasonal cycles of phytoplankton species in Arctic and Antarctic waters. What little information exists seems to indicate that diatoms form a significant fraction of the total flora. In a transect between 43° South and 62° South, Jacques *et al.* (1979) showed that diatoms became increasingly dominant with increasing South latitude. Species of *Chaetoceros, Rhizosolenia, Thalassiosira* and *Fragilariopsis* were the most important. This pattern of distribution would be consistent with the high $P{:}R$ ratios of diatoms and their high sinking velocities.

Diatoms are characteristically abundant in polar lakes and in temperate lakes in winter when low insolation and deep mixing leads to a requirement for efficient photosynthesis. The Great Lakes, for example, have abundant populations of *Fragilaria*, *Tabellaria*, *Stephanodiscus*, *Melosira*, *Diatoma* and *Asterionella* in winter when mixing occurs to about 200 m. Diatoms accounted for 80 % of the total algal biomass in winter and spring in Lake Ontario (Nalewajko, 1966; Munawar and Nauwerk, 1971). A number of smaller species of phytoplankton occurred in winter also including species of *Cryptomonas* and *Rhodomonas* which are suspected of partial heterotrophy (Haffner *et al.*, 1980). Kalff (1970) reviewed the seasonal cycle of algal populations in Arctic lakes and showed that species of small flagellates (*Cryptomonas* and *Rhodomonas*) were important in polar lakes and that diatoms were frequently, but not invariably, the dominant fraction of the total biomass. Small flagellates and nanoplankton were generally an important group. Blue-green algae, he noted, were not abundant in Arctic lakes. Kalff (1970) decided that the small size of many species in Arctic lakes was consistent with the extremely oligotrophic character of many Arctic lake waters. The species composition of Arctic waters is therefore consistent with the foregoing arguments. In oligotrophic waters small species are favoured, while in more eutrophic, deep mixed water, diatoms are better able to survive.

9.8.2. TEMPERATE WATERS

As Melack (1979) indicated, the amount of variability in both photosynthetic rates and algal biomass increased with latitude. Variability in both parameters is associated with shifts in the species composition so the phytoplankton assemblage in temperate and high latitude waters shows strong seasonality (Round, 1971; Kalff and Knoechel, 1978). The major changes in the vertical mixing regime induced by the formation and disruption of the seasonal thermocline induce massive changes in the populations in surface waters. Round (1971) identified such periods as 'Cardinal points' in the seasonal succession. The classic sequence of events in spring in temperate waters was worked out by Riley (1949) and Sverdrup (1953). Deep mixing in winter causes energy limitations to all populations but the lengthening days and the increased insolation in spring allow diatoms to grow under low Z_{eu}/Z_m ratios and low temperatures. As a broad generalization diatoms exhibit high $P{:}R$ ratios, high sedimentation velocities and rapid photoinhibition (Harris, 1978). Reynolds labelled these diatoms 'W' forms (Fig. 9.8).

The spring diatom bloom in freshwaters may terminate in a number of ways but a combination of grazing, Si depletion (Lund, 1950; Jewson *et al.*,

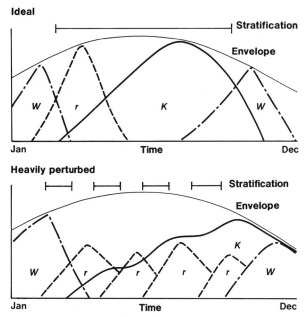

Fig. 9.8 The ideal and 'perturbed' seasonal successions of phytoplankton in temperate waters. Winter diatoms *W* are replaced by *r* flagellates as stratification sets in spring. Large *K* species succeed the *r* types in summer. *W* species may recur in autumn. The upper limit of biomass (the envelope) is set by the availability of light and the stocks of *N* and *P*. In the perturbed state *W*, *r* and *K* may occur at various times as the stratification comes and goes. Reynolds *et al* (1983) first suggested this scenario.

1981), thermal stratification and sedimentation (Lund, 1954, 1955, 1971a) and C depletion (Jaworski *et al.*, 1981) usually brings the bloom to an end. C depletion arises from photosynthetic activity and a resulting rise in the pH of surface waters. Lund's (1949, 1950a,b) early work on *Asterionella* was a clear demonstration of the lack of density dependent controls on the growth of the spring diatom bloom as the bloom was terminated by different factors each year depending on the precise sequence of physical events in that year. The fate of the spring diatom bloom appears to be total sedimentation. There is good evidence (Jewson *et al.*, 1981, Reynolds *et al.*, 1982a,b; Reynolds and Wiseman, 1982) that almost the entire standing crop of diatoms may be recovered from sediment traps or on the sediment surface. Reynolds *et al.* (1982b) noted that, in Windermere, the algal remains permanently contributed to the sediments, were dominated by diatoms and that the sedimentary fluxes of certain heavy metals were almost entirely accounted for by this process. The spring diatom bloom may also dominate the sedimentary flux of P to the sediments. Edmondson and Lehman (1981) showed that the removal of P to the sediments in Lake Washington occurred mostly during a brief period in spring and that marked depletion of P occurred in surface waters as a result. The spring diatom bloom is mostly new production. Much of this new production ends up in benthic food chains. Hence the requirement for processes which speed nutrient regeneration in summer and the increased proportion of nutrients derived from such processes later in the year. The classic sequence of events in the oceans also involves the growth of a spring diatom bloom which is terminated in a similar fashion.

The spring diatom bloom is highly dependent on the physical structure of the water column and hence on the vagaries of the climate from year to year. The size of the spring bloom of diatoms makes a useful marker of physical conditions in spring and surveys over a number of years reveal marked year to year variations which are linked to climatological events. The precise sequence of events in spring will determine the timing of the bloom and which of the potential limiting factors comes into play. Physical conditions in summer may lead to the recommencement of vertical mixing so that resuspension and regrowth of diatoms may occur once more. Harris *et al.* (1980b) noted that *Stephanodiscus* only grew in summer when vertical mixing was particularly active and Lund (1971) experimentally altered the seasonal cycle of *Melosira* by artificial mixing in mid summer. For these organisms, a benthic innoculum of cells is important so that vertical mixing will lead to resuspension and growth. The benthic innoculum preadapts the community by providing a ready made reservoir of colonizing individuals. Even though only 1% of the cells which sink out in spring may survive (Jewson *et al.*, 1981) this is sufficient to seed the water column when vertical mixing returns. In deep oceanic water the use of a benthic innoculum is not possible as the turnover time of deep ocean water is far too long. Diatoms in such habitats must persist in surface waters until conditions become favourable for growth.

The sequence of events in spring in many freshwaters is now relatively well understood and the growth and loss terms for the phytoplankton community can be apportioned between the various size classes. Once the seasonal thermocline has been formed in temperate waters the first species to appear are often small flagellates or other '*r*' strategists. These small forms have high rates of growth but seem to require high energy inputs (Harris *et al.*, 1983); hence the requirement for longer days and shallow mixed layers. The conditions of warming of the water and the growth of small phytoplankton also favour the growth of the zooplankton.

There are two major effects of zooplankton grazing worthy of note: the reduction in the phytoplankton biomass and the size selectivity of the animals. The maximum intrinsic rates of increase of zooplankton are a function of temperature. At reasonable surface water temperatures (10–

25°C) values of k (ln units) for Cladocera vary from 0.2 d^{-1} to 0.5 d^{-1} with some higher values (up to 0.7 d^{-1}). The rates of increase of copepods are somewhat less, being mostly in the range 0.05 d^{-1} to 0.35 d^{-1} (Allen and Goulden, 1980). Other marine grazers, including Salps show similar rates of increase (Heron, 1972). Inspection of the rates of increase for the phytoplankton in size classes which are consumed ($10–10^4$ μm^3) indicates that under optimum conditions the zooplankton are growing at much the same rate as their food supply, if not faster. As might be expected the growth rates of the microzooplankton (50–200 μm) are much higher than those of the larger animals (> 200 μm). Le Borgne (1982) recorded growth rates of (ln units) 0.3 d^{-1} to 1.2 d^{-1} for microzooplankton in the tropical ocean while the larger animals only increased at rates between 0.13 d^{-1} and 0.48 d^{-1}.

Examination of the size fractions grazed by the zooplankton indicates that phytoplankton larger than about 10^4 μm^3 are infrequently grazed by Cladocerans such as *Daphnia* whereas rotifers seem to favour phytoplankton smaller than 10^3 μm^3. Size selective grazing is clearly important in controlling the sizes of phytoplankton in the water at certain times of the year, both in marine and freshwater habitats (Porter, 1973). The mechanism of zooplankton grazing is clearly not simply filter feeding (Kerfoot, 1980). Copepods catch algal cells by both filtering smaller cells and by individually grasping larger cells. They are also able to switch the size classes upon which they feed depending on the sizes of food available. Cladocerans filter feed on a restricted range of size classes but the threshold volume of 10^4 μm^3 is a reasonable upper limit. Zooplankton grazing can therefore suppress the populations of nanoplankton, leave the populations of larger phytoplankton largely unaffected and even increase the populations of some algae which pass through the gut unaffected (Porter, 1973). Passage through the gut of zooplankton without digestion may allow these species to take up nutrients while passing through, and grow more rapidly when voided into the water. Porter (1973) gave the results of the grazing experiments using species of *Daphnia*, *Diaptomus* and *Cyclops*. She found that small algae, Primarily small flagellates such as *Cryptomonas* and *Rhodomonas* and nanoplankton were suppressed by the grazers along with some of the larger diatoms (*Asterionella*). Large green algae such *Sphaerocystis* increased as a result of the grazing. Desmids, dinoflagellates and chrysophytes as well as the lager colonial blue-green algae were unaffected.

Frost (1980a) showed that size selective grazing is as important in the oceans as it is in freshwater by using examples from the model of Steele and Frost (1977). Using a variety of size selection curves Steele and Frost (1977) were able to examine the interactions betweeen grazing copepods and a range of phytoplankton size classes. Their major conclusion was that size was probably more significant than total biomass in understanding the flow of energy and materials between trophic levels. Failures of herbivore populations occurred when there is an absence of the right sized food at a particular stage of development. The critical stages appeared to be the early stages in the life cycle. The control of herbivore populations by the phytoplankton was found to be at least as important as the converse, if not more so. Steele and Frost (1977) found that zooplankton could be grouped by size (hence growth rate and size selective characteristics) into three classes. The organisms fell into small (*Oithona*, 1 μg C), medium (*Pseudocalanus*, 10 μg C) and large (*Calanus*, 100 μg C) categories with a number of species in each category. This is highly reminiscent of Schaffer's (1981) arguments concerning interactions between species, time scales, holons and community structure. The overall situation with a spectrum of sizes of both phytoplankton and zooplankton is therefore highly complex. There is both 'bottom up' and 'top down' control in the food chain and the precise mechanism may differ in different size classes. The complexity of the interactions between nutrients, phytoplankton and zooplankton in a trophic hierarchy has been elegantly illustrated by Lane and Levins (1977).

The events of early summer, therefore, lead to the establishment of the seasonal thermocline, the rapid growth (> 1.0 d^{-1} ln units) of small flagellates and nanoplankton and, after a delay, the rapid growth (<1.0 d^{-1}) of zooplankton populations. This sequence of events leads to a pulse of *r*-strategist phytoplankton followed by a pulse of zooplankton and a brief period during which the impact of copepod and Cladoceran grazing is considerable. For brief periods (a few days) the grazing rate by all zooplankton may exceed 1000 ml l^{-1} d^{-1} (Thompson *et al.*, 1982) but this rate cannot be maintained and the collapse of the zooplankton population usually ensues. Precisely why the zooplankton population rapidly declines is unclear, but as the concentration of food in edible size classes falls, egg production falls also (Ferguson *et al.*, 1982). There is clear evidence that there is a threshold concentration of food below which filter feeding zooplankton find it difficult to reproduce (Lampert and Schober, 1980). At POC concentrations below 0.2–0.3 mg l^{-1} in algal size classes less than 10^4 μm^3, the reproduction rate of *Daphnia* is severely reduced. After the water has been cleared by the brief period of severe grazing (the 'clear water' phase) reproduction rates and population densities are severely depressed. Evidence is now at hand to suggest that food limitation also occurs in the oceans (Checkley, 1980; Durbin *et al.*, 1983). Zooplankton populations may also be depressed by interference from larger size classes of phytoplankton (Gliwicz, 1977) or by direct density-dependent interactions between the animals.

Oscillations and resurgences of the edible size classes of phytoplankton and of zooplankton may occur throughout the summer if physical conditions permit the entrainment of nutrients into surface waters or if external loadings are significant. The disparity in the rates of increase of phytoplankton and grazing animals means that overshoots and crashes will always occur. Porter *et al.* (1983) demonstrated that the growth strategy of *Daphnia* is in many respects sub-optimal but is well-suited to tracking brief pulses of suitable food species. Because of their high threshold food requirements, Cladocerans appear to be energetically suited to growth in mesotrophic and eutrophic waters. This may explain the relative rarity of Cladocerans in the oceans where oligotrophic waters with low food concentrations predominate.

Cushing (1959) and Heinrich (1962) discussed the time delay between the spring phytoplankton bloom and the zooplankton population increase and showed that in polar and temperate waters the delay was longer than in tropical waters. The difference is likely to result from a disparity in zooplankton growth rates brought about by a difference in water temperatures and systematic size differences. Zooplankton in tropical waters tend to be smaller than those in temperate waters (Taniguchi, 1973). The shorter time-lag in tropical waters ensures that the coupling between primary and secondary production is much tighter. Time-lags are reduced and the whole system is much more dependent on internally regenerated nutrients. There is also evidence that the ecological efficiency of tropical oceanic herbivores is greater than that of temperate species (Taniguchi, 1973). In the terminology of Margalef (1963) and of holons, the tropical situation may be regarded as much more orderly than the unbalanced situation in temperate waters. The food chain in temperate waters at scales of days and weeks is disrupted by perturbations at higher levels.

The nanoplankton on which the zooplankton feed, not only have the highest rates of increase but also the greatest loss rates (Sommer, 1981). Unlike diatoms which sink, the fate of the nanoplankton is to be grazed (Reynolds *et al.*, 1982). This means that by mid-summer the phytoplankton community in oligotrophic and mesotrophic waters becomes dominated both by very small plankton (picoplankton) and by the large *K*-strategists greater than 10^4 μm^3 in size. This leads to very disjunctive size distributions in the phytoplankton with size classes < 50 μm^3 and $> 10^4$ μm^3 predominating. In oligotrophic and mesotrophic waters, once the spring bloom has reduced the nutrient concentrations in surface waters, the community metabolism switches over from new production to regenerated sources of nutrients. Lehman (1980) demonstrated that during the summer months in Lake Washington the epilimnetic zooplankton contributed ten times the P and three times the N contributed from all other sources. The regulation of nutrients by the zooplankton was sufficient to supply a sizeable fraction of the daily requirements of the phytoplankton. The regenerated P and N were rapidly taken up by the phytoplankton so that the ambient concentration was small and the turnover was rapid. I have already discussed the importance of microscale patchiness and the effects this will have on the phytoplankton present.

As the soluble nutrient pools in surface waters decline and the phytoplankton community becomes more and more dependent on the small-scale, patchy, recycled flux, the regeneration of N and P at a scale of a day is insufficient to supply the growth of r-strategist phytoplankton at rates of $1.0 \, d^{-1}$ or so. The only viable possibility is to shift scales, either up to larger cells, slower growth rates and low loss rates (K-strategy) or down to very small cells, rapid regeneration and tight coupling between production and consumption (picoplankton). The picoplankton and the K-strategists have quite different nutrient uptake characteristics. Picoplankton have small storage capacities and require frequent small pulses of nutrient. The larger phytoplankton can survive on fewer, larger pulses. Both in the way they are grazed and the way nutrients are utilized, cell-size categories make better functional units than taxonomic categories.

In summer with stratified water and lower stocks of nutrients in surface waters the stage is set for either tightly coupled (orderly) nutrient regeneration by picoplankton and microheterotrophs or for the large, slow growing K-strategists. The precise sequence of events in summer depends to a degree on trophic state. At low TP concentrations and low biomass in oligotrophic waters, the orderly coupling of production and consumption results in a predominance of small forms. Tropical and subtropical oligotrophic waters with low seasonality may be regarded as equivalent to permanent summer (Fig. 9.5). In such waters the rapid regeneration of nutrients is essential and the whole system is dependent on tight coupling. The production process may be likened to a small rapidly spinning wheel (Harris, 1980). The physiological strategies of the small phytoplankton are well-suited to the small scale nutrient regeneration processes in that they are energy inefficient but nutrient efficient. This is not to say that a few large K-strategists will not be present in such waters. Such cells may either migrate vertically to obtain nutrients or may have generation times long enough to utilize the irregular pulses of nutrients brought into surface waters by the occasional entrainment of deeper water (10–15 days scales). In order to survive, such species must have long generation times and low loss rates. Grazing on such forms must be minimal. Goldman (1984) has suggested that the major food source of zooplankton in the oligotrophic oceans is formed from aggregates of small cells and that the major portion of the nutrient regeneration takes place in such aggregates.

At high TP and high algal biomass the sequence of events is different as nutrient stocks may be so high in spring that severe depletion does not occur. Alternatively the external loading of nutrients may be high and continuous and may be high enough to supply the growth requirements of the algae throughout the summer. In such circumstances the phytoplankton become much less dependent on internally recycled resources and the grazing pressure may be negligible. A number of factors favour the growth of large K-strategists under these circumstances; these include stable water columns, temporally persistent habitats, high nutrient concentrations and light limitation brought about by self shading. Light limitation favours large cells because the self-shading coefficients of large cells or colonies (10^6 μm^3) are smaller than those of smaller (10^3 μm^3) units (Harris, 1978) and thus more biomass may be accommodated when the mixed layer is 'filled in'. This 'package effect' (Talling, 1971) extends to functional units or clumps of cells as well as to individual large cells. Thus the characteristic growth form of many summer blue-green algae, where many small cells are aggregated into large clumps up to mm in size (*Microcystis*), represents a solution to the problem of light limitation as well as a means of preventing grazing. Observation on species of *Microcystis*, *Aphanizomenon* and *Anabaena* support the suggestion that the growth form is a way of minimizing loss rates in late summer (Shapiro, 1980). The clumping of the cells also influences the buoyancy regulation of the populations and allows more rapid repositioning of the colonies after a mixing of the water column (Walsby and Reynolds, 1980).

Many mesotrophic and eutrophic successional sequences show a shift to larger, slower growing species in late summer (Margalef, 1963, 1978; Reynolds, 1980; Sommer, 1980) and these species are indeed characterized by low loss rates as well as low growth rates (Reynolds, 1982; Reynolds *et al.*, 1982; Sommer, 1981; Gliwicz and Hillbricht-Ilkowska, 1975). The late summer species, typified by blue-green algae which regulate their buoyancy and by dinoflagellates, are too large to be grazed effectively so that the ultimate fate of these species is decomposition (Gliwicz and Hillbricht-Ilkowska, 1975). The slow growth of these larger species (Fig. 5.9) dictates that these organisms grow in stable water columns, fronts or thermoclines – habitats with slow mixing times (Table 3.1) which allow sufficient time for a significant biomass to build up. These organisms must also be good competitors because, unlike the transient environments of the mixed layer, these more stable environments may persist for periods sufficient for competitive exclusion (Harris, 1983). Many of these buoyancy regulating organisms move vertically to exploit the resources of the whole water column. An excellent example of this is the vertical migrations of *Ceratium hirundinella* (Heaney and Talling, 1980). The vertical migrations of *Ceratium* allow it to exploit hypolimnetic sources of nutrients which are not available to other non-migratory species and thus become the dominant organism (Harris, 1983). Because the growth rate of the organism is low, numerical dominance takes up to ten weeks, so it requires environments where the vertical stability of the water column is matched by the temporal stability of the stratification. Dinoflagellate blooms in the oceans are also associated with fronts and pycnoclines; temporally stable habitats (Holligan and Harbour, 1977).

Many of the blue-green algae which exist in eutrophic waters in summer are N-fixers. The advantage of such a strategy in waters rich in P but low in N is obvious, particularly when hypolimnial anoxia favours denitrification and the removal of N from the water. The cycling of N speeds up in eutrophic waters (Chapters 4 and 7). The habit of cell aggregation also provides a suitable microenvironment for N-fixation.

In shallow waters, where such a strategy is viable, many of the K–strategists have perennation mechanisms. Dinoflagellates overwinter as cysts (Heaney *et al.*, 1983), blue-green algae overwinter either as vegetative cells or as akinetes and diatoms overwinter as vegetative cells. All species must be resuspended by vertical turbulence or, in the case of blue-greens, by positive buoyancy provided by gas vacuoles. What the perennation mechanism effectively does is to give these slower growing species an innoculum. This is particularly noticeable in the case of *Ceratium* (Heaney *et al.*, 1983; Harris, 1983) in Esthwaite where there is an abrupt rise in the population in March as free-swimming cells hatch from the cysts. Trimbee and Harris (1984) have shown that the innoculum of blue-greens may be considerable, with as much as 2– 4% of the summer population rising from the sediments to seed the planktonic population. Lund's (1954, 1955, 1971) work on *Melosira* was an early demonstration of the importance of perennation by vegetative cells on the sediment surface. Even if only a few percent of the overwintering stages survive (Jewson *et al.*, 1981) it is sufficient to ensure a significant population the following summer. These perennation mechanisms preadapt the summer community and, if the species are good competitors in a stable environment, the course of the summer succession may be largely determined at spring turnover. Anomalous summer weather may still, of course, deflect the course of the summer succession if vertical mixing is more prevalent than usual.

Parasitism

While the late summer K-strategists are, to a large extent, free from grazing they are not totally free from attack. It is now becoming clear that many of these large species are susceptible to attack from a wide range of parasites. The list of species known to be vulnerable includes species of *Volvox* (Ganf *et al.*, 1983), *Microcystis* and *Ceratium* (Heaney, personal communication), *Cosmarium* (Canter, 1979) and a number of the larger spring diatoms (*Asterionella, Stephanodiscus, Fragilaria*, Canter, 1979). The parasitic organisms are an equally wide range of types; viruses, ciliates, amoeboid protozoa and chytrid fungi have been observed to infect natural populations. The impact of these parasites can, on occasion, be considerable as more than 50 % of the phytoplankton population may be affected at any one time and numbers may drop dramatically as a result. It is interesting to speculate on the role that such parasitic organisms may play in freshwater and marine phytoplankton population dynamics. It is entirely possible that parasitic attacks may influence the competitive interactions between the late summer *K*-strategists in the warm, stable environments of late summer by reducing the populations of affected species. A recent parasitic attack on *Ceratium* in Esthwaite drastically reduced the population in the lake in mid summer, although the population subsequently recovered. By influencing the formation of cycts such infections could have a great impact on the year to year variations in the summer assemblage.

9.8.3 TROPICAL AND SUBTROPICAL WATERS

The main features of the seasonal succession are clearly set by the seasonal readjustments of water column stability. The form of the variance spectrum of energy or thermocline depth will determine the overall form of the phytoplankton response. Tropical waters may or may not show seasonality in their physical regimes and in many respects mimic permanent summer. Little or no seasonal succession of species may result (Ganf, 1969, 1974; Lewis, 1978a). The seasonal succession of species in subtropical and tropical oceanic waters is noticeably weaker than that in temperate latitudes (Smayda, 1980). For example, analysis of rates of succession in Lake Lanao, (Lewis, 1978b) showed little seasonal pattern and indicated that the phytoplankton community responded to irregular overturns which occurred as a result of the passage of atmospheric weather systems. In the case of Lake George, Uganda, the greatest scale of variability was found to be a diurnal scale with as much as a 10°C temperature change from morning to afternoon (Ganf, 1974). Overturn took place each morning. The lake was dominated by species of buoyancy regulating blue-green algae (*Microcystis*) which showed little seasonality. As noted above (Chapter 3) tropical lakes may show seasonality in relation to wet seasons and rainfall and, while most species of phytoplankton in Lake George showed little seasonality, two species (*Anabaena* and *Melosira*) did show a marked outburst in abundance during the study period and therefore indicated some alteration in the hydrographic regime. Melack (1979) reviewed the patterns of variability in lakes as a function of latitude and showed that tropical lakes varied little in productivity or biomass at a scale of weeks or months. Melack identified three patterns of behaviour in tropical lakes; a seasonal pattern associated with fluctuations in rainfall, river flow or vertical mixing; a total lack of seasonality and very low biological variability; and a pattern of abrupt shifts from one stable assemblage to another. The first two types were explained by the relative importance of daily versus seasonal fluctuations. No explanation was offered for the third pattern of behaviour.

The subtropical oceanic gyres are also very stable environments which show little physical variability from one month or year to the next. They contain a stable phytoplankton assemblage (Venrick, 1982) which is highly diverse. Little is known about the relative importance of different physical processes operating at different scales so the only explanation of the phytoplankton diversity is the equilibrium explanation of Venrick (1982). At a macroscopic level the system appears to be at steady state but little is known about the scales of variability at the scales of the organisms and there may therefore be more variability at small scales than is presently suspected.

9.9 Nutrient ratios and the role of competition

I have already shown (Chapter 7) that there is little evidence for reductions in growth rate brought about by low nutrient concentrations. The theoretical models of Tilman and Kilham (1976) (Chapter 5) are based on the simple Michaelis–Menton relationship between concentration and growth rate. All the evidence points to the fact that rates of nutrient regeneration in surface waters are sufficient to support rapid growth by the species which match the scales of regeneration. Periods of strong nutrient limitation of growth must be brief and infrequent or else there would be more direct evidence for their existence. There appears to be strong selection for species with varying minimum cell quotas over the range of nutrient availability found in natural waters. The dynamic relationships between production and nutrient regeneration, and the essentially non-equilibrium relationship between the two at a variety of scales, points to the importance of nutrient flux rates, uptake parameters (K_t) and cell quotas (q) rather than the equilibrium approach of concentrations and K_s values.

Tilman and Kilham's models were extended to include the use of nutrient concentration ratios. The ratios in themselves are of little use unless there are some regular relationships between flux rates and concentrations. After all, the use of ratios removes an absolute concentration component and a ratio of 10 µg N:1 µg P (oligotrophic) would work out the same as 1000 µg N:100 µg P (eutrophic). The species composition, trophic state and nutrient flux characteristics of the two water types would be completely different. As is indeed shown by Tilman and others the nutrient ratio approach may be useful with *K*-strategists if, and only if, waters of one trophic state are chosen and if one of the nutrient components varies widely and the other varies little (i.e. Si:P ratios in Lake Michigan and N:P ratios in eutrophic lakes, Tilman *et al.*, 1982). Then, if there is a monotonic relationship between fluxes and concentrations (as appears to be the case) the nutrient ratio approach may be a simple surrogate for more realistic ecological variables.

There is a further problem with competition in that equilibrium theory assumes coexistence during the period of interaction. While some have studied the competitive interactions between diatoms and blue-greens in culture, these species rarely coexist in nature. There are undoubtedly some long range autogenic interactions between species in that the growth of the spring diatom bloom changes the nutrient regime in surface waters and sets the scene for the blue-greens in late summer. They cannot strictly be said to be competing in the normally accepted use of the word as they do not coexist.

For many species, particularly the extreme '*r*' strategists, the non-equilibrium approach dictates that competition will be very weak or non-existent in a fluctuating environment. Density dependent effects will also be absent. Exponential growth to a ceiling set by high level resource parameters seems to be the rule. High cropping by grazers will also minimize competition. The only group of organisms for which density dependent effects should be observed should be those *K*-strategists which grow in temporally stable habitats and show low loss rates. In the midst of a blue-green or dinoflagellate bloom most of the nutrient is tied up in the algal biomass and regeneration rates are minimal. Concentrations may then be more important than fluxes. The long time required to reach maximal biomass means that the bloom must persist for times equivalent to the time required for competitive exclusion. The low loss rates are important. Kalff and Knoechel (1978) showed that competitive exclusion could not occur based solely on a difference in growth rates: loss rates must be invoked to explain the result. Species with low loss rates have an advantage over species with higher loss rates. Harris (1983) examined the growth responses of *Ceratium* in Esthwaite and showed that there were indeed strong density dependent effects. The population growth could be fitted to a logistic model which indicated a 10–12 week period of exponential growth followed by a plateau. The initial population increase due to the hatching of cysts was clearly visible. The precise growth rate varied from year to year but the doubling time was of the order of 7–10 days. As the environment was stable for much longer time periods than this, the argument for equilibrium conditions could be sustained. Further evidence for density dependent effects comes from C:N:P ratios. The only C:N:P ratios which can be found in natural populations which support the idea of reduced relative growth rates come from late summer dinoflagellate populations (Heaney, personal communication; Serruya and Berman, 1975). C:P ratios in late summer populations of *Peridinium* in Lake Kinneret reached 600:1.

If the competitive exclusion models of Tilman and Kilham are to be used at all, then a good case can only be made for their use for species which are known to coexist in late summer in stable waters. Tilman (1982) proposed that disturbance was 'a process that influences the relative supply rates of resources for which competition occurs'. If disturbances do not influence the supply rates of the resources, equilibrium interactions ensue. It is not surprising that equilibrium conditions can be found; the real question is how often are such conditions encountered in nature? It is difficult to see competition operating in phytoplankton communities, or indeed any community in which the environmental fluctuations override the tendency for internal order and stability. I have already shown that competitive exclusion takes 20–50 days in culture and even then cannot be complete if growth processes are the sole determinant of the outcome (Kalff and Knoechel, 1978). Loss rates are important. It is necessary for the environment to be stable for twenty generations or so for exclusion to occur. The surface waters of lakes and the oceans are rarely stable for these time scales. Few other ecosystems are stable for equivalent multiples of the generation times of the dominant organisms. The rapid growth of the phytoplankton in relation to the seasonal changes which occur ensures that coexistence between the dominant organisms rarely occurs for such time periods. Species may persist in low numbers but little interaction occurs.

CHAPTER 10

Community structure and function in turbulent environments

10.1 Processes in phytoplankton successions

Margalef (1963) identified the following major features which show regular trends in the course of phytoplankton successions.

(1) Reduction in mixing and turbulence. Margalef considered that only in more stable water columns could the organization accumulated by previous populations be preserved and built upon.
(2) Resources are progressively depleted so that the succession is driven in a fixed direction by changes in the availability of energy or nutrients.
(3) P/B ratios and efficiencies change with time so that in the initial stages there is a higher flow of energy per unit biomass.
(4) Changes in plant pigments accompany both changes in the physiology of individual populations and changes in the species composition.
(5) Grazing efficiency increases as selective grazing becomes important.
(6) Levels of dissolved organic carbon and detritus increase.

These six major features are themselves interrelated through the reaction of the organisms on their environment and through the effect of external perturbations on the whole system. Plant successions are influenced by both autogenic and allogenic factors.

Reduction in mixing and turbulence may be regarded as one of the main driving variables, as the development of the seasonal thermocline is of major significance to the organisms. The development of the thermocline reduces Z_m, the mixing depth and allows the surface layers to become more stable. The increased stability in surface waters leads to the development of vertical heterogeneity as the mixing rates decrease (Chapter 3) and it becomes possible for the organisms to migrate vertically and/or buoyancy regulate. The development of the seasonal thermocline, by reducing vertical diffusivity, also allows the biomass of phytoplankton to increase in surface waters and may lead to the light limitation of summer populations. At the same time the thermocline prevents the regeneration of nutrients from bottom waters during the summer and thus leads to a reduction in total nutrient concentrations in surface waters. As the stratified period progresses there are, therefore, a number of factors which change in concert. The trophic status of the water body will determine which particular pathway is followed, as oligotrophic waters tend to become nutrient-limited during the stratified period whereas eutrophic waters tend to become light-limited as the biomass in surface waters is higher.

The nature of the limiting resource (or resources) determines the phytoplankton strategies in the water column in conjunction with other non-limiting resources. Life history strategies and efficiencies change in response to the ecological niches available. The succession of communities is therefore quite different in oligotrophic and eutrophic water. The changes in strategies and ecological efficiencies may be monitored by pigment changes, by photosynthetic parameters or by nutrient uptake characteristics. As we have already seen cell size is a useful scale for many of these functions. Cell-size distributions interact with grazing efficiencies in that certain size fractions in the phytoplankton community are grazed more effectively than others. There is therefore a strong feedback between light limitation, nutrient regeneration, cell-size distributions and grazing. As the seasonal succession progresses there is a tendency for the dissolved organic carbon and detritus component to increase in importance as products of decomposition build up in the water. In eutrophic waters the food chain is highly dependent on such carbon sources and the food chain may become essentially a detrital food chain in such waters. This will be especially true in waters where the phytoplankton are large and the zooplankton small.

I have already demonstrated that the planktonic environment is far from homogeneous (Chapters 3 and 5) and that patchiness in space and time is widespread. Changes in the variance components of resource availability are as important as changes in the mean level of those resources. Niche diversification was supposed to be difficult to demonstrate in plants (cf. Venrick, 1982), but I have identified a number of examples of significant niche diversification (Chapter 8). The reason that such diversification was underestimated in the past appears to lie in the fact that there is significant diversification in the way different species use the variance components of the environmental fluctuation (rather than the mean resource levels) and the fact that cells of different sizes appear to show quite different photosynthetic and nutrient uptake strategies in environments with different periodicities. Thus size may be more important than taxonomy in some circumstances. The work of Turpin and Harrison (1979) (Chapter 8) illustrated very clearly the influence of patchiness on the physiology and competitive ability of phytoplankton and illustrated the tracking ability of phytoplankton in fluctuating environments. The ability of cells of different size to track different resource fluctuation spectra appears to be a crucial factor in determining the species composition of natural phytoplankton communities. Thus the niche diversification between species contains components of both the mean level of the resource and the variance. While Tilman's models of competition apply in continuous culture, I have shown that they can only be applied in the field with difficulty as the relationship between nutrient concentration ratios and nutrient flux ratios is complex (Chapter 7). The small scale variance in the recycled nutrient flux appears to drive a number of significant non-equilibrium processes and competitive exclusion may be avoided because of the variance of the flux in space and time. The universality of the nutrient turnover relationships in lakes and the oceans and the apparent lack of suppression of growth at low nutrient concentrations is worthy of note (Chapters 6 and 7).

The other area of significant niche diversification lies in the temperature optima of different species (Chapter 8). As the temperature signal in the surface waters of lakes and the oceans is a slow seasonal trend then there is a greater probability that such diversification leads to an equilibrium response as the trend changes over scales of a number of generations; scales sufficient for competitive exclusion. The temperature optima of phytoplankton are very broad however, (Chapter 8) and are unlikely to lead to the generation of the observed diversity. Temperature changes do lead to successions of species in mass cultures (Chapter 8) but the overall diversity of the assemblages in such cultures is less than in nature. Shugart *et al.* (1980) showed by means of a simulation that temperature effects on succession may produce discontinuities by influencing the competitive ability of species. Furthermore the effect of temperature displayed an hysteretic effect dependent on the direction of temperature change. Increases and decreases in temperature produced discontinuities at different points. While temperature may act as a coarse filter for the selection of species at different times of the year, it appears that the variance components of other factors play an important role in determining the species composition and diversity. Thus the various environmental factors which determine the species composition in surface waters may operate in different ways and at different scales.

10.2 Structural trends in succession

Margalef (1963) envisaged that all phytoplankton successions begin with a period of intense mixing and progress as the water column stabilizes. This will produce one, two or many successions during the year depending on the particular physical regime. As the water column stabilizes, random turbulent movements cease to dominate the organisms and structure develops in the vertical dimension (cf. vertical mixing time scales, Chapter 3). Margalef (1963) called this the passage from random flow to structured 'viscosity'. In less turbulent waters sinking velocities become critical and loss processes crucial for the survival of populations. I have shown that the sinking velocities of natural phytoplankton populations are well matched to the time scales of physical processes in surface waters during stratified periods (Chapter 5). Margalef (1963) considered that as the succession proceeded and the rate of succession declined, the decline was due to both a reduction in the net rate of increase of the species and to an increase in grazing.

As the species composition of the communities changes and the 'viscosity' in the system increases, the order in the community increases. As the order increases the transition probabilities become much more limited and to quote Margalef again: 'Structures that accumulate or copy information (that is preserve it with time) at a small thermodynamic expense have greater probability of survival and influencing events' (Margalef, 1963). Order and diversity are not the same, so that order may increase as the diversity first increases and then decreases in the expected fashion. Margalef (1963) noted this trend fifteen years before the publication of Connell's (1978) paper. Order (predictability) may be increased by tight feedback loops, increased efficiency of grazing and reduction in time lags between pools. This represents adaptation to and incorporation of the perturbing holons. Also a reduction in reproduction rates and buffered population fluctuations (adaptation of physiological holons) make changes more predictable. A decrease in the rate of succession over time is therefore correlated with an increase in order.

The increase in order gives the phytoplankton community a 'feedforward' or preadapted capacity as the information carried by the constituent species limits the transition probabilities when the community is faced with perturbations. If these perturbations are predictable at seasonal scales then the species may (for example) adopt encystment and emergence strategies which exploit equally predictable environmental cues. A number of species appear to do this. This, once again, represents the incorporation of predictable signals into a functional holon.

Margalef (1980) explicitly recognizes the difference between processes which structure diversity and processes which structure food chains. Diversity is brought about by individual species playing the 'game' whereas food chains are structured by thermodynamic constraints. This is quite a different view from the traditional equilibrium viewpoint and is one which leads to different models and predictions. I have already argued (Harris, 1980a) that the usual form of ecological model does not adequately describe the dynamics of the real world. A model which is composed of sets of differential equations solved at equilibrium only represents a subset of the real spectrum of ecological possibilities. It is worth repeating though that the success of such models will depend on the time scales of processes and the variances of the fluxes involved. Over the time scale of the model it is a matter of 'conservative' versus 'nonconservative' behaviour. Instead of relying on equilibrium theory as a representation of the real world it is essential to demonstrate that the theory is truly applicable before it can be applied to a given problem.

Dissipative structures can clearly only exist in situations where the environment is stable for a large number of generations and sufficient time exists for close coupling between production and consumption. The development of order takes time. Such situations seem to be prevalent in oligotrophic waters where the generation times of the picoplankton are short. Close coupling between production and consumption appears to exist only in summer in temperate oligotrophic waters and in the subtropical oligotrophic oceans. For most organisms in surface waters the seasonal cycle of stratification is sufficient to disturb the relationships between production and consumption and to make the relationships essentially non-steady state.

Planktonic food chains are clearly structured in part by the size of the organisms. The predator/prey size ratios brought about by hydrodynamic constraints result in trophic levels which act as holons with quite different time constants. The biomass at each level may be easily calculated from size distributions and ecological efficiencies. Furthermore, it may also be postulated that the yield per unit of nutrient is inversely proportional to the ecological efficiency and directly proportional to the size of the organisms. In ultra-oligotrophic waters the preponderance of picoplankton with rapid growth rates leads to high ecological efficiencies and close coupling between production and consumption. The vast majority of primary production on a global scale is regenerated within surface waters. In such waters the tight coupling between production and consumption requires high energy inputs. Little 'new' production is available. As the nutrient status of the water increases and more larger cells occur, the ecological efficiency drops and the yield of cells per unit of nutrient increases. In eutrophic waters the energy requirement is less but the production and consumption are coupled at larger scales. On a seasonal scale there is a burst of 'new' production available in spring. 'Displaced metabolism' of the carbon and other major nutrients regenerates nutrients at seasonal scales or after intermittent overturns (cf. Chapter 4). In the extreme case the nutrients are returned by upwelling after a long sojourn in deep ocean water. Sustained 'new' production may then be available.

Parallels may be drawn between aquatic and terrestrial systems. In tropical rain forests the stock of nutrients in the soil is small and the velocity of cycling is high. There is little 'new' production available. In temperate ecosystems the energy input is less and the stocks of nutrients in the soil are larger. Large amounts of nutrients are recycled at seasonal scales and more 'new' production is available on a seasonal basis. In the same way as in aquatic systems, the seasonality of production and the lower temperatures lead to time-lags between production, consumption and regeneration. A system like that can only operate with large reservoirs of nutrients.

There appears to be a particular group of organisms which occur in situations where there is a pool of soluble nutrients to be exploited. These organisms might be regarded as the classic extreme 'r-' strategists because they grow opportunistically in situations where resources are plentiful and energy inputs are high. These flagellates and diatoms are characteristic of spring blooms in temperate waters and tend to be intermediate in size between picoplankton and the large net plankton. Picoplankton may have both high growth and mortality rates (birth rate (b) and death rate (d) = 2.0 d^{-1}) but the close coupling of production and consumption leads to ordered systems ($|b - d| \simeq 0.0$). In the case of the large, slow growing 'K-' strategists growth and death rates are small (b and $d = 0.1$ d^{-1}) and the system is also ordered ($|b - d| \simeq 0.0$). In the small flagellates and diatoms characteristic of spring blooms and areas of 'new' production growth is rapid ($b > 1.0$ d^{-1}) and mortality is initially small. As the grazing organisms catch up and the soluble nutrient pool is depleted the reverse situation occurs and mortality from grazing exceeds 1.0 d^{-1}. In either case $|b - d| > 1.0$ d^{-1}. It is interesting to note that high yields of biomass arise from poor coupling, low efficiencies and poor ordering of systems. These occur in highly seasonal environments. Human consumption is also dependent on high biomass yields in similar circumstances. Highly ordered and closely coupled systems yield little and attempts to manage such systems seem to be doomed to failure. Enforced uncoupling, as in the case of upwelling and eutrophication produce biomass yields. Only in the case of upwelling is the yield of use to man. Eutrophication leads to blooms of nuisance algae.

Margalef pointed out that the resilience of a system should be a function of $|b - d|$ as environmental fluctuations should interact with b and d through the ability of organisms to integrate over generation times. The generation times of the picoplankton and the large net plankton differ by an order of magnitude. There is a dichotomy in the transition from 'r' to 'K' in phytoplankton successions as, while $|b - d|$ is reduced and the system becomes more ordered by both picoplankton and the larger net phytoplankton in late summer, the absolute values of b and d must also differ by an order of magnitude. Also, the ecological efficiencies of the cells of different sizes and the degree of coupling between production and consumption differs markedly. The presence of a distribution of periodicities in the regenerated nutrient flux means that the picoplankton can use the small scale flux while the larger, slow growing forms can use the longer term (days) scales which arise from atmospheric interactions. In late summer, and in persistently nutrient poor waters this may be the reason for the coexistence of picoplankton and large dinoflagellates.

The velocity of turnover of nutrients is clearly a feature of considerable importance in determining the structure of food chains. The fact that there are clear trends in cell-size distributions and in turnover velocities in different water types is evidence of structured processes. Cell size is also related to turnover velocities through the inverse relationship between cell size and growth rate. The dynamic view of food chain structure is that of flux rates and efficiencies rather than concentrations and growth rates. That concentrations and growth rates do explain some features of the real world is indicative of the fact that there are monotonic relationships between nutrient concentration ratios and flux ratios, and between growth rates and efficiencies. These relationships are not necessarily linear as demonstrated by the non-linear relationships between TP and algal biomass in lakes.

Margalef (1980) proposed an index of ecosystem stability of the form

$$dL/dt = 2 \sum_{i=1}^{n} a_i N_i^2 \qquad\qquad 10.1$$

where

$$a_i = |b - d|$$

and dL/dt is the velocity of change which Margalef supposed would tend to a minimum over time. This is indeed the case apparently as long as the effects of environmental fluctuations do not override this tendency. The trends in succession therefore depend on the balance between the trend towards internal order (equilibrium) and the disruptive forces of external change (non-equilibrium). The final result is most frequently a wobbly compromise between the two extremes.

The distribution of different sized organisms in different water types can therefore be related to both the inherent constraints of predator/prey size ratios and efficiencies as well as the external influences of environmental fluctuations. Organisms of different sizes have differing capacities to track and integrate different environmental signals. There is a complex set of links between the intrinsic and extrinsic controls of food chain structure. The control is by no means entirely 'top down' as many standard food chain models would have it. With built in time-lags and large differences in the sizes of organisms and in growth rates there will be considerable 'bottom up' control. Nutrient fluxes and the supply of energy are determined by physical factors as well as biological processes and, as the size classes of phytoplankton are under strong direct physical control, the consumers will likewise be influenced, if less directly, by extrinsic forces. In the terminology of holons and hierarchies; the holons are not completely nested. External perturbations frequently occur.

The non-equilibrium view recognizes the importance of the interplay between structure, function and the fluctuations in both extrinsic and intrinsic functions and relationships. Extrinsic fluctuations may be scaled by the $-5/3$ power function of the turbulence spectrum. Intrinsic fluctuations will be brought about by the time scales of materials recycling within the organisms as well as by fluctuations in generation time, cell volume and migration. Further fluctuations (Turing behaviour) are likely to arise from a combination of extrinsic (diffusion) and intrinsic (growth) factors which will keep the entire system in sustained motion. As Prigogine (1978) pointed out there is a complex interplay between structure, function and fluctuations. Diversity arises from the opportunistic growth of organisms within thermodynamic constraints (Margalef, 1980). In such a system high level (large number) properties will be more predictable than low level properties. Cell-size distributions are more predictable than species compositions: biomass is more predictable than population size. It is possible to identify all of Tilman's (1982) spectrum of possibilities; from situations where diversity is limited by a lack of fluctuations (equilibrium), through the point where the fluctuations become a resource in themselves, to the point where the fluctuations become destructive. As might be expected the middle ground seems to be the norm. The organisms have had plenty of time to adapt to the normal spectrum of environmental perturbations.

10.3 Rates of succession

Shugart and Hett (1973) reviewed the similarities in the species turnover rates of a number of successional sequences. They concluded that as the successional communities approached some equilibrium state the proportion of species lost per unit time did indeed decrease. The rate coefficients for species turnover decreased with the age of the community for all types of succession studied. The deceleration was most pronounced early in the successional sequence. This conclusion was borne out by the work of Jassby and Goldman (1974) who calculated a succession rate from the rate of change of species composition in natural phytoplankton communities. Jassby and Goldman used biomass (volumes) to weight the species abundances. In Castle Lake, a dimictic, mesotrophic lake, the rate of succession declined with time during the stratified period in each of three years (Fig. 10.1). Lewis (1978a) revised the calculations of Jassby and

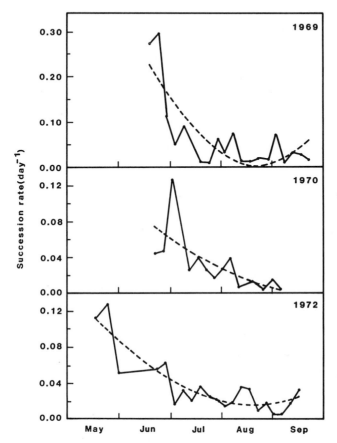

Fig. 10.1 Succession rate as a function of time in the summers of three years in Castle Lake (Jassby and Goldman, 1974). Rates calculated from volume weighted abundances over periods of five days.

Goldman and introduced a new measure of succession rate: Lewis did not like the weighting technique that Jassby and Goldman had used. Lewis (1978a,b) achieved a very different result from Jassby and Goldman however, not so much because his index produced widely differing statistics, but because he applied his index to the succession rate of Lake Lanao in the Phillipines. That lake showed no distinct seasonal pattern but merely showed a series of fluctuations in rate throughout the year.

Reynolds (1980) also devised a modification of the Jassby and Goldman (1974) index which was used to describe the rate of succession in a number of English lakes. Reynolds used a volume weighted index but based it on the relative size of the species present. The results (Fig. 10.2) indicated that the

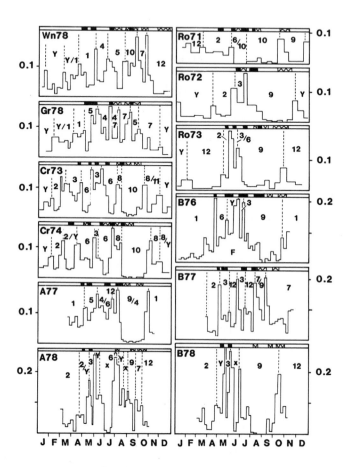

Fig. 10.2 Rates of community change observed in each of a number of lakes over the stratified period. 1, *Asterionella, Melosira italica*; 2, *Asterionella, Stephanodiscus astraea*; 3, *Eudorina, Volvox*; 4, *Sphaerocystis*; 5, *Chrysophytes*; 6, *Anabaena, Aphanizomenon*; 7, *Tabellaria, Fragilaria, Staurastrum*; 8, *Melosira granulata, Fragilaria, Closterium*; 9, *Microcystis*; 10, *Ceratium*; 11, *Pediastrum, Coelastrum*; 12, *Oscillatoria agardhii*; X, *Opportunists*; Y, *Cryptomonads*. Note the periodic rise and fall of the rate of change of community structure. Bars indicate periods of mixing. Lakes and years as follows: Gr: Grasmere, Cr: Crose Mere, Ro: Rostherne Mere, Wn: Windermere, A and B: Lund Tubes in Blelham Tarn (from Reynolds, 1980).

seasonal succession was frequently interrupted by mixing events in surface waters. These events occurred at scales of a few days to a month and served to perturb the sequences. Reynolds noted that the peaks in the plot of community change over time were skewed to the left. This was taken as an indication that the rate of change after the perturbation was initially rapid but slowed as new dominant species became important. Two thirds of the peaks in the rate of community change that occurred during the stratified period were the direct result of wind mixing while one third occurred independently of changes in the physical conditions. Reynolds (1980) concluded that these latter shifts in species composition were due in large part to the depletion of nutrient stocks in the epilimnion and were therefore due to internal interactions rather than external perturbations.

There is a problem with all these indices of rate of change however in that they are dependent on the time interval of sampling. Trimbee (1983) used both daily samples and weekly samples to calculate Lewis' (1978a) index of succession rate. There was very little similarity between the two measures. Daily rates of change were smaller than weekly rates, but more importantly the temporal sequence of changes in rate was quite different. One of the reasons for the discrepancy between the daily and weekly rate estimates was probably the redistribution of species within the reservoir studied. The layering of species at different depths led to strong advection effects, with species moving in different directions at different depths (Trimbee and Harris, 1983). Indices of rates of successional change can therefore be affected by the temporal and spatial scales of sampling, by horizontal and vertical heterogeneity, by basin mixing scales and the growth and loss of populations. Sampling schemes should reflect the fundamental physical and biological scales rather than the convenience of the researcher (Trimbee and Harris, 1983).

10.4 Succession in turbulent mixed layers

Both Margalef (1978) and Reynolds (1980) showed that the changing physical conditions in surface waters were the major determinants of community change. Harris and Piccinin (1980) showed that quantitative estimates of the rate of community change were correlated with changes in the mixing depth in Hamilton Harbour and this was illustrated by an ordination of the seasonal abundances of the major species. In a situation where nutrient limitation played no major role, Harris and Piccinin (1980) showed that there were critical Z_{eu}/Z_m ratios for certain groups of species. Wofsy (1983) showed that many phytoplankton assemblages in waters where nutrients are not limiting increase in biomass until a set Z_{eu}/Z_m ratio is achieved. Z_{eu} is reduced by self-shading until the mixed layer is 'filled' with biomass. The P:R ratio of the populations making up the biomass is the major determinant of the final ratio (Chapter 8). In the case of Hamilton Harbour the mean Z_{eu}/Z_m ratio lay close to Wofsy's (1983) mean ratio but massive, intermittent changes in Z_m (driven by storm activity and exchange with Lake Ontario) led to severe perturbations in the Z_{eu}/Z_m ratio. While the populations were not light limited in the classic sense they were affected by the variance of the Z_{eu}/Z_m ratio and the interaction of the time scales of the Z_m perturbations and the time scales of the algal response. The variance in the ratio over time was more important than the mean. All the evidence points to the importance of these allogenic processes in the natural state. The fluctuations in the turbulence in surface waters at a scale of days have an important effect by determining which species can grow. The normal situation is therefore not a simple winter, spring, summer sequence with a shift from r- to K-strategists, but a more complex sequence of changes driven not only by the mean resource levels but also by the variance components at a variety of scales.

Harris (1983) used spectral analysis to examine the time series relationships within and between data sets for biomass, phytoplankton diversity and water column stability in Hamilton Harbour. The data sets were not stationary and therefore needed to be transformed before analysis. The biomass data (as chlorophyll) was log-normally distributed because it was the sum of a number of exponential growth processes (Equation 11.2, MacArthur, 1960). There was an overall negative relationship between biomass and diversity with the peak biomass occurring early in the stratified period and the peak diversity occurring at the end of the summer stratification, 6–8 weeks later (Fig. 10.3). Surface biomass increased whenever there was a reduction in vertical mixing. The biomass increase was usually the result of growth by a single species and the effect of such growth on the rest of the assemblage was minimal (Harris and Smith, 1977). The calculated diversity index (H') was sensitive to numerical dominance of one, or a few species, as described by Harris and Smith (1977). At a scale of a few days to two weeks, therefore, a reduction in vertical mixing led to increased biomass and reduced diversity. The 6–8 week lag between maximum biomass and maximum diversity was the result of the slow accumulation of species in the mixed layer during the stratified period.

The slow rise in diversity can be explained by the alternation of heating and cooling cycles in surface waters; in other words by the variance in physical structure. Periods of vertical stability allowed vertical differentiation to occur (Haffner *et al.*, 1980; Wall and Briand, 1980). Breakup of the vertical structure by cooling and/or current shear dispersed the cells through the mixed layer again. The slow accumulation arose from the persistence of species between their individual growth periods so that, over time, the total number of species in the water column rose until autumn overturn. The total number of species encountered in the usual counting proceedure doubled between May and early September. Others (Lund, 1964; Margalef, 1958, 1968; Reynolds, 1976a,b) have noted the same effect in other water bodies so the effect is widespread. The persistence of species between growth periods points again to the importance of loss processes in determining the course of the summer succession.

Principal component analysis of the phytoplankton assemblage in Hamilton Harbour (Harris, 1983) showed that there were three distinct periods of community change throughout the year which were tightly coupled to changes in the physical structure of the water column. The week by week changes in the phytoplankton community illustrated in Fig. 10.4 (Harris, 1983) showed that the largest changes in the structure of the assemblage took place in spring and autumn, as noted by Round (1971). At other times the week to week changes in the assemblage were smaller. There was, in Hamilton Harbour, a correlation between the diversity of the assemblage and changes in the physical mixing regime. Periods of constant overturn led to almost monospecific blooms whereas periods of intermittent mixing led to high phytoplankton diversity. This is consistent with the intermediate disturbance hypothesis of Connell (1978), and with the foregoing arguments about the role of disturbance at a variety of scales.

Cross-covariance analysis of water column stability and species abundances showed that each species had a unique covariance relationship and transfer function indicating that each species was tracking the environmental fluctuations in a different manner. While the changes in community structure were driven by the changes in the Z_{eu}/Z_m ratio, the response was not immediate and the periods of growth and persistence lagged the environmental changes. Different species exhibited different lags. Not only that, but some species responded positively to changes which affected others negatively (Harris, 1983). The amplitude of the transfer functions increased at longer time scales; a clear indication of the integrative nature of cellular physiology. Some time-lags were of the order of 4–8 weeks indicating that long time-lags may occur between the initiation of growth and the appearance of the species as a dominant form in the assemblage. This bears out Vollenweider's (1953) assertion that it is the factors which cause the bloom that are important, not the conditions at the time of the bloom. Such time-lags doom to failure any attempt to correlate the species composition with ecological conditions at any point in time.

In a recent study of the day-to-day relationships between population dynamics and fluctuations in water column stability, Sephton and Harris (1984) examined the cross-correlations between species abundances and both N^2 and Z_{eu}/Z_m. As expected, different species showed quite different covariance relationships with significant lags in the 1–10 day range. As expected also (Chapter 8), the apparent changes in physiological parameters

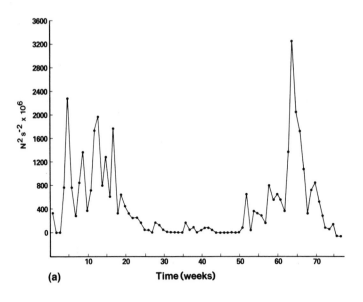

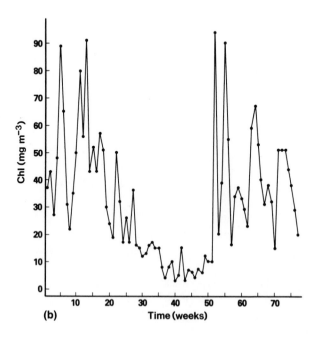

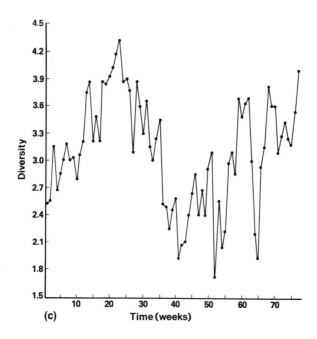

Fig. 10.3 A time series of data from Hamilton Harbour in which are plotted weekly values of (a) water column stability (N^2), (b) algal biomass and (c) diversity.

were produced by changes in the species composition. Small *r*-strategists such as *Rhodomonas* showed rapid growth in response to changes in N^2 at lags of 1 and 3 days whereas the larger *Coelastrum*, a *K*-strategist, showed a similar response but with lags of 4 and 9 days. *Oocystis* showed a negative covariance with N^2 at a scale of 5 days. It grew in response to increased vertical turbulence. The large, heavy, *Stephanodiscus* also increased in abundance with increased vertical turbulence but the lags were 4–5 and 9 days. Thus Levin's (1979) predictions are borne out in that different species utilize the variance in N^2 and Z_{eu}/Z_m in different ways. The conditions for non-equilibrium interactions are satisfied as the lag-times are of the order of 1–9 days; time scales which interact strongly with the dominant internal periodicities of 5–6 days (Chapter 5). This is no more than a formal statement of Hutchinson's (1941, 1953, 1961, 1967) ideas about the importance of temporal fluctuations in the control of community diversity. The characteristic time lags of different species bear a close resemblance to the best estimates of the generation times of the organisms and it is possible that such techniques could be used to obtain estimates of growth rates *in situ*.

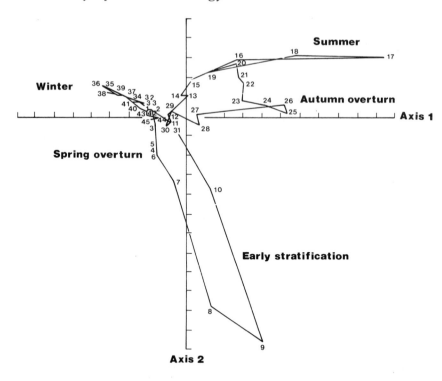

Fig. 10.4 A plot of the results of a week-by-week ordination of the abundance of phytoplankton species in Hamilton Harbour. The sequential weeks are joined and numbered. Week-to-week changes are small in winter but large changes occur when the seasonal thermocline breaks down and reforms (Harris, 1983).

Harris *et al.* (1983) were able to throw more light on the underlying mechanisms of the changes in species composition by a study of cell-size distributions in Hamilton Harbour. The overall cell-size distribution in surface waters was characterized by the presence of 'small', 'medium' and 'large' size classes which could be identified with particular species or groups of species (Fig. 8.11). As noted by Sheldon and Parsons (1967) the spectrum of cell sizes showed few peaks: never more than three. The 'small' size class was dominated by species of *Rhodomonas* and *Cryptomonas*, the 'medium' size class was dominated by a species of *Oocystis* and the 'large' size class was dominated by a species of *Stephanodiscus*. (For taxonomic details see Harris and Piccinin, 1980.) Marked differences in the importance of the three size classes occurred in different years, depending on the particular physical regime in that year. There were times when it looked as though the 'small' and 'medium' size classes were mutually exclusive (Harris *et al.*, 1983). It was not possible to invoke competition to explain the observed pattern as the growth periods of the two size classes were out of phase at a scale of days. *Oocystis* ceased growth 6–8 days after a major heating event in 1979 when the *Rhodomonas* population was very low. *Rhodomonas* reached its peak abundance 4 days after the growth of *Oocystis* had ceased by which time *Oocystis* was growing again. Because of the persistence of species between growth periods the three cell-size classes frequently coexisted. Coexistence was possible because of the differences in cell size, growth rates and characteristic lag-times: much as predicted by Schaffer (1981). The physiological basis of this niche diversification based on cell size was discussed in Chapter 8. The coexisting species are not physiologically plastic and the differing degrees of energy and nutrient efficiencies shown by cells of differing size can be used to explain the distribution of natural populations (Laws, 1975). Not only is there a seasonal cycle of cell size, but day-to-day fluctuations in the environment control shorter term responses and lead to coexistence by a variety of species.

In a general sense, it is possible to identify the conditions which lead to the growth of certain algal assemblages and the factors which cause shifts from one assemblage to another. These results have been summarized in a series of diagrams published by Margalef (1978) and Reynolds (1980a, 1984b). These diagrams relate the observed patterns of phytoplankton occurrences to nutrient availability (trophic state) and indices of turbulence (Fig. 10.5). On a seasonal scale, it can be argued, the day to day variance in the environment can be averaged out and broad 'equilibrium' generalizations become possible (Reynolds, 1980a). These diagrams bring together much of the foregoing discussion but it is clear that indices of turbulence are not sufficient in themselves. What is required is a measure of the temporal fluctuations and the persistence of habitats. However, just as broad patterns in the behaviour of $P:R$ ratios can be discerned it is possible to identify broad taxonomic patterns of behaviour. The demonstration of non-equilibrium conditions at a scale of days does not render this approach invalid, even though the factors leading to the growth of certain assemblages must be sought in short-term events which preceed the appearance of the organisms in abundance (Vollenweider, 1950, 1953). The original appearance of the species in the water column depends on seasonal strategies such as survival and perennation mechanisms or encystment. Subsequent growth of the organism depends on the maintainance of a particular set of conditions for long enough to enable sufficient doublings to occur. The fact that repeatable seasonal patterns occur in the face of considerable environmental variability is a good indication that the real world is a wobbly compromise between equilibrium and chaos.

The occurrence of lags of up to 6–8 weeks has important management implications. Management models will have to be statistical time series

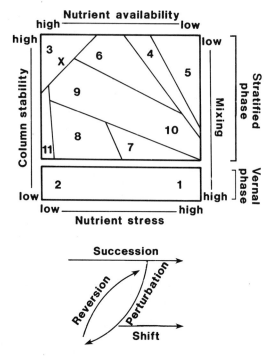

Fig. 10.5 Reynolds' possibility matrix showing the most likely phytoplankton assemblages as a function of nutrients and mixing. The lower figure indicates the direction of motion in response to perturbation and loosely indicates the possible sequences of events (from Reynolds, 1980a). Communities classified as follows: 1, *Asterionella, Melosira italica*; 2, *Asterionella, Stephanodiscus astraea*; 3, *Eudorina, Volvox*; 4, *Sphaerocystis*; 5, *Chrysophytes*; 6, *Anabaena, Aphanizomenon*; 7, *Tabellaria, Fragilaria, Staurastrum*; 8, *Melosira granulata, Fragilaria, Closterium*; 9, *Microcystis*; 10, *Ceratium*; 11, *Pediastrum, Coelastrum*; 12, *Oscillatoria agardhii*; X, *Opportunists*; Y, *Cryptomonads.*.

models rather than the present type of equilibrium plankton model (Harris, 1980a). The utility of such models was discussed by Roughgarden (1975a,b) and Poole (1976). The predictions from these models will never by 100 % sure because there will always be a degree of uncertainty associated with the vagaries of the physical environment. The relationships between physical perturbations, algal biomass, diversity and species composition appear to be ripe for exploitation as a management tool (Harris, 1983; Reynolds *et al.*, 1983). By artificially increasing the frequency of physical perturbations it appears that it is possible to depress the biomass below the 'ceiling' by not giving species time to fully exploit the resources available. Fig. 9.6 demonstrates the effect which was produced naturally in Hamilton Harbour and Fig. 9.8 diagrammatically illustrates the underlying processes (Reynolds *et al.*, 1983). The effect relies on the interaction between the frequency of perturbation and the growth responses of the species in the mixed layer. The result of intermittent perturbation is the production of a more diverse community and the suppression of some nuisance blooms.

In confirming the theoretical work of Levins (1979) this view of phytoplankton assemblages demonstrates the importance of environmental variability in determining the strategies of the organisms. The niche for these species must be defined both by the environmental variance components and by the organisms' perception of those components. The environmental non-linearities behave as resources. The organisms appear to conform to Levin's (1968) best strategy of not attempting perfect tracking. Survival and the avoidance of competition arise from a set of species-specific tracking strategies which lead to a community structure determined by the mean and the temporal components of the variance of the resources. Phytoplankton show a hierarchy of responses (Slobodkin and Rapoport, 1974) from encystment and perennation at seasonal scales, to structural changes associated with growth and persistence at scales of a few days to weeks, to physiological integration mechanisms at scales of less than a few days.

A planktonic ecosystem which is in a state of dynamic disequilibrium can be expected to show long term fluctuations such as those observed by Cushing (1978), Cushing and Dickson (1976), Russell *et al.* (1971), Russell (1973), Southward (1980), and Gieskes and Kraay (1977). Given the time and space scales of oceanic and climatic fluctuations (Lamb, 1972, 1977; Monin *et al.*, 1977; Smith, 1978) an ecosystem that responds to the frequency components of the environmental spectrum at scales from days to years will rarely, if ever, achieve equilibrium (Connell, 1978). Equilibrium conditions may occur in physical regimes which are stable in time but non-equilibrium conditions will prevail if temporal variability is predominant. The effects of physical regimes on the growth of phytoplankton are profound (Margalef, 1978) and have been underestimated in the past.

10.5 Manipulation of the species composition

It is possible to manipulate the species composition of natural waters by altering both the mean and the variance of some of the critical variables which control the seasonal succession. In Chapter 7 it was shown that the $TN:TP$ ratio in lakes broadly controls the species composition by the effect of N and P turnover on the replacement of species. Low TN concentrations (rapid N turnover), and $TN:TP$ ratios less than 25, favour the growth of N-fixing blue-greens. Reductions in average P loads will therefore influence the species composition by altering the relative rates on N and P turnover as well as reducing the total biomass. Lund and Reynolds (1982) studied the effects of changing the periodicity of the P loadings with experiments in the Lund Tubes (Fig. 10.6). Lund and Reynolds fertilized the Blelham Tubes weekly and stimulated the growth of r-strategists throughout the summer. This experiment copied the situation in the eutrophic waters of the Tarn (Fig. 10.7) remarkably well and demonstrated the importance of the distribution of loadings over time for the determination of species composition. Variation in the timing of P loadings caused variations in both seasonal biomass cycles and species composition. r-strategists tended to grow throughout the summer when stocks of nutrients were available for 'new' production.

In the same way it is possible to manipulate the species composition of lakes and reservoirs by altering both the mean Z_{eu}/Z_m ratio and the variance of that ratio. The relationship between the Z_{eu}/Z_m ratio and critical depth calculations was discussed in Chapter 8. The seasonal cycle in the Z_{eu}/Z_m ratio leads directly to the seasonal sequence of W, r to K species in the seasonal succession. There is also an interaction between the cycle of stratification and the cycle of nutrients as the sedimentation of nutrients in summer leads to reduced biomass and increases in Z_{eu}. Vertical mixing will not only increase Z_m but will also tend to decrease Z_{eu} if the growth of phytoplankton is stimulated. Growth of W diatoms (Fig. 9.8) may result if an increase in vertical mixing serves to increase the availability of P and Si and to decrease the pH in summer (Lund, 1954, 1955, 1971).

Reynolds *et al.* (1983) performed a series of artificial mixings in one of the Lund Tubes in Blelham Tarn in 1981. They induced vertical mixing by means of an air lift pump and over a period of a few days Z_m was increased. Between mixings the tube was allowed to restratify; a process assisted by the flux of heat sideways through the walls of the tube. Artificial mixing was induced four times during the summer and on each occasion Z_m was increased by about 1 m d^{-1} to a maximum of about 8 m. Reynolds *et al.* (1984) also performed a similar sequence of artificial mixing experiments in 1982 and the artificially induced changes in Z_{eu}/Z_m are shown in Fig. 10.8. These fluctuations in the Z_{eu}/Z_m ratio were induced at scales of between 10 and 20 days which produced six peaks in the ratio between June and the end of September. While these fluctuations in the Z_{eu}/Z_m ratio are reminiscent of the fluctuations in N^2 seen in previous discussions of natural situations (Chapter 5) they are both more severe and more frequent than those normally encountered. In other words Reynolds *et al.* (1983, 1984) intensified an otherwise normal sequence of events. It must be pointed out that these fluctuations in the Z_{eu}/Z_m ratio can only be achieved if both Z_{eu} and Z_m are of the same order of magnitude. Artificially mixing a very clear water column where $Z_{eu} \gg Z_m$ will have little effect, so the manipulation is most effective when optically deep water columns are perturbed. Water columns with high biomass are optically deep and manipulation is both desirable and feasible.

The effects of the artificial manipulation experiments on community structure were clear and could be explained by the sequence of events outlined above. Under stable conditions of high Z_{eu}/Z_m ratios, growth of r-strategist flagellates was possible whereas, when vertical mixing was induced, a reversion to W forms was noted. The K-strategists persisted throughout the summer but reached a lower, later peak of abundance. This was apparently due to the fact that the fluctuations in the Z_{eu}/Z_m ratio slowed their growth. Growth continued through the stable periods but ceased under mixed periods. As the loss rates of these species were low the populations slowly accumulated. Lund (1954, 1955, 1971) showed that persistent mixing of the water column of Blelham Tarn in mid summer produced a total reversion to W species. A bloom of *Melosira* resulted.

The frequency of perturbation is crucial. No mixing in summer leads to domination by K species and possible nuisance blooms of blue-greens or dinoflagellates (Reynolds and Walsby, 1975). Total turnover may lead to blooms of diatoms which may clog water intakes if sufficient Si is present. Strong perturbations in the Z_{eu}/Z_m ratio (from > 4 to < 1) at 10 to 20 day scales result in a diverse flora with W, r and K species coexisting at reduced biomass. The reduction in biomass results from the fact that no species has time to grow up to the 'ceiling' set by light or TP before the conditions change. Indirectly this demonstrates that phytoplankton are well adapted to cope with the normal frequency components in the planktonic environment; the 'envelope' is normally filled (Chapter 8).

The effects on two species of a changing Z_{eu}/Z_m ratio have been demonstrated by Reynolds (1983a,b). Reynolds (1983a) looked in detail at *Fragilaria crotonensis*, a W species, and showed that during mixing the species grew at net rates between 0.4 and 0.8 d^{-1}. Reynolds was able to derive a photosynthetic model for the growth of *Fragilaria* based on a modification of the critical depth model. *Fragilaria* grew at rates approaching μ_{max} during periods of mixing, an observation which supports earlier arguments. During stable periods growth ceased abruptly and sedimentation velocities increased due to photoinhibition or CO_2 limitation (Jaworski *et al.*, 1981). The cessation of growth and the increased sedimentation velocity produced abrupt population declines. *Volvox*, on the other hand, responded by growing during stable periods and declining during the mixed periods (Reynolds, 1983b). *Volvox* had a complex strategy which was a mixture of r (rapid growth under stable, shallow Z_m) and K-strategies (the ability to migrate vertically to achieve self regulation). This mixture is brought about by the grouping of small flagellated cells into colonies. The colonial habit frees the cells from grazing and makes rapid vertical migration possible.

Fig. 10.6 Blelham Tarn and the Lund Tubes, English Lake District. By courtesy of Trevor Furnass, Freshwater Biological Association.

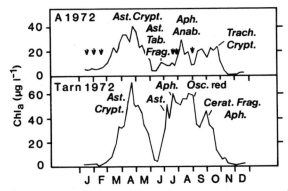

Fig. 10.7 The effects of continuous nutrient additions (↓) to the Lund Tube (A) of Blelham Tarn (from Lund and Reynolds 1980). Phytoplankton species as follows; *Ast, Asterionella; Crypt, Cryptomonas; Tab, Tabellaria; Frag, Fragilaria; Aph, Aphanizomenon; Anab, Anabaena; Osc, Oscillatoria; Cerat, Ceratium; Trach, Trachelomonas.*

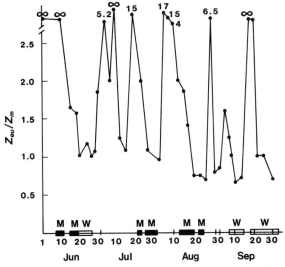

Fig. 10.8 The time series of fluctuations in the Z_{eu}/Z_m ratio in the Lund Tube of Blelham Tarn in 1982. M means artificially mixed and W means wind mixed.

Despite the colonial habit the *Volvox* in Reynolds' (1983b) experiments displayed the ability to grow rapidly ($k' = 0.4$ d^{-1}) when nutrients were available in shallow mixed layers.

The experience from the manipulation experiments is therefore entirely consistent with the notion that opportunistic growth during brief favourable periods is a predominant feature of the ecology of phytoplankton. Species switch from periods of exponential growth to periods of persistence and the interaction of environmental perturbations with growth rates brings about coexistence and diversity. The predominant role of allogenic processes in the seasonal succession of species may be exploited by those wishing to manipulate the species composition in surface waters. The fact that experimental manipulations have not only proved to be possible, but have been so when periodic fluctuations were imposed at the expected scales, has provided further evidence for the basic ecological mechanisms involved.

10.6 Biomanipulation

There have been a number of recent suggestions that it is possible to manipulate the biomass and species composition of phytoplankton communities by changing the food chain of lakes. This implies that there is considerable 'top down' control, that there is predator limitation of prey numbers and that there is some long-term equilibrium in numbers. This raises, once again, the whole question of the ability of predators to control the numbers of their prey. Lane and Levins (1977) and Briand and McCauley (1978) have used loop analysis to show that, at equilibrium, there are some complex interactions between food chain components and that enrichment of the soluble nutrient pool can have unexpected effects. For example, increase in soluble nutrients may theoretically have no effect on the phytoplankton biomass but may increase the biomass at higher levels. These analyses assumed equilibrium and equally tight links between all elements in the food chain.

All the arguments assembled in this book so far deny the reality of the equilibrium approach, except in special circumstances, and so render these approaches invalid as models of the real world. Equilibrium between grazers and phytoplankton simply does not exist frequently, if at all. One of the major pieces of evidence for the non-equilibrium nature of food chain dynamics lies in the size differences between predator and prey and the evident disparity in growth rates. Phytoplankton in spring can grow at rates in excess of $1.0 \, d^{-1}$ and so can easily outstrip the zooplankton. Grazing pressure by zooplankton is only capable of clearing the water for brief periods at a time. Seasonal perturbations in the resource base of the phytoplankton lead to seasonal cycles characterized by outbursts followed by crashes (Chapter 9). The degree of uncoupling between production and consumption depends on the latitude and the seasonality in primary production, on growth rates and on temperature. Only at annual scales are there any good empirical relationships between primary production and production at higher trophic levels (Chapter 7).

A further problem lies in the different rates of growth exhibited by the different algal size classes and the differential effects of size selective grazing. The seasonal cycle of phytoplankton shifts from easily grazed *r* species to large, inedible *K* species. There is evidence that the larger *K* species actually interfere with zooplankton grazing on the smaller species (Gliwicz, 1980). The 'top down' control of food chains assumes that each trophic level is strongly resource limited. Given the marked fluctuations in the abundance of prey numbers over the season, one is led to wonder how the predators survive when prey is scarce unless alternative sources of food are available. There is now good evidence of food limitation at certain times of the year in both marine and freshwater zooplankton populations (Lampert and Schober, 1980; Durbin *et al.*, 1983).

The origins of much of the biomanipulation research lie in the paper by Brooks and Dodson (1965) in which changes in the size classes of zooplankton were noted when planktivorous fish were introduced to New England lakes. There was much explicit equilibrium ecology in the original statement and the paper has become something of a classic in the field. Brooks and Dodson noted that in the presence of the Alewife (*Alsoa pseudharengus* (Wilson)) the larger species of zooplankton were replaced by smaller species. They suggested a size-efficiency hypothesis which stated that the removal of the larger zooplankton species by predation would favour smaller species because of the removal of competition. It was assumed that large zooplankton would always be able to out-compete the smaller species because of their greater food gathering and growth efficiencies. Brooks and Dodson (1965) suggested that if predation pressure on the zooplankton was high then the biomass of phytoplankton would also be high because of the inefficient utilization of the phytoplankton biomass. They quoted the work of Hrbacek *et al.* (1961) in support of this suggestion.

The size efficiency hypothesis itself has not stood up to detailed scrutiny (Hall *et al.*, 1976; Gliwicz, 1980). There is considerable overlap in the size classes of phytoplankton taken by the various zooplankton species, and there is still some controversy over the mechanisms of grazing in different animals. Body size is not a good indicator of niche dimensions (Frost, 1980b) and it has been difficult to demonstrate the role of competition in natural populations (Kerfoot and DeMott, 1980). Some recent examinations of zooplankton community structure indicate that non-equilibrium models may be more satisfactory than equilibrium models (Poulet, 1978). Diversity in resource availability and utilization, growth rates, the ability to track environmental fluctuations (Porter *et al.*, 1983) leads to coexistence.

A number of workers have published accounts of experiments in which fish populations have been naturally or artificially manipulated and which have resulted in changes in the abundance of different zooplankton size classes. This effect appears to be well documented (cf. review by Hall *et al.*,

1976) as the planktivorous fish feed on the size classes of zooplankton that can most easily be seen. Lynch (1983) however, measured the size specific mortality of zooplankton in such an experiment and showed that the mortality did not fall in the expected size classes. Hall *et al.* (1976) did note that the design and execution of these fish manipulation experiments required great care. 'Results of such experiments can be difficult to evaluate because of possible confounding effects of toxins used to sacrifice the fish, insufficient duration of the experiments, unnatural species composition or densities of introduced fish, and inadequate baseline data' (Hall *et al.*, 1976).

Most of the reports of successful manipulation of phytoplankton and zooplankton populations come from very shallow lakes or bag experiments (Lynch, 1979; Lynch and Shapiro, 1981; Andersson *et al.*, 1978; Leah *et al.*, 1980) and/or with extremely high fish population densities (Hrbacek *et al.*, 1961). In shallow water and in bags not only will the fish be able to search out most of the larger zooplankton and eat them, but the effects of nutrient regeneration by the fish themselves will be maximized. Thus the growth of phytoplankton in the presence of planktivorous fish may be as much due to the altered nutrient flux as to the removal of predators by zooplankton. Fish have a remarkable ability to change growth rates in response to crowding and may, in fact, lose weight if overstocked (Weatherly, 1972). As a general rule Weatherly noted that fish appear to have forsaken density dependent mortality in favour of density dependent growth. In deeper water the zooplankton will be able to migrate vertically and hence escape visual predators by descending below the photic zone during the day. Koslow (1983) noted that food chain models were inadequate in that they did not describe the true interaction between fish predators and prey and that vertical migration served to minimize the effects of fish predation in natural systems.

Invertebrate predators have the opposite effect to vertebrate predators (Hall *et al.*, 1976) in that they remove the smallest zooplankton species. Smyly (1976, 1978) noted that the exclusion of *Chaoborus* larvae from the Blelham Tubes resulted in changes in the zooplankton assemblages that were consistent with predation in certain size classes. Smyly (1976) also noted that the changes in the enclosed zooplankton populations were poorly correlated with changes in the phytoplankton. Hall *et al.* (1976) noted the difficulties in the interpretation of experiments involving the manipulation of invertebrate predators and the problems were similar to those listed above for fish manipulation experiments. Hall *et al.* (1976) observed that the effectiveness of invertebrate predators depended very much on the food limitation of prey populations. This is also true of fish predation. Grazing is not a static process and the size distribution of the zooplankton species present in the water column depends as much on the growth and longevity of cohorts of animals as it does on the size selective predation (Lynch, 1983).

The effects of fish predation on zooplankton in enclosures is by no means clear. The responses of phytoplankton and zooplankton to fish predation in shallow waters and in bags must be seen in the light of the seasonal succession of phytoplankton species and the seasonal development of cohorts of animals. A reduction in longevity either by a reduction in food supply or by an increase in grazing will lead to smaller animals. As the summer progresses, the phytoplankton species shift towards less edible forms as the spring burst of 'new' production passes and K species come to dominate the biomass. This is particularly true in eutrophic waters which tend to be dominated by smaller zooplankton which feed mainly on bacteria and detritus (Gliwicz, 1969). There is, therefore, an interaction between phytoplankton and zooplankton size classes which depends on trophic state. The model of Hall *et al.* (1976) displayed the interactions between body size, growth, reproduction and predation. There is a food concentration below which reproduction of zooplankton ceases (Lampert and Schober, 1980). Short periods of starvation (the clear water phase) may cause little or no mortality among adult zooplankton but may exclude some species because of severe juvenile mortality.

In shallow waters and in bags the zooplankton are not only vulnerable to predation because of the restricted volume of water. Their vulnerability because of their inability to migrate vertically may explain the observed relationships between phytoplankton, zooplankton and fish in shallow waters. In shallow lakes the zooplankton (particularly species of *Daphnia*) may utilize detrital sources of food to tide themselves over periods when phytoplankton are scarce. This will allow them to maintain a sufficient biomass to control successive phytoplankton blooms. As a growth response to algal outbursts is not required there will be no time-lag between the growth of the algae and the onset of effective grazing. So in the absence of fish, large populations of zooplankton may persist all summer and be able to effectively control the algal biomass. In the presence of planktivorous fish the zooplankton cannot escape predation by vertical migration and may be severely depleted in numbers. Predation pressure on the zooplankton may be increased if the zooplankton are concentrated close to the bottom, as the fish may concentrate their activity in what is essentially a two dimensional environment.

In deeper waters ($\geqslant 2$ m) the planktivorous fish have little effect on the zooplankton populations. The mortality of the zooplankton is far higher than the grazing pressure (Gliwicz and Prejs, 1977). Furthermore the temporal patterns of zooplankton mortality and grazing pressure during the summer do not match. In two lakes with different fish stocks, Gliwicz and Prejs (1977) found that the sequence of peak numbers of Cladocerans was the same in both lakes, that the smaller species actually decreased in the lake with higher predation and that the largest numerical abundance of the largest Cladoceran occurred in the lake with the more intense zooplankton predation. In an analysis of data from 30 lakes Gliwicz and Prejs (1977) showed that by far the most important determinants of zooplankton community structure were morphometric indices (influence of littoral zone and bottom area in reach of epilimnion, both a function of mean depth) and trophic state indices (oxygen depletion). Sprules (1977) also showed that mean depth was the most important correlate of zooplankton community structure in a study of a number of lakes in Ontario. Gliwicz and Prejs (1977) concluded that it was necessary to reach a planktivorous fish biomass of in excess of 900 kg ha^{-1} before a strong effect on zooplankton community structure was noted. This in fact was in agreement with Hrbacek *et al.* (1961). More recently Koslow (1983) tried to model the food chain in the North Sea and concluded that a similar biomass of planktivorous fish was required to effectively graze the zooplankton in that system. Direct observation and models point to the same conclusion. Zooplankton biomass and size structure can only be influenced by unusually high levels of fish biomass; levels that are rarely seen in nature.

What are the effects of zooplankton grazing on phytoplankton? The equilibrium model predicts that extensively grazed communities should be more diverse than less highly cropped systems. In short, coexistence through predation. McCauley and Briand (1979) attempted to test this hypothesis by using 800 litre enclosures. Surprisingly the removal of zooplankton resulted in no change in the diversity of edible species and a decrease in the number of larger, inedible species. Unfortunately McCauley and Briand (1979) gave no information on the biomass or the abundances of the populations so it is not possible to say what the precise sequence of events was. Remembering the links between abundance, counting effort and measures of diversity, it is possible that the change in the apparent number of species was really a change in abundance. Lynch and Shapiro (1981) also tested the hypothesis that increased predation increased diversity. In experiments where planktivorous fish were removed they also noted the number of phytoplankton in a standard volume of water. When zooplankton were removed the number of phytoplankton species actually increased. Again these results may have been biased by the total number of cells counted and the abundance of the phytoplankton, but the results do not support the initial hypothesis. In many cases the total number of species in enclosures declined with time and enriching the enclosures with nutrients strengthened this trend.

A decline in the number of species in enclosures with time is to be expected as the physical regime in the enclosures will be much less dynamic than that in the lake outside. The addition of nutrients will also serve to impoverish the spectrum of disturbances within the enclosure and, over time, the diversity of phytoplankton will decrease. The concept of time in all these enclosure experiments is crucial. Experiments with the removal of fish may lead to an initial decrease in the phytoplankton biomass, but over seasonal time scales the growth of large populations of blue–greens would lead to the opposite effect. In other words there does not yet appear to have been a proper test of the assertion that phytoplanton populations can be manipulated by 'top down' control. No one has yet fully taken the interaction of the spectrum of turbulence, the spectrum of cell sizes and time into account.

Perhaps the most complete and satisfying demonstration of the lack of 'top down' control in lakes comes from the experiments of Lund and Reynolds (1982) in the Lund Tubes in Blelham Tarn. These Tubes have been operated for years with almost no fish in them. Despite this there have been repeatable patterns of biomass, seasonal succession and phytoplankton assemblages year after year. I have already discussed many of the results previously. None of these results (including the manipulation of species composition by Reynolds *et al.*, 1983) would have been achieved if there had been large populations of herbivorous zooplankton consistently consuming large amounts of phytoplankton biomass. An analysis of the loss rates in the phytoplankton assemblage by Reynolds *et al.* (1982) showed that grazing was only an important process for some species and size classes at certain times of the year. Zooplankton pulses and crashes followed the growth of edible species of phytoplankton (Ferguson *et al.*, 1982). In 1979 and 1980 large numbers of fish were added to Tube B in an attempt to influence the zooplankton community and, indirectly, the phytoplankton community in the Tube. The attempt failed. Another series of arguments which point to 'bottom up' control may be found in the long term studies. Cushing's 'match-mismatch' hypothesis which was used to explain the long-term changes in the English Lakes, the Great Lakes and the North Sea, depends on food chain control by the seasonal cycle of primary production. The obvious effect of simple climatological variables on zooplankton and fish points to the same conclusion.

The large scale empirical data sets from many lakes also bolster the case for 'bottom up' control because of the correlations between trophic state and the various food chain components. The correlations between zooplankton and fish yields and such high level variables as *TP* and the MEI indicate that it is the flow of energy from the primary producers which determines the size of the consumer pools. Overshoots and crashes do occur but at annual basin scales there is a regular flow of energy from bottom to top determined by ecological efficiencies, body size and the turnover of particulate pools. Throughout the food chain heuristic processes lead to much the same result as that envisaged by equilibrium theory, but without the restrictions on realism. As Margalef (1980) envisaged, ecology is the result of thermodynamics and a game.

CHAPTER 11

Large number systems: empiricism

Large number systems are those in which there are a sufficient number of interacting processes or species for broad statistical properties to appear. These may be likened to physical gas laws. As was seen in Chapter 7, the C:N:P ratios of phytoplankton communities and detritus revealed regular statistical properties and patterns of behaviour that were explicable in terms of small scale nutrient dynamics and the growth rates of the constituent species. In precisely the same way, statistical distributions of biomass distributions may be sought that are explicable in terms of the underlying population dynamics. Having introduced the concept of opportunistic growth within an envelope set by high level resources, we are about to define the effective size of the envelope in different waters and to look at other high level descriptors of community structure.

Empirical (statistical) studies of large scale ecological processes are widely accepted in limnology. There are evidently two reasons for this. First, lake basins are well defined units in space and time with clear structural and functional boundaries so that averaged properties may be studied with ease. Secondly, planktonic organisms in lakes are so small and the generation times so short that annual basin averages reveal the properties of very large number systems. Similar studies in forestry would require averaging at continental scales over hundreds of years or millennia. There is no reason to believe that similar relationships do not exist in the oceans. The only real difference between the lakes and the oceans lies in the ease with which boundaries may be defined and nutrient inputs determined.

It is wrong to assume that empirical correlations between 'state variables' such as algal biomass and TP are bad ecology. Rigler (1982a,b) and Kerr (1982) pointed out that a totally reductionist approach to the dynamics of multispecies communities will never reveal the emergent properties of community level processes. (Just as it is impossible to study the quantity of noise or speed contained in the piston rings of a motor cycle.) Rigler (1982a) showed that because of the complexity of the interactions, the unpredictability of many events and the evolutionary changes within populations, it is impossible to accurately forecast the abundance of individual species. While it is true that empirical predictions of algal biomass say little about the dynamics of individual species it is possible to use an improved knowledge of population dynamics to explain the observed statistical relationships (Rigler, 1982b).

A combination of research at a number of levels may yield empirical relationships between high level state variables and at the same time may provide explanations for such behaviour. Rigler (1982a) may have been unnecessarily pessimistic in his assessment of the ability to predict the future size of individual populations. A non-equilibrium, opportunistic model of population growth within an envelope or boundary such as been introduced above will be easier to work with in that the relationships and competitive interactions between species are weak and the interactions between species and the environment are dominant. In such a framework the relevant questions are 'is it there?', 'is it growing?', and 'how long a period has it in which to grow?'. As we have already seen there is evidence that there are times in the seasonal succession when species are absent from apparently suitable environments, when species are present but not growing and when the environment is correct and the species are present but the time for growth is short. These points will be raised again below in a different context but with similar results.

11.1 Observations of ecological diversity

Much ecological research in the field begins with the acquisition of a list of species present in a specific area together with some indication of relative abundance. The two measures combined give some indication of ecological diversity. This is a high level descriptor as one number contains all the information about both the number of species in the community and their relative abundances. Preston (1948) plotted the relative abundance of species as a histogram in which the relative abundance was expressed as functions of $\log_2$-class intervals which he called octaves. For many different groups of organisms he consistently found that the resulting plot was symmetrical and Gaussian, in fact a lognormal plot. Furthermore he discovered that the statistics of the curve were consistent from group to group and were deemed to be a basic function of the organization of community structure. May (1975) reviewed the patterns of species abundance and diversity and showed that Preston's lognormal was one of a number of types of distribution in which the relative abundances of the species were either more or less even than those of the lognormal. So is it possible to say anything about community structure if the total number of species and their relative abundances are known?

MacArthur (1960) used the equations for exponential growth as a simple means to discuss the relative abundance of the ith species at time t, with $r_i(t)$ defined as the intrinsic growth rate of the ith species at time t, then:

$$r_i(t) = \frac{1}{N(t)} \cdot \frac{dN_i(t)}{dt} \qquad\qquad 11.1$$

and by integration:

$$\log N_i(t) = \log N_i(o) + \int_o^t [r_i(t)\, dt] \qquad\qquad 11.2$$

Note that in this model r is not assumed to be constant over time. The school of ecology concerned with density dependent processes and the competitive equilibrium relegates the integral portion of Equation 11.2 to relative unimportance as sustained exponential growth was not regarded as usual or possible. (Indeed MacArthur (1960) suggested two models of relative abundance based on this equation. The model based on the integral portion of Equation 11.2 was dismissed in one sentence. Remember that r strategists are regarded as unimportant in the received version of the theory.) Thus, following Darwin, the log $N_i(o)$ term is regarded as the most important term as equilibrium is assumed and species are therefore close to their carrying capacities. MacArthur (1960) developed a model for the distribution of relative abundances of such equilibrium species which he called the 'broken stick' model. This assumes that the basic pattern of distribution is due to the even partitioning of some major resource and results in a distribution of relative abundances that is more even than the lognormal. MacArthur (1960) and others since have presented data from animal communities to support the existence of this type of distribution. It should be stressed that the mere existence of data which fits the model cannot be construed as direct evidence of the existence of such a mechanism.

May (1974) reviewed the statistics of the lognormal and showed that the statistics of the curve were a very weak function of species number and that by the Central Limit Theorem the curve should be asymptotically lognormal if the sample was taken from a large enough area and from a large enough number of species. Thus May concluded that the fact that ecological data frequently fitted a lognormal distribution was not, in itself, a significant result. May (1975) wrote 'the lognormal distribution reflects the statistical

Central Limit Theorem; conversely, in those special circumstances where broken stick, geometric series, or logseries distributions are observed, they reflect features of community biology'. Preston's (1962) work revealed that the lognormal distributions found in ecological data were remarkably consistent, with a particular shape and spread. The curves were found to be canonical lognormals which assume a particular relationship between the distribution of relative abundances of the species and the distribution of the total number of individuals (May, 1975). Hutchinson (1953) observed that, for ecological data, a measure of the width of the lognormal distributions was remarkably constant and that this might mean something significant. May (1974, 1975) reviewed the mathematics of the lognormal distribution and showed that the spread of the curve was a very weak function of species number when the distribution was canonical; hence the apparent constancy in spread. This conclusion was recently denied by Sugihara (1980) who showed that while the ecological data does conform to a particular type of lognormal, the classes of lognormal exhibited are only a subset of a much larger group of possible distributions. Thus there is some evidence that the results of sampling large numbers of species do indeed reflect an underlying organization.

Whatever the outcome of this debate, it is clear that if ecological data conforms to an approximation of a canonical lognormal distribution then the species area relationships of MacArthur and Wilson (1967) follow directly from this. May showed that because of the statistical properties of the lognormal distribution the species–area relationships of MacArthur and Wilson are a robust function of any reasonable lognormal distribution of the relative abundance of species. May (1975) came to two major conclusions in his review of the three types of distribution. Firstly, small samples taken from particular situations may show either broken stick or other distributions but the distribution becomes increasing lognormal as the sample size and the total number of species is increased. Secondly, many common measures of species diversity do not distinguish between the different types of distributions if the total number of species is small.

It should be clear by now that the indices of diversity are sensitive to the relative distributions of relative abundances, as they all contain expressions for the proportional abundance of the ith species in the community. The measurement of diversity is by no means as easy as it looks (Peet, 1974) as it assumes an adequate census method; but May (1975) has shown how the two most common measures of diversity are related to both the total number of species and to their proportional abundances. The two most common measures of diversity are H', the Shannon–Weaver diversity index and D, Simpson's index (Simpson, 1949). These two indices may be calculated as follows:

$$H' = -\sum_{i=1}^{S_t} p_i \ln p_i \qquad 11.3$$

where:

$$p_i = N_i/N_t \qquad 11.4$$

the proportional abundance of the ith species and S_t is the total number of species.

$$C = \sum_{i=1}^{S_t} p_i^2 \qquad 11.5$$

where Simpson's index

$$D = 1/C \qquad 11.6$$

Hurlbert (1971) has also suggested the use of 1–C. These various indices have also been compared by Peet (1975) and De Jong (1975). May (1975) calculated the effect of changing the distribution of relative abundances from geometric or logseries, to lognormal, to broken stick on the relationship between S_t and both H' and D. As noted above, if S_t is small, it is difficult to see much of an effect, but as S_t rises above ten species, the relationships are discernibly different. Thus, if the counting techniques are adequate, then a plot of log S_t against H' or D should give some information about the form of the distribution and should give some indication of whether or not there was significant resource partitioning (Fig. 11.1). As

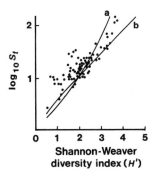
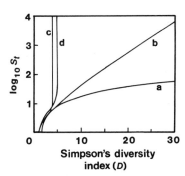

Fig. 11.1 Plots of the relationship between diversity and the total number of species (May, 1975). Left: the theoretical curves for (a) lognormal distribution and (b) the broken stick distribution compared to a variety of natural communities. Right: the value of Simpson's D compared to the total number of species assuming (a) broken stick (b) lognormal (c) geometric series and (d) logseries distributions.

observed by May (1975) most of the existing data indicates a reasonable fit to the lognormal, which in itself does not say much, but there are also data sets that indicate the existence of communities in which the distributions of relative abundance are significantly more even than the lognormal (broken stick) or significantly less even (logseries).

Non-equilibrium communities would be expected to show lognormal distributions as the exponential term of Equation 11.2 would be expected to be dominant and if $r_i(t)$ is normally distributed then $N_i(t)$ will be lognormally distributed. Thus there may be some confusion between equilibrium and non-equilibrium communities if the test of S_t and the diversity index is used. This will be particularly true in a patchy spatial distribution (the norm?) as each patch may be a closed system but the overall system may be open by virtue of migration between patches. Thus sampling data from one patch may well show a logseries distribution but as the sample size is increased the data will more and more come to be lognormal as more and more different patches are included. Thus some previous knowledge of the scales of spatial patchiness and the dispersal abilities of the organisms is needed in order to correctly design a sampling programme.

There have been a number of previous reviews of phytoplankton diversity, notably those of Hutchinson (1967) and of Margalef (1958, 1963a,b, 1967, 1968). In general, using an index like H', phytoplankton diversity varies from 1.0–4.5 bits with a strong seasonal component as already discussed (Fig. 11.2). Many values of phytoplankton diversity calculated as H' fall in the range of 2.4 to 2.6 bits. Margalef (1980) has provided a more

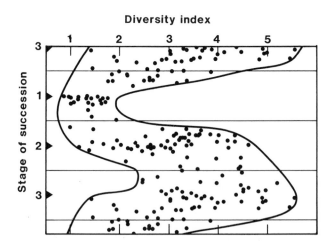

Fig. 11.2 The changes in diversity $(d = S_t - 1/\log_e N)$ during the seasonal succession in the Bay of Vigo (Margalef, 1958). The successional stages began in May and finished in October.

recent comparison of aquatic and terrestrial data. The diversity of phytoplankton is higher in oligotrophic than in eutrophic waters and there is some evidence that marine phytoplankton are more diverse than freshwater assemblages. This may be explained by the widespread occurrence of oligotrophic oceanic waters.

So is it possible to tell anything about the nature of biological interactions from an examination of diversity data? In an earlier chapter, I showed that there was little evidence for competition in the Hamilton Harbour phytoplankton community. Further evidence for a lack of competition was obtained from an examination of the diversity data. Harris *et al.* (1983) showed that a plot of $\log_{10}S_t$ versus H' for the Harbour data revealed a distribution of species abundances that was very even: more even in fact than the broken stick model of MacArthur (1960). So if the distribution of relative abundances was so even why was this not evidence for competition? The answer lay in the comparison of the actual data with a 'null' model constructed on a computer with a random number generator. This model was similar to that used by Moss (1973d) and was constructed as follows. Up to 100 species were randomly assigned a realistic growth rate in the range 0.1 d^{-1} to 1.0 d^{-1} and were allowed to grow for a random number of days. Growth began on a randomly chosen day and continued either for the chosen growth period or until the end of 'summer' (100 days) whichever came first. After the growth period the cells were removed from the water column at a realistic loss rate. At the end of the growth period the species abundances were ranked and the diversity calculated. A number of runs of the model were carried out at each of a number of chosen loss rates.

The results of the model compared well to the real Harbour data (Fig. 11.3) and the loss rates proved to be the most important driving variable.

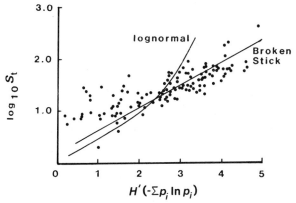

Fig. 11.3 The relationship between H' and $\log_{10}S_t$ for communities in Hamilton Harbour (Harris *et al*, 1983). The data are compared to the lognormal and broken stick distributions.

Decreased loss rates increased both the total number of species and the diversity in a fashion similar to the field data. The diversity of species in surface waters can therefore be seen to arise from the growth and persistence of species in surface waters as suggested previously. The tendency for natural phytoplankton assemblages to have similar diversities says as much about the similarity of loss rates in different habitats as it does about other biological interactions. We have already seen that there is a remarkable similarity between the settling velocities of phytoplankton and the vertical mixing rates in surface waters (Chapter 5). This is a restatement of that relationship. As Wofsy (1983) noted, loss rates by sedimentation in summer are usually minimal in relation to growth rates (cf. Chapter 6). The increase in diversity during the summer stratified period is a result of reduced loss rates in surface waters and the consequent persistence of species. Persistence leads to apparent coexistence, but as many of the persistent cells are apparently inactive, cooccurrence would be a better term. The presence of inactive cells in the water column is a frequent observation when autoradiographic techniques are used (Harris, 1978).

Harris and Smith (1977) noted a similar lack of competition in natural populations of phytoplankton. They showed that species fluctuated in abundance more or less independently of the other species in the assemblage and that the changing value of the diversity index (H') was more a feature of bias in the calculation of the index than a real change in diversity. The apparently paradoxical diversity of phytoplankton can be accounted for by niche diversification (based on cell size and efficiencies for example), temporal fluctuations in the habitat and persistence of temporally inactive cells. When the known spatial heterogeneity of the planktonic habitat is added to the habitat variance (Richerson et al., 1970; Harris and Smith, 1977) the diversity of phytoplankton is not unexpected (Harris, 1980a).

Hutchinson (1967) suggested that at least a part of the paradoxical diversity of phytoplankton could be explained by the presence of meroplankters: organisms that were not truly planktonic but spent only a part of their life history in surface waters. There is some evidence that meroplankters are important in small, shallow lakes as the relationship between the surface area of the lake and the total number of phytoplankton species found is usually weakly negative (Ruttner, 1952; Jarnefelt, 1956; Hutchinson, 1967). Harris et al. (1983) produced a similar result from an analysis of data from 125 North American lakes and reservoirs (Fig. 11.4). The increasing number of species encountered in small, shallow lakes indicates that some of the planktonic diversity is due to the presence of species resuspended from the bottom but the weak inverse relationship indicates that the effect is small. In deep waters the presence of meroplankters is unlikely.

The lack of a standard species/area relationship is further evidence of a lack of equilibrium in surface waters (but see Colinvaux and Steinitz, 1980).

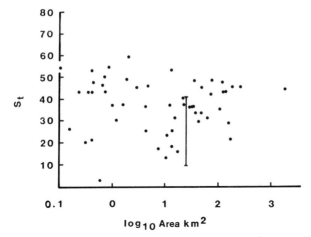

Fig. 11.4 The species/area relationship for a number of North American lakes taken from US EPA data. The vertical bar shows the range of species numbers found in Hamilton Harbour. Note the lack of an increasing trend with increased lake area.

The lack of a species/area relationship is also an indication that it is small-scale turbulence that controls the diversity of the phytoplankton as increasing lake area has little effect on community diversity.

Diversity therefore arises from a balance between growth processes and loss processes for all the species in the assemblage. In an equilibrium community predation may actually increase diversity if predation falls mainly on the numerically dominant organism. In a non-equilibrium community, however, the greatest diversity will arise when many species grow rapidly one after the other and persist between growth periods. Persistence implies maximizing the growth terms and minimizing the loss terms for each species. In such a system increasing any of the loss terms will reduce diversity by making the relative abundances of the species more uneven and ensuring that the diversity calculation is numerically dominated by the species that grew most recently. During periods of rapid growth by one species diversity will decline (Fig. 11.5). In an opportunistic universe diversity arises from the individual growth strategies of the species playing the 'game' of growth whenever possible (Margalef, 1980). This is not an optimal universe but a heuristic one. The ensemble does however show repeatable statistical properties arising from the growth of many species. There is a chance component to the species composition and not all resources may be utilized. Equally the diversity may be reduced in circumstances of excess environmental variance when the growth periods of many species are too short to achieve significant increases in abundance. It is not possible to predict the precise species composition or the precise relative abundances (Margalef, 1980). The only good predictors are higher level parameters such as biomass, C:N:P ratios and TP.

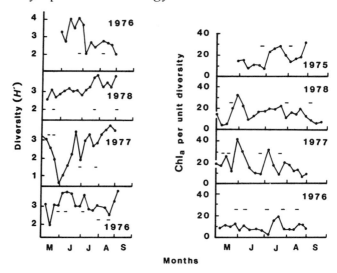

Fig. 11.5 The changes in diversity and biomass per unit diversity in Hamilton Harbour in 1975 to 1979. Bars indicate periods of stabilization. (From Harris *et al*, 1980.)

The 'games' which are the underlying components of the overall diversity are constrained by thermodynamic considerations such as efficiencies (Margalef, 1980), cell size and food chain relationships. Also the individual species that make up the assemblage at each trophic level differ in their abilities to integrate, filter and lag the external environmental fluctuations. While there may be a large number of species in the overall phytoplankton assemblage when they are classified on the basis of efficiencies and cell size the number of species in each group is quite small. Taxonomy is not a good indicator of ecological diversity as the functional role of species is not explicitly stated.

11.2 Predictions of the distribution and occurrence of species

If biological interactions between individual species are weak then it may be possible to predict the occurrence of individual species in lakes and reservoirs if the correct ecological driving variables are chosen. This may be particularly true in eutrophic waters where the external nutrient loading is a major forcing function and many of the species are large enough to escape grazing. As I have shown many of these species are K-strategists with low growth and loss rates in late summer and many are dependent on the stability of the water column for buoyancy regulation. Furthermore, many of the species which dominate eutrophic waters in summer form nuisance blooms and some predictive capacity would be useful for management.

There are two ways to attempt to predict the occurrence of nuisance blooms of phytoplankton in lakes. The analytic approach attempts to understand the interactions between the environment and the species present, and the interactions between the species. This knowledge could then be used to make predictions. I have already shown that this is not possible in a non-equilibrium universe because of chance effects and the vagaries of the weather. The other approach is to take a large number of lakes and reservoirs and to examine them to see if statistical properties emerge. This is essentially the search for large number properties of many bodies of water. Vollenweider was the first to pioneer this approach (1968, 1969, 1975).

Data from a large number of United States lakes and reservoirs has been published by the United States Environmental Protection Agency (US EPA National Eutrophication Survey, 1975, 1978a,b,c). These reports give morphometric, chemical and biological information as well as phytoplankton species composition for over 800 lakes and reservoirs. In this analysis, data from July, August and early September will be used to demonstrate that it is indeed possible to use this data to predict the occurrence of summer nuisance blooms of blue-green algae. The two driving ecological variables that were identified were, as expected, measures of the prevailing physical and chemical conditions. Physical conditions in summer were estimated from the ratio of the estimated thermocline depth (from Ragotzkie's (1978) relationship, Chapter 3) to the mean depth of the lake:

$$M = D_{tb}/\bar{z} \qquad\qquad 11.5$$

$M < 1$ indicated stable conditions in summer as the mean depth of the lake exceeded the estimated depth of the late summer thermocline. $M \gg 1$ indicated strong vertical mixing as the estimated thermocline depth far exceeded the mean depth of the lake. The chemical conditions in the lakes were estimated from the ratio of $TN:TP$. This ratio is a useful guide to the relative turnover times of N and P in the system (Chapter 7). The differing $TN:TP$ ratios in lakes are mostly accounted for by changes in TP although at high TP there is evidence for changes in the metabolism of N in lakes (Chapter 7).

Inspection of the US EPA Reports revealed that 435 lakes satisfied the criteria for analysis by having complete physical, chemical and biological data at the correct time of year. The species composition of the phytoplankton in the lakes was taken from the EPA reports. Taxonomic problems limited the usefulness of some records as not all organisms were identified to the species level. Nevertheless the dominant organisms were identifiable in most cases. Organisms identified as *Aphanizomenon* were characteristically *A. flos-aquae*; the *Melosira* records were dominated by *M. granulata*; *Microcystis* by *M. aeruginosa*; *Lyngbya* by *L. limnetica*; *Fragilaria* by *F.*

crotonensis and *Stephanodiscus* by *S. astraea*. The other genera (*Anabaena, Oscillatoria, Raphidiopsis* and *Synedra*) presented greater taxonomic difficulties and were not consistently attributed to a single species. A measure of the relative size of the blooms of the various species was developed. The abundance of the various species was divided by the TP in the water in each lake to obtain a measure of the fraction of the TP tied up in a given species in each case. Thus lakes of different types developed *Microcystis* blooms but the lakes in which there was a higher *Microcystis*/TP ratio formed a more restricted subset of the total indicating those cases in which more of the TP was tied up in relatively fewer species, indicating a more concentrated bloom.

When the data plots were examined the importance of the physical and chemical driving variables became clear (Figs 11.6 and 11.7) There was a

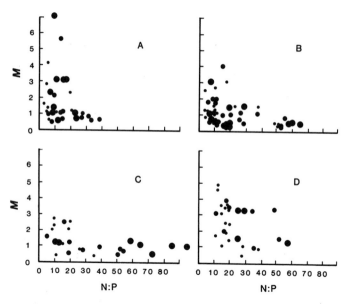

Fig. 11.6 The occurrence of blooms of four common species of algae in the US EPA lakes as functions of *M*, the stratification parameter and the ratio of total N:P (see text for details). Species as follows: A, *Anabaena*; B, *Aphanizomenon*; C, *Microcystis*; D, *Oscillatoria*.

clear dividing line at a TN:TP of 25–30. At ratios above this value nitrogen fixing blue-green algae rarely formed blooms, whereas below this value nuisance blooms became common (Smith, 1983). Thus *Anabaena* and *Aphanizomenon* occurred most frequently at TN:TP < 30. TN:TP < 30 clearly favoured the growth of N-fixing organisms and indicated that the

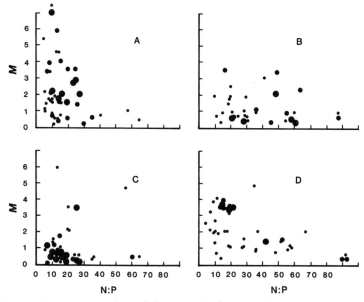

Fig. 11.7 As Fig. 11.6, species as follows: A, *Melosira*; B, *Synedra*; C *Fragilaria*; D, *Stephanodiscus*.

flux of N in the water was possibly limiting. The correspondence between the critical ratio of $TN:TP < 30$, the frequent presence of blue-greens and the dynamics of the whole basin fluxes of N and P was discussed in Chapter 7. Buoyancy regulating species such as *Aphanizomenon* and *Microcystis* occurred most frequently at $M < 2$; the latter species was rarely found at $M > 1.5$ indicating a strong preference for stable water columns. The species of *Oscillatoria* and *Lyngbya* were most frequently found in vertically mixed water bodies $(M > 1)$ at $TN:TP$ ratios in excess of 15–20. The major species of diatoms had equally diversified and characteristic distributions. Species of *Synedra* rarely occurred at $TN:TP < 20$ whereas *Stephanodiscus*, *Melosira* and *Fragilaria* rarely occurred at $TN:TP > 30$. *Fragilaria* and *Synedra* occurred frequently in stable water columns whereas the heavier species of *Melosira* and *Stephanodiscus* required more vertical turbulence to keep them in the water column.

There is clear evidence of niche diversification between the major groups of algae from this analysis but what confidence may be placed on these conclusions? Can such data be used to predict when certain species might occur? To determine how often each species occurred in apparently suitable conditions the whole set of data from 435 lakes was scored for presence or absence of the major species. The percentage occurrences are given in Table 11.1. There was a 67% probability that *Aphanizomenon* would occur in

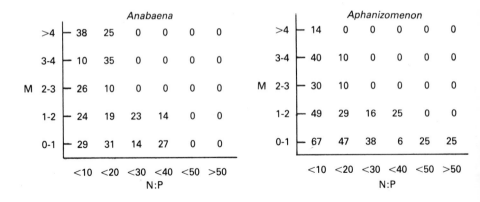

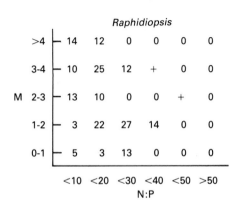

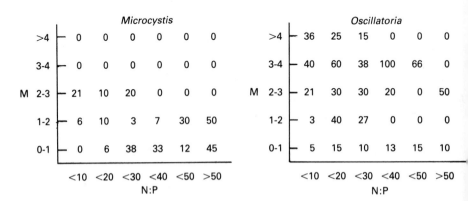

Table 11.1 Probabilities of occurrence of the given species under the stated conditions M, the stability parameter and N:P the ratio of total N:P (US EPA data).

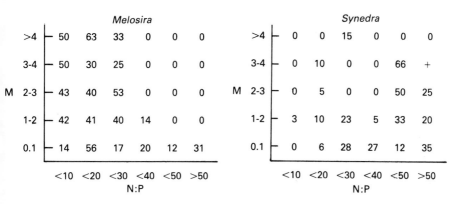

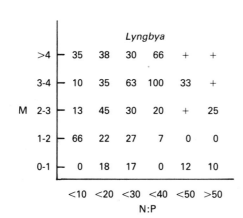

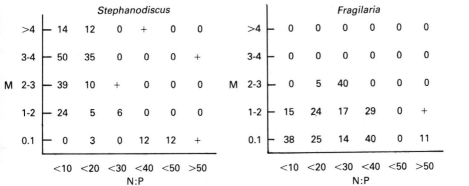

lakes with $M < 1$ and $TN:TP < 10$, a 40–50% probability that *Microcystis* would occur with $M < 2$ and $TN:TP > 50$ and a 60–70% probability that *Lyngbya* or *Oscillatoria* would occur when $M > 3$ and $TN:TP > 20$–30. For diatoms, there was at least a 50% probability that species of *Melosira* would occur at $TN:TP < 30$. Similarly there was a 50% probability that *Stephanodiscus* would occur at $M > 3$ and $TN:TP < 10$. Surprisingly, *Fragilaria* occurred most frequently in stable water columns ($M < 2$).

There was an indication of interaction between the two variables in that species such as *Microcystis* showed a pattern of distribution that curved upwards at low $TN:TP$ ratios (Fig. 11.6). This indicated that some vertical mixing may havé compensated for low $TN:TP$ ratios, presumably by influencing the turnover of N and P in the lake. Vertical mixing makes nutrients available from the sediment surface in a way which would not be available in a stable, stratified lake with a well-developed hypolimnion.

The element of chance in the occurrence of species in lakes and reservoirs was due to two sources of uncertainty: (1) the uncertainty associated with the processes of immigration and establishment, as each lake is an ecological 'island' (Talling, 1951; Maguire, 1963, 1971) and (2) the uncertainty surrounding the environmental conditions in the lake in the days and weeks prior to the date of sampling. The ratio M was only a predictor of the average end of summer conditions and could not take the vagaries of the previous weather into account. Given the two sources of uncertainty it was still remarkable that the presence of the most common species could be predicted with at least 50% confidence if the conditions in the lake were known. The fact that the high biomass/TP ratios (blooms) formed a small subset of the overall species occurrences (Fig. 11.6) and that the percentage occurrences of these blooms was also high, meant that severe bloom conditions could be predicted with the most confidence.

The data set only included mesotrophic and eutrophic lakes so the overall picture is somewhat biased. It was evident from an examination of the data that at $TN:TP > 60$ the diversity of the phytoplankton rose and the predictable occurrence of these species became rarer. The nuisance species occurred most predictably at $TN:TP$ ratios below 50 and this appears to be associated with the changing patterns of metabolism of N and P in these lakes (Vollenweider and Kerekes, 1981; Janus and Vollenweider, 1981). This is further evidence for previous suggestions (Harris, 1980) that in oligotrophic lakes the diversity is maintained in part by the spatial and temporal patchiness of the recycled nutrient flux. When this recycled flux is overridden by external loadings and significant nutrient concentrations occur in summer, blooms of nuisance species occur more frequently. The dependence of the nuisance species on the $TN:TP$ ratio (Smith, 1983; Tilman *et al.*, 1983) and the sensitivity of the ratio to TP concentration

(Vollenweider and Kerekes, 1981) means that control of *TP* loadings will not only control the mean annual biomass (Chapter 9) but will also directly influence the species composition and diversity. These results have, in fact, been tested by whole lake fertilization studies (Schindler *et al.*, 1973; Schindler, 1977) and have been confirmed. For example Schindler (1977), in one of the Experimental Lakes Area lakes, produced a bloom of *Aphanizomenon* by reducing the *TN*:*TP* loading ratio from 15 to 5.

The consistent dependence of the species on physical and chemical parameters means that such determinants of community structure override the biological interactions between species and are further evidence for allogenic controls. The consistent patterns of occurrence between lakes indicate consistent patterns of behaviour of the species and indicate that the seasonal cycles of species are largely determined by seasonal cycles of physical and chemical perturbations. The dependence of species on physical parameters raises the possibility that nuisance blooms of *Microcystis* and *Aphanizomenon* may be controlled by manipulation of the stratification of the lake or reservoir. Intermittent mixing of the water column should eliminate those species which depend on $M < 1$ if the scale of the mixing is such as to interfere with the growth strategy of the organism. I have already shown (Chapter 8) that perturbations in Z_m at a scale of a few days have a strong effect on diversity in surface waters by virtue of the interaction between physiology, growth rates and competition. Such management techniques would be important in situations where the nutrient loading is not easy to control due to extensive diffuse inputs.

11.3 Phosphorus loadings and algal biomass

In a series of important papers Vollenweider (1968, 1969, 1975, 1976; Vollenweider and Kerekes, 1980; Vollenweider *et al.*, 1980) has studied the large number properties of the relationships between P loadings and algal biomass. This topic was introduced in an earlier chapter in the discussion of nutrient cycling in lakes. Vollenweider deliberately looked for general high level properties of whole lake basins at a scale of years. By plotting both P loadings and *TP* against algal biomass Vollenweider showed that it was possible to predict algal biomass from nutrient loadings and that these relationships could be used for management. The assumptions of the models and the basic formulations were reviewed in Chapter 7. When algal biomass relationships are added to the nutrient mass balance models of Chapter 7 it is necessary to assume, in addition, that macrophyte growth has no serious effect on the cycling of P in the lake and that light is not limiting. At high algal biomass light does become limiting because of self-shading but, from the point of view of the loading models, the contribution of suspended detritus and sediment to light attenuation should not be too great (Canfield and Bachmann, 1981; Wofsy, 1983).

If algal biomass (as chlorophyll) is plotted against *TP* in lakes then a clear correlation is observed. The form of the plot has to be a plot of the logarithm of the biomass versus *TP* as the biomass data is lognormally distributed. The explanation for this was described in Equation 11.2. Thus plots of biomass versus *TP* should be either log/linear, or log/log if it is desirable to express the *TP* data on a logarithmic scale also. It should be noted here that these plots also provide evidence for alterations in the way nutrients are processed in lakes of different trophic state. Evidence for such changes can be found in the plot of mean annual algal biomass (as chlorophyll, the photosynthetic pigment) against *TP* given in Janus and Vollenweider (1982). This plot gives data for the OECD lakes with the addition of a number of oligotrophic Canadian lakes. The inclusion of these oligotrophic lakes reveals that the biomass versus *TP* plot is not linear (Fig. 11.8). If *TP* is added to an oligotrophic lake the relationship predicts that the biomass

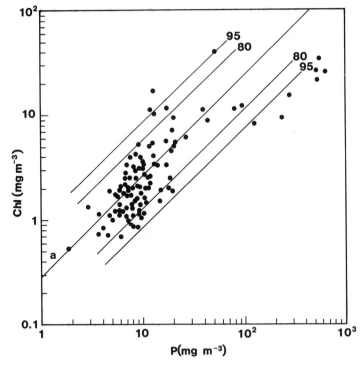

Fig. 11.8 Annual mean chlorophyll concentrations in relation to annual mean total phosphorus concentrations. Data from a number of lakes in an OECD study (Janus and Vollenweider 1981). The 80% and 95% confidence limits are given for the line of best fit.

would rise more steeply than it would if the same amount of TP were added to an already eutrophic lake. The steepest shift up occurs at about 10 mg m^{-3} TP. At TP levels above about 50–80 mg m^{-3} the curve levels off somewhat, indicating that light is probably limiting the overall biomass.

The shift in lake metabolism has been noted before and Reynolds (1978) showed that the use of added P was more efficient in oligotrophic lakes. The ratio of chlorophyll produced to DIP available varied by a factor of almost four and was highest in the oligotrophic lake. Golterman (1973) pointed out that it is only possible to assess the impact of external P inputs if the magnitude of the internally recycled flux is known. He estimated that the turnover rate of the TP pool could reach 20–40 y^{-1} but this would appear to be a conservative estimate. Certainly the changing relationship between the external and internally recycled fluxes has an important impact on the expression of the TP as algal biomass. The relationships between the turnover times of TN and TP, the partitioning of the TN and TP pools and the expression of biomass are all consistent with a marked change in the processing of nutrients in eutrophic as compared to oligotrophic water. Thus the empirical models which relate in lake TP concentrations to physical and morphometric variables tell only a part of the story.

From a physiological point of view the relationship between the increase in algal biomass (as chlorophyll) and P-uptake should be between 1 and 2 mg Chl mg P^{-1}. Harris (1983) showed a relationship of 1.68 mg Chl mg P^{-1} for a data set from Hamilton Harbour and showed that P-uptake was linearly related to the algal biomass in the photic zone. The Chl:P ratio changes with trophic state (Chapter 7). We have already seen that there is a systematic change in the minimum cell quotas of P in cells of different sizes (Chapter 8; Shuter, 1978). The changing ratios of P:Chl in lakes result in a changing slope of the Chl:TP plot (Fig. 11.8) which are associated with systematic shifts in cell size. As might be expected nanoplankton dominate oligotrophic waters and the proportion of nanoplankton in the total plankton falls as the TP and Chl rise (Watson and Kalff, 1981).

Empirical studies of large number properties are in accord with the mechanisms discussed earlier. The relationships between these high level 'state variables' are best studied as annual or basin averages. Because of the opportunist growth strategies of the phytoplankton and the strong seasonal cycle of water column stability in many waters, seasonal events strongly perturb these empirical relationships. Averages over smaller time and space intervals therefore introduce more scatter as the pool sizes of soluble and particulate nutrients fluctuate over time. Production estimates are also less well-correlated with TP than the more conservative biomass estimates (Schindler, 1978) and there is evidence for an upper limit to production brought about by self-shading and light limitation (Schindler, 1978; Vollenweider and Kerekes, 1980). There is a similar inflection in the Chl:TP plot (Fig. 11.8) at high biomass.

Averages at scales of years and whole basins may break down at smaller time intervals if the loading relationships are not in equilibrium. Some lakes do not fit the loading plots if the load is increasing or decreasing over time scales much less than the water residence times (Chapter 4). If the time scales are congruent (i.e. loads change slowly in relation to water residence times) then data from single lakes over time match the data from a series of lakes at one point in time quite well. This is equivalent to saying that the P:Chl ratio is stable in time and space and reductions in P-loadings can be used to manage algal biomass in lakes. This is the basic reason for the success of the Vollenweider approach. The classic case of P removal and a consequent reduction in algal biomass is Lake Washington (Edmondson, 1972; Edmondson and Lehman, 1981) but many other cases have been equally successful (Jones and Lee, 1982). The paper by Edmondson and Lehman (1981) gives some excellent examples of the relationships between P loadings, algal growth, sedimentation and seasonal cycles of stratification. It nicely demonstrates the biological processes of production and nutrient regeneration which underlie the empirical models and demonstrates the need for annual, whole basin averages.

The fact that there is a clear empirical relationship between TP and Chl in lakes indicates that it is the pool size of TP which limits the pool sizes of particulate P in the biomass. Indeed, the entire pool of TP in surface waters is partitioned into a number of soluble, particulate, organic and inorganic pools and the Chl:TP plots reveal some regularity in the partitioning across lakes of differing trophic state. Clearly all the TP cannot be sequestered by the algae, neither can it be permanently bound in any other pool. Life in the planktonic ecosystem depends on a continuous flow of energy and materials. Fluctuations and irregularities in the flows and pool sizes occur constantly but regeneration processes at a number of scales ensure that, over the year, the activities of the many organisms result in broad statistical distributions.

The fact that the reduction in P-loadings may be used to control algal biomass and that the response of one lake to changing loads is the same as that of many lakes with different loads, illustrates an important property of planktonic ecosystems. The response of the many opportunist growth strategies of the species in the mixed layer produces an overall heuristic behaviour that responds to high level forcing by nutrients and light. The fact that the algal biomass in one lake may be changed in a predictable fashion indicates that the Chl:TP plots are not simply a Central Limit phenomenon, but a true reflection of the partitioning of P between soluble and particulate pools in the food chain. The non-linearities in the relationships, which are explicable in terms of known biological properties further reinforce this view.

11.4 Manipulation of algal biomass

Vollenweider's empirical models which relate TP to annual average algal biomass can be used as an empirical framework for the manipulation of algal biomass. A high level holon such as TP, which has a residence time equivalent to the water residence time, is used to control another holon at the same level. The annual average algal biomass responds only weakly to strong low level perturbations. In Hamilton Harbour the average summer chlorophyll dropped from 45 μg l^{-1} to 30 μg l^{-1} under the influence of strong perturbations in mixed layer depth. On an annual basis the decrease would be about half that amount. Low-level physical perturbations have been shown (Chapter 9) to affect the variance in biomass (and hence the maximum expected value, Harris, 1983) but leave the average biomass little affected. By influencing TP through changes in the P-loadings, the manager is influencing the 'size of the envelope' within which the species grow. Studies have shown that the tracks of individual lakes through time as P-loadings are reduced, follow the regression of TP versus chlorophyll for many lakes (Jones and Lee, 1982; fast *et al.*, 1983). This will only be true if the TP and chlorophyll interact through processes which ensure that the two variables are functionally related at time scales less than the averaging period (< 1 y).

Even though there is only good data for a restricted number of lakes (< 20) the manipulation of algal biomass by means of TP loadings is well demonstrated. Essentially the technique rests on chlorophyll:P ratios (Chapter 7); or the ratios of two high level variables. The systematic change in the Chl:P ratio with trophic state (Chapter 7) means that as the TP in the water is reduced the algal biomass will drop most quickly, as oligotrophic conditions are reached. I have shown that this change in the Chl:P ratio is associated with systematic changes in phytoplankton cell size and the structure and efficiency of the food chain.

11.5 Phosphorus loadings and hypolimnial anoxia

The links between inputs, production, stratification and sedimentation are further displayed by the correlations between P loadings and hypolimnial anoxia. As the nutrient loading is increased the tight coupling between production, consumption and nutrient regeneration in oligotrophic waters is broken and the system begins to show large excursions in pool sizes over space and time. Rather than rapid nutrient regeneration in surface waters the system begins to show 'displaced metabolism', sedimentation of nutrients into bottom waters and regeneration at predominantly seasonal scales. The depletion of oxygen in bottom waters in both lakes and the oceans is a function of the decomposition of particles raining down into bottom waters. There is therefore a straightforward relationship between production in surface waters and the areal hypolimnetic oxygen deficit (AHOD) as long as the stock of oxygen in bottom waters is taken into account. Jones (1972) presented data from the English Lakes which show a variety of trophic states. The data showed clear relationships between *TP*, algal biomass and hypolimnia anoxia and were a confirmation of the original Pearsall series (Macan, 1970). These were the same lakes that Lund (1954–5) had used for his correlations between major ions and phytoplankton biomass (Chapter 4) so links can be made which extend from conservative major ions to nutrients to primary productivity to hypolimnial anoxia. Through a series of stronger or weaker correlations all can be related to biological activity in the water column.

Vollenweider and Kerekes (1980) and Jones and Lee (1982) displayed the results of comparisons between the P loading and AHOD (Fig. 11.9) but the

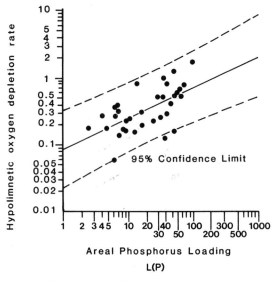

Fig. 11.9 The observed relationship between areal phosphorus loading and the aerial hypolimnetic oxygen deficit as summarized by Jones and Lee (1982).

correlation showed considerable scatter. Cornett and Rigler (1979) showed that there was a clear relationship between the P retention coefficient and AHOD, and were able to improve the predictions of AHOD by including the average hypolimnetic temperature and the average hypolimnetic thickness. Thus biological and morphometric factors influence the coupling between P-loading and AHOD through the metabolism of P in surface waters. Charlton (1980a) successfully developed a model which predicted AHOD on the basis of a multiplicative relationship between Chl, hypolimnial thickness and temperature. Shallow lakes with warm, shallow hypolimnia are more prone to anoxia than lakes with large, cold hypolimnia. Vollenweider and Kerekes (1980) commented 'The higher uncertainties in predicting the hypolimnetic depletion rates are due to complex interactions between epilimnetic and hypolimnetic conditions as depending on lake morphometry, length and type of thermal stratification, vertical entrainment and oxygen transfer and the interaction between sediments and the overlying water'.

11.6 Phosphorus loadings and the biomass of other organisms in the food chain

The TP loadings to lakes influence the envelope dynamics within which the opportunistic phytoplankton grow. There is little or no evidence of growth limitation by TP but an upper limit to biomass is set by the partitioning of the TP between the various pools of the food chain. In the short-run there is much temporal and spatial variation in the various pools of nutrients as growth, grazing, sedimentation and regeneration occur. What these empirical approaches reveal is that there is, over large scales, a regular partitioning of the major nutrients between the particulate pools. There is no need to invoke optimization to achieve the observed result. Heuristic processes by many species produce smoothed patterns at the community level, be they phytoplankton, zooplankton or fish.

McCauley and Kalff (1981) examined the relationships between phytoplankton and zooplankton biomass and showed that there was indeed a positive relationship. This, in itself, is some indication of 'bottom up' control. The ratio of zooplankton to phytoplankton biomass decreased as the total phytoplankton biomass increased. This was because the zooplankton biomass was correlated only with the nanoplankton fraction of the total algal biomass. This fraction declines in relative abundance as the total algal biomass increases (Watson and Kalff, 1981) and, as it is the edible fraction, zooplankton biomass declined proportionally, in concert. There was considerable scatter in the individual phytoplankton to zooplankton ratios. This would be expected from the essentially non-equilibrium nature of the species interactions in the many water types.

The same considerations apply to the correlations between phytoplankton biomass, primary production or zooplankton biomass and fish production. Ryder (1965) was the first to note that a morpho-edaphic index (MEI) made up from the ratio of $N/\bar{z}$ (where N is a nutrient parameter (or a close correlate) and $\bar{z}$ is the mean depth) could be used to predict fish yields in lakes. The basic philosophy of the use of the MEI is similar to that of Vollenweider (Ryder, 1982) and, in the same way, spans the gap between ecological theory and management practice. There have been a number of arguments about the use of the MEI and, indeed, some claims that the correlation is spurious (Youngs and Heimbuch, 1982). Ryder *et al.* (1974) provided an historical review of the use of the MEI and explained the background to the use of total dissolved solids (TDS) as a surrogate for nutrient concentration. I have already shown that TDS is well correlated with a number of indices of water quality and is therefore as good a master chemical variable as any (Chapter 4). The use of mean depth in the MEI is similar to Vollenweider's use in that it determines the relationships between area and volume, loadings, production and sedimentation. Larger lakes are deeper and do indeed tend to produce more fish (Youngs and Heimbuch, 1982; Prepas, 1983b) but the fish production of larger lakes *per unit area* is less than that in smaller (shallower) lakes. This probably reflects the influence of the relative proportions of littoral and profundal benthos in lakes of different size and the differences in productivity between shallow and deep lakes.

As might be expected there is a strong link between primary production, algal biomass and fish production. Guillen and Calienes (1981) showed that it was possible to accurately predict the biomass of anchovy in the Peruvian upwelling from estimates of algal biomass as chlorophyll. Jones and Hoyer (1982) reported a strong correlation between summer chlorophyll values and fish harvests while Melack (1976) showed a strong logarithmic relationship between primary production and fish yields in tropical lakes. Melack pointed out that dynamic biological correlates are preferable to consideration of static (biomass) concentrations. Oglesby (1977b) examined the relationships between primary production, algal biomass and the MEI and showed that the expected trends existed. Productivity and biomass were indeed higher in shallow lakes with high TDS. (This had already been demonstrated by Lund (1954–5) in a largely neglected paper.) The relationships between algal biomass and the MEI were not as strong as those with TP however and the use of P loading models for the estimation of fish production is probably preferable. Matuszek (1978), Hanson and Leggett (1982) and Prepas (1983b) have all reviewed aspects of the relationships between MEI, algal biomass and production and fish biomass and have reached similar conclusions.

The effects of latitude on primary productivity and fish production are very similar (Brylinsky and Mann, 1973; Schlesinger and Regier, 1983) and are probably mediated by a similar master variable, that of mean temperature. Thus such indices as MEI are better confined to regional comparisons of lakes with similar climate (Jenkins, 1982) so that broad climatological differences and effects on temperature and stratification are avoided (Ryder *et al.*, 1974; Ryder, 1972). After all, Vollenweider's models of phytoplankton biomass in lakes are almost totally confined to temperate lakes with a period of summer stratification. *TP* and MEI models therefore reveal broad statistical properties of planktonic food chains and the data from many lakes is consistent. There is no need to assume that the oceans are different. What little data there is, is consistent with the freshwater models. The lack of clear boundaries to water masses makes the loading approach extremely difficult in oceanic waters but the same relationships between *TP* and Chl should exist. It would be instructive to measure *TP* and *TN* in the oceans and to correlate the results with Chl in surface waters.

11.7 Size distributions in food chains

There is another way in which broad statistical properties of food chains may be examined. In Fig. 8.11 I showed that there were characteristic cell-size distributions in the waters of Hamilton Harbour. These cell-size distributions were seen to be directly related to environmental factors such as nutrient fluxes, Z_{eu}/Z_m ratios and vertical turbulence. Elsewhere in the text I have shown that cell-size distributions were influenced by nutrient availability and grazing pressures (Chapters 6, 7, 8 and 9). Only certain phytoplankton cell-size classes may be grazed effectively so the transfer of energy along food chains is not simply a function of nutrient loadings and primary productivity, but of primary productivity in certain size classes.

Kerr (1974) developed a theory for size distributions and food chain relationships in ecological communities based on the observations of Sheldon *et al.* (1972, 1973). Sheldon and his coworkers measured the amount of biomass in a wide range of size classes spanning a number of trophic levels. The phytoplankton size distributions were similar to those found in Hamilton Harbour (Fig. 8.11) and were doubtless generated by similar mechanisms. More importantly Sheldon *et al.* (1972) noted that the biomass in a wide range of size classes (from 1 to 10^6 µm equivalent spherical diameter) was more or less constant. In other words the biomass of whales was only slightly less than the biomass of phytoplankton. Kerr (1974) explained these observations by recourse to a size dependent food chain model in which growth and metabolism were a function of body weight. Kerr's model showed that the biomasses of adjacent trophic levels existed in a simple proportionality to one another depending not so much on body size, as on the efficiency of transfer of energy from one trophic level to another and on the ratio of the sizes of predator and prey. Using an ecological efficiency of 20% and estimating that the predators were, on average, twenty times the body length of the predators, Kerr (1974) obtained the result that the standing stock of phytoplankton should be about 1.2 times the standing stock of zooplankton and so on up the food chain. This was in close agreement with the observations of Sheldon *et al.* (1972) in the oceans.

Sheldon *et al.* (1977) examined the two bases of Kerr's (1974) model in a little more detail and refined the model somewhat, using more data on the power relationships between metabolism, growth and body weight (Fenchel, 1974; Peters, 1983). Sheldon *et al.* (1977) were able to show that the size ratio of predator to prey in the oceans was close to 14 and that this resulted in a significant disparity in maximum growth rate. They also showed that 15% was a better estimate of the ecological efficiency, assuming high prey utilization. By plotting the theoretical relationships between the ratios of the standing stocks and predator–prey size ratios, and by assuming that the product of the ecological efficiency and predation efficiency lay between 0.10 and 0.20, Sheldon *et al*, (1977) were able to estimate the range of likely observed values. They found that the observed ratios of standing stocks of predators and prey should vary between 0.3 and 3.0 but should show a mean ratio of about 1.0. In the previous discussion it was clear that grazing (at least in temperate waters) was not a steady-state process, so variation in standing stocks is to be expected. At a scale of years steady state may be assumed but over time periods encompassed by the normal types of sampling schemes great variability is to be expected. Papers by Cushing (1959) and Heinrich (1962), and the review by Melack (1979), indicate that there are strong latitudinal components to the seasonal cycles of phytoplankton and zooplankton in lakes and the oceans and that the entire food chain is driven by high level, seasonal perturbations in physical and chemical processes. Energetic considerations determine the limits of the biomass fluctuations that might be expected.

Sprules *et al.* (1983) sampled 37 lakes in Ontario and estimated the biomass in a number of size classes spanning the range from phytoplankton to fish. Their data showed that the ratios of phytoplankton to zooplankton varied from 0.04 in an oligotrophic lake in the clear water phase following the spring bloom to 597 in a small, shallow, eutrophic lake with a dense bloom of filamentous algae. The mean ratio of standing stocks was 1.63 and 70% of the values fell in the range from 0.3–3.5, much as expected. In accordance with these results, when McCauley and Kalff (1981) summarized the relationships between the standing stocks of phytoplankton and zooplankton in seventeen lakes from a variety of trophic types, they found that the ratios of the mean standing stocks in the lakes varied in the range from 0.6–3.5. The range of individual values was greater. In oligotrophic waters the mean ratio of the standing stocks was close to unity, but as the total biomass increased the ratios of the standing stocks declined. McCauley and Kalff (1981) showed that the ratio of zooplankton biomass to nanoplankton biomass remained constant (as might be expected from the size dependency of grazing) so that the apparent reduction in conversion efficiency as waters became eutrophic was caused by the changing relationships between phytoplankton cell-size distributions and trophic state. McCauley and Kalff (1981) observed therefore, that it was not necessary to invoke changes in ecological efficiency to explain the proportionally reduced biomass of zooplankton in eutrophic waters.

This size analysis of food chains can be taken one step further to predict fish biomass and production from phytoplankton biomass and production. Sheldon *et al.* (1977) were able to estimate fish production in the Peruvian upwelling quite accurately from phytoplankton biomass and production estimates. Their prediction of the total annual fish production corresponded remarkably well with the estimates of the annual catch when almost all the fish production was harvested. Observation indicated a close agreement between model predictions and fish biomass and production (Sheldon *et al.*, 1972, 1977). Sprules *et al.* (1983) compared fish biomass to phytoplankton and zooplankton biomass and obtained similar biomass ratios and variances to the expected values. The mean ratio was again slightly above unity with results falling between 0.9 and 2.3. When the phytoplankton and fish biomass was compared in nine lakes, a more or less linear relationship was revealed.

Borgmann (1982) used a similar model to estimate total animal production in pelagic ecosystems. Whereas other models were models of the efficiency of energy transfer between trophic levels, Borgmann used the clear size distinctions between trophic levels to calculate size dependent conversions rather than trophic level conversions. The advantage of using size rather than trophic level lies in the problems in assigning trophic levels to animals and the fact that fish change trophic level as they increase in size. Borgmann (1982) found that particle size conversion efficiencies were similar to, and varied no more than, the ecological efficiencies used above. What Borgmann was in fact doing was to include the effects of the predator–prey size ratio and the ecological efficiency in one factor. This formulation allows the calculation, for example, of total fish production from zooplankton production. Cushing (1975) showed that apparent ecological efficiencies in planktonic ecosystems were a declining function of total production and biomass. The apparent ecological efficiency between trophic levels will be a function of thermodynamic considerations and the distributions of different sized organisms in different waters types. Borgmann and Whittle (1983) have recently used the model to calculate the distribution of contaminants in food chains. The concentration of contaminants in the various food chain components is a function of the particle size conversion efficiencies and the turnover times of the various particulate pools.

11.8 Phytoplankton biomass and fish production

Fish populations are remarkable sensitive to production at lower trophic levels and there is a good deal of evidence which points to a strong link between phytoplankton production and fish production (Larkin and

Northcote, 1969). This holds both within and between lakes. Empirical correlations of phytoplankton biomass and fish biomass have been successfully sought in a number of waters (Matuszek, 1978; Hanson and Leggett, 1982; Prepas, 1983). Jones and Lee (1983) have demonstrated a significant relationship between Vollenweider's P loading model and fish yield in a number of United States lakes and reservoirs (Fig. 11.10). In Lake Memph-

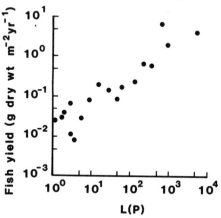

Fig. 11.10 The empirical relationship between areal phosphorus loading L(P) and the fish yield in a number of lakes (Jones and Lee, 1982).

remagog, Quebec, there is a strong gradient in water quality, primary production and phytoplankton biomass from one end of the lake to another. This is reflected in ratios (from south to north basins) of 1.6:1 in algal biomass; 1.25:1 in primary production (Ross and Kalff, 1975); 3:1 in benthic invertebrate biomass (Dermott, 1974); 2.5:1 in perch biomass (Nakashima and Leggett, 1975), and 2.5:1 in number and 3:1 in biomass of littoral fish (Gascon and Leggett, 1977).

A number of authors have noted a strong link between primary production and fish production (cf. references in Gascon and Leggett, 1977; Cushing, 1982). There is an increasing amount of evidence which would lead one to believe that recruitment is more important than growth in determining fish production. As Weatherley (1972) observed, fish seem to have evolved a strategy of density dependent growth rather than density dependent mortality. In other words the survival of the larval stage is more important in determining adult biomass than density dependent mortality later in life. Survivorship in the earliest weeks and months of life is low in fish populations and small changes in larval fish survival have dramatic effects on the subsequent adult population. The final production and biomass of many fish and invertebrates with planktonic larvae are determined by the success of juvenile stages feeding low in the food chain; either on zooplankton or on phytoplankton directly. Kerr and Martin (1970) observed that the production of lake trout was proportional to an index of primary production irrespective of the trophic level at which the adult fish were feeding. Ryder's (1965) regression of fish production against the MEI for a set of Northern temperate lakes also suggests a close link with primary production.

Primary production and fish production are also strongly correlated in the oceans, both in different ocean areas and over time. Ryther (1969) reviewed the relationships between photosynthesis and fish production in the sea and showed clearly that the areas of higher primary production and fish production (shelf areas < 200 m deep and upwellings) were only a small proportion (< 10%) of the total ocean surface area. Size distributions are also important (Parsons and Le Brasseur, 1970). The oligotrophic waters of the central oceans are dominated by nanoplankton so the food chain leading to fish is longer in those waters. Because of the low biomass and small size of the nano- and picoplankton in oligotrophic ocean water there is insufficient food to support large populations of herbivorous zooplankton. Primary production is largely consumed by microheterotrophs and much of the primary production is associated with particles (Goldman, 1984). I have already shown that in oligotrophic waters much of the primary production and nutrient regeneration occurs at very small scales (Chapters 6 and 7). Crustacean zooplankton become predominantly carnivorous in such waters. For example, the proportion of carnivorous zooplankton species increased from 16% to 39% in a transect from North American coastal waters to the Sargasso Sea (Ryther, 1969). In temperate coastal waters the phytoplankton are larger and juvenile fish may feed directly on the large populations of herbivorous zooplankton which occur seasonally. Spawning periods tend to be linked to periods of high 'new' production in such systems (Cushing, 1982). In upwelling areas the adult fish may themselves be herbivorous filter feeders which feed directly on the large phytoplankton present in the eutrophic waters. Guillen and Calienes (1981) estimated anchovy biomass directly from phytoplankton biomass in the Peruvian upwelling. The net result of all this is that the largest fish production occurs in upwelling areas despite the fact these account for only 0.1% of the ocean surface area (Ryther, 1969). Upwelling areas and coastal areas account for nearly all of the world's commercial fisheries. Thus nutrient status, phytoplankton cell size and primary production set an upper limit to fish production in oceanic areas.

Within this framework, the strong relationship between primary production and fish production has been exploited by Cushing (1982) to explain the long-term fluctuations in fish stocks which occur in many waters. Energetic considerations may set an upper limit to fish production, but noticeable changes in fish populations occur with irregular declines and periods of recovery. Cushing (1982) suggested that the recruitment of commercial fish stocks was determined partly by major differences in the ratio of carnivorous to herbivorous zooplankton and in the species composition of fish; in other words by changes in water mass distributions at scales of oceans basins and years. Cushing (1982) also noted that a number of changes in the recruitment of commercial fish stocks were correlated with simple physical factors such as temperature and wind stress on a year to year basis. These factors are the major determinants of the timing and magnitude of the spring outburst of primary production by their control over the relationship between the mixing depth and the critical depth (Chapter 9). Many important commercial fish stocks spawn over a relatively brief period and if the period of spawning does not closely match the period of the spring outburst of primary production then the 'new' production of spring cannot be harvested and passed up the food chain. Cushing and Dickson (1976) and Cushing (1982) used this 'match–mismatch' hypothesis to explain the fluctuations in recruitment that occur in commercial fish populations. Year-to-year climate variations therefore drive year-to-year variations in the *timing* of the spring pulse of primary production, and similar variations in the *magnitude* of fish populations in subsequent years.

The close coupling between primary production and fish production is also indicated in the close spatial correspondence between the two processes. In order to sustain high fish production there must be sustained 'new' production and a sustained transfer of nutrients from soluble or other reserve pools to algal biomass. The brief pulse of 'new' production in spring in temperate waters occurs throughout surface waters but dies away quickly. In coastal waters where run-off from the land supplies a larger stock of nutrients, the spring pulse may last longer but sustained 'new' production only occurs in upwelling areas. Sustained primary production may also occur at oceanic fronts and ring margins where physical processes serve not only to concentrate biological activity by virtue of reduced horizontal and vertical diffusivities, but also to produce a sustained flow of nutrients from the mixing of water masses (Yentsch, 1980). The primary productivity of fronts is well illustrated by the satellite image of the eastern seaboard of the United States, and the occurrence of fish at these fronts is elegantly illustrated by the maps of fishing effort for similar areas. Fournier (1978) reproduced maps of the positions of Canadian and other trawlers off the Grand Banks of Newfoundland and the trawlers can be seen to be fishing, not on the Banks, but at the position of the shelf-break front (Fig. 11.11). The same is true of the fishing grounds around Cape Cod and Georges Bank. The occurrence of demersal fish stocks in frontal areas

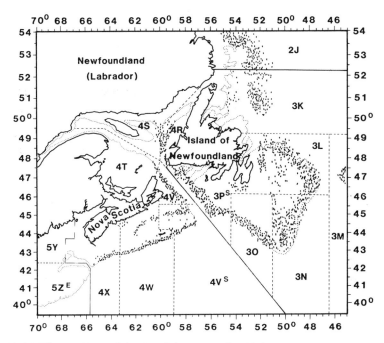

Fig. 11.11 The position of foreign fishing vessels off the east coast of Canada in 1972. The dotted lines indicate the 200 m contour. From Fournier (1978), standard fishing zones are also indicated.

indicates a close coupling between primary production in surface waters and secondary production in bottom waters. High primary production in surface waters associated with the more or less permanent occurrence of the front in a zone over the shelf break (Fig. 3.6) will ensure a continuous rain of particles into deep water at that point. This is a good example of physical/biological coupling in the water column and of the importance of vertical fluxes of materials.

One aspect of the coupling of planktonic and benthic food chains that is of interest is the fact that the size spectra of benthic organisms appear to be the mirror image of planktonic spectra (Schwinghamer, 1981, 1983). Given the necessary size ratios between predator and prey it is perhaps not surprising that the benthic organisms fill the gaps in the planktonic spectra. Herbivorous zooplankton graze the nanoplankton; the fate of the larger phytoplankton is sedimentation (Chapters 9 and 10). Thus the benthic consumers of the larger phytoplankton should be larger than their planktonic counterparts and the peaks in the size spectra should be staggered. Kerr (1982) has noted that even though the physical nature of the benthic substrate is quite different, these size dependent producer/consumer relationships confirm the ubiquity of the phenomenon. Both planktonic and benthic organisms are feeding in a viscous environment.

Size spectra and envelope dynamics are therefore statistical properties of the nutrient and energetic relationships of planktonic food chains. The statistical distributions are both lognormal for size and for biomass but are not simply Central Limit phenomena arising out of the dynamics of the many species involved. The characteristics of the distributions are biologically significant. The mean annual algal biomass in lakes may be changed by altering the P-loading, and the track of one lake over time is the same as that of many lakes at one point in time. The standard deviations of the lognormal biomass distributions are a function of physical perturbations both within (Harris, 1983) and between (Vollenweider and Kerekes, 1980) lakes. Variation in the physical regime drives variation in algal biomass through a number of biological mechanisms (Chapter 5). There is some indication that the size spectra of physically perturbed water columns are broader than temporally stable water columns. Temporal variability in the mixing regime increases diversity and this diversity is reflected in the coexistence of a number of peaks in the size spectrum (Harris *et al.*, 1983). All the evidence points to a relative lack of interactions between organisms at any given trophic (or size) level so that lognormal distributions arise from overlapping, non-equilibrium, opportunistic growth strategies. Organisms are uncoupled by differences in size and growth rate (Schaffer, 1981). This is the essence of Margalef's (1980) 'game'. Where biological order does abound is in the food chain interactions between size classes or trophic levels. Energetic (thermodynamic) considerations determine both the size classes and the biomass in each as well as the standard deviations of both.

This is not to say that vertical coupling in the food chain is necessarily very tight. Phytoplankton appear to play an important role in determining the events above them in the food chain. Coupling between phytoplankton and zooplankton appears to be determined by the ability of the consumers to track the availability of certain size classes of producers. The disparity in growth rates results in non-steady state interactions. In the fluid environment coupling between the trophic levels by nutrient regeneration is necessarily weak as regenerated nutrients diffuse away rapidly (Jackson, 1980). The phytoplankton must therefore be at least as dependent on external (physical/chemical) cues as on the internal (biological) recycling of nutrients and other materials. This appears to be increasingly true in waters with a strong seasonal nutrient pulse or as the water column becomes more and more eutrophic. Only in continuously oligotrophic waters is there tight coupling between trophic levels and then aggregation on particles is necessary to defeat the forces of diffusion (Goldman, 1983).

CHAPTER 12

Interannual variability

12.1 Long-term changes in phytoplankton abundance

There is a relative dearth of good long-term phytoplankton data as it is difficult to sustain the necessary effort using consistent methods of sampling, counting and identification. Many of the earlier sampling and counting methods did not adequately enumerate the smaller flagellates and nanoplankton but merely concentrated on the net phytoplankton fraction. In other cases inconsistencies in the siting of sampling stations, analytical methods and changes in personnel have rendered the interpretation of long-term data sets difficult or impossible. Also, regrettably, there seems to be a general lack of appreciation of the value of long-term routine monitoring which, in a time of financial stringency, has made it difficult to sustain the effort in recent years. It has, perhaps, also been difficult to justify long-term monitoring because the direct results of such activities were not apparent until recently. In recent years there have been a number of published accounts of long-term monitoring and of similar long-term climatological data. Such results have, no doubt, contributed to the unease with equilibrium theory, as it is clear when a number of years of data are analysed that the world is an inconstant and inconsistent place. Time cannot be ignored. The best long-term records are probably those from fishing activities as catch records have existed for centuries. Direct, quantitative sampling of phytoplankton and zooplankton populations has only been carried on on a continuous basis since the 1930s and there was much disruption during the war. Despite the problems and the paucity of data there are now a sufficient number of accounts in existence from both marine and freshwater environments to demonstrate both the nature of the biological processes involved and the necessity of long-term data collection. For example, long-term data records exist for certain English Lakes, the Laurentian Great Lakes, the North Sea and the English Channel. Conveniently these data span a wide range of basin scales from small lakes to oceanic basins. Some common mechanisms and properties will however be identified.

As might be expected the factors influencing long-term fluctuations include climatological effects, nutrient loadings and availability, and food chain interactions. In order to describe the climatological fluctuations which drive the year to year variations in the biomass of phytoplankton in different waters it is necessary to have very long data sets. Lamb (1977) has extensively reviewed the long-term (20–50 y) cycles of climatic change and has shown that evidence for such climatic change can be obtained from biological as well as climatological data. The existence of cyclic climatic changes over periods of 20–30 years means that any short-term (1–5 y) data set suffers from problems of interannual variability. This problem was particularly apparent in my own five year study of Hamilton Harbour. Each year is different, and without a long-term baseline reference it is difficult to tell just how different. Nutrient loadings also differ with time due to climatological effects, biological and geochemical processes on land and in the water, and on human activities. Historical estimates are again hard to come by and few estimates go back more than 20 or 30 years. This has led to severe difficulties in the interpretation of the effects of pollution control in the last decade as climatological cycles interfere strongly with the apparent effects of pollution abatement.

It is clear that there is a strong link between the biomass of phytoplankton, zooplankton and fish in all waters. The links between trophic levels are sufficiently strong to ensure that if physical processes drive fluctuations in phytoplankton biomass from year to year then there will also be similar fluctuations in the other food chain components. The examination of long-term data can throw light on the nature of the controls on food chain structure; the 'top down' versus 'bottom up' argument. The existence of strong climatological effects on the biomass of zooplankton and fish is a good argument for some degree of 'bottom up' control. This is not to say that some 'top down' density dependence does not occur; it is a matter of degree. The strong empirical relationships presented in the previous chapter are, in themselves, arguments for 'bottom up' control. Even so some (Shapiro, 1977, 1978; Shapiro *et al.*, 1975, 1982) have argued strongly that it is possible to manipulate phytoplankton populations by manipulating fish populations. In a series of recent books and papers Cushing (1959, 1966, 1969, 1975, 1978, 1982, 1983, 1984) has argued strongly for 'bottom up' control of fish stocks in the sea and has postulated that a sequence of physical events in spring which control the spring bloom of phytoplankton is the key to understanding recruitment to commercial fish stocks. This 'match–mismatch' hypothesis can be extended to freshwater environments and will be used to explain the sequence of long-term changes in the English Lakes. As indicated in the previous chapter, the timing of the spring bloom determines the magnitude of the recruitment to the fishery. From year to year the climate in spring determines the timing of the spring phytoplankton bloom. The transition from mixed to stratified water in spring, which is controlled by increasing daylength, sunshine hours and temperature, controls the onset of the seasonal phytoplankton succession (Chapter 9). Growth of the *r*-strategist phytoplankton in early summer provides food for the zooplankton and hence for the fish larvae (Chapter 9), many of which feed planktonically in the early weeks of life. As the survival of the larvae depends on the relative timing of spawning and the spring outburst then the emphasis has shifted from mean values to phase relationships. In a fluctuating environment much of what occurs depends on the match of phase relationships rather than on average values. The changing populations of phytoplankton to be found in the oceans and in lakes are strongly influenced by the sequence of events in certain years and mean annual statistics may hide much biological information at the population level. Diversity depends very much on the temporal sequence of events (Chapters 10 and 11).

12.1.1 THE ENGLISH LAKES

Estimates of phytoplankton, zooplankton and fish populations as well as chemical and physical measurements have been carried out by members of the Freshwater Biological Association since the 1930s. Weekly or biweekly sampling has been concentrated in Windermere and nearby lakes but other less frequent data records exist for most of the English lakes and many of the tarns. While much of the data remains unpublished, enough has now appeared to make it possible to discern a broad sequence of events and to suggest some important processes. The total data record is unique and important in many respects, not the least of which is the fact that changes in sampling stations and methods, analytical techniques and counting methods have been kept to a minimum. The most extensive data comes from the North and South Basins of Windermere, from Esthwaite Water and from Blelham Tarn. Details of these, and all the English lakes, may be found in Macan (1970) and many of the papers referred to below.

(a) *Chemistry*

Sutcliffe *et al.* (1982) have discussed the long-term changes in the chemical composition of the surface waters of these lakes and Figs 12.1 and 12.2 show that both DIN and DIP have increased markedly in some basins in recent years. In all basins except the North Basin of Windermere there has been a marked rise in winter DIP levels in the 1960s and 1970s. The total population around Blelham Tarn has remained quite stable but Lund (1978) related the changes in the Tarn to the changes in farming practice in the drainage basin (1954–5), to the installation of piped water supplies (1953) and to the installation of a sewage system in the village (1962). Lund also

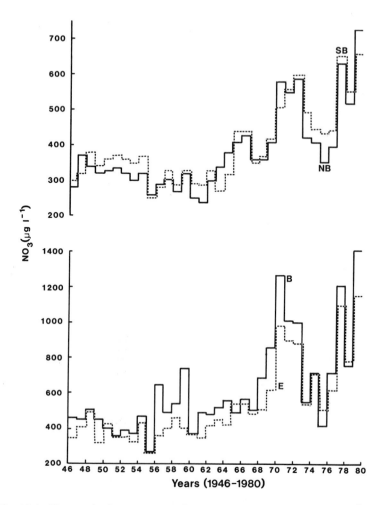

Fig. 12.1 Changes in the mean annual nitrate concentration in the North Basin (NB) and the South Basin (SB) of Windermere and in Blelham Tarn (B) and Esthwaite (E) since 1946. Data from Sutcliffe *et al* (1982).

documented the steep rise in the use of agricultural fertilizers in this same period. These factors together led to a sharp increase in DIP in the mid 1950s in the Tarn and the highest values were recorded around 1970. The pronounced rise in DIP in Esthwaite Water was delayed until the 1970s and the rise is coincident in time with the introduction of a sewage plant in the village of Hawkeshead which lies in the drainage basin (Lund, 1981). The increase in DIP levels in the North Basin of Windermere was small and hardly significant despite the slight increase in P-loadings from upstream.

The steep rise in the DIP concentration in the South Basin of Windermere, however, was a function of the sewage input from one of the major tourist centres in these lakes, the towns of Bowness and Windermere.

Nitrate levels in all four basins have also risen markedly in the last 30 years. Significantly, there were similar long-term fluctuations in the DIN in all four basins which Sutcliffe *et al.* (1982) interpreted as being due to climatological effects. Quite what it was which drove the DIN fluctuations is unclear. Possible candidates are air temperature and rainfall which could influence both nitrification and run-off in the drainage basin as well as lake levels and water residence times. There is no simple explanation available at present. The DIN peaks are associated in time with a statistically significant ten-year cycle in June temperatures (Kipling and Roscoe, 1977). This cyclic pattern in mean June temperatures (Fig. 12.3) had peaks around 1940, 1949–50, 1959–60 and 1968–9 with troughs in between. The periods of warm springs and warm summers around 1960 and 1968 were associated with noticeable peaks in DIN in all lake basins. Whatever the explanation the fluctuations in DIN and DIP over time were a function of basin geochemistry, human activity and climate.

(b) *Spring phytoplankton blooms*

The history of the phytoplankton in these lakes may be found in a number of publications most noticeably those of Lund (1949, 1950, 1954, 1955, 1954–5, 1961, 1969, 1971a,b, 1972a,b, 1979, 1981). The year-to-year variability in the magnitude and the timing of the phytoplankton peaks is considerable and the underlying factors are complex, but clear influences of nutrient loadings and climatological factors are discernible. The year-to-year variability in the timing and magnitude of the spring *Asterionella* bloom depends on physical and chemical conditions as outlined in the chapter on seasonal cycles (Chapter 9). Talling (1971) elegantly demonstrated the relationships between basin morphometry, water residence times, stratification and surface irradiance which controlled the growth of *Asterionella* over seven years. Growth began earlier in the shallow basins of Esthwaite and Blelham as compared to the deeper basins of Windermere. *Asterionella* should show the earliest growth in Blelham as it is the shallowest of the basins but it can be flushed out by spring rainfall and this may delay the onset of the spring bloom. As first noted by Lund (1949, 1950), the *Asterionella* bloom is terminated by strong stratification and sedimentation, by silica depletion or by sharp rises in pH (Jaworski *et al.*, 1981). Different mechanisms come into play in different years and they result in strong year-to-year variations in the magnitude of the spring *Asterionella* bloom and the timing of the silica depletion in surface waters (Lund, 1972b). The rate of silica depletion is a direct function of the growth of the *Asterionella* crop.

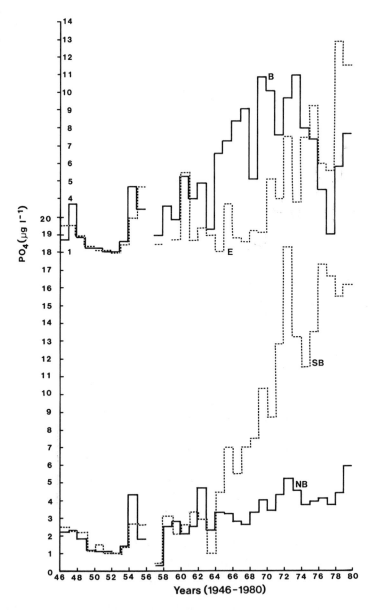

Fig. 12.2 Changes in the mean annual phosphate in the North (NB) and South (SB) Basins of Windermere and in Blelham Tarn (B) and Esthwaite (E) since 1946. Data from Sutcliffe *et al* (1982).

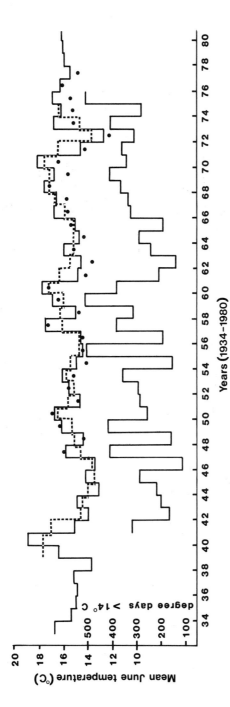

Fig. 12.3 Changes in the June temperature of the North Basin of Windermere since 1934. Data from the North Basin buoy (dots), from daily data (averaged) from Back Bay (solid lines). Dotted line: three point moving average. Degree day data from Kipling and Roscoe (1977).

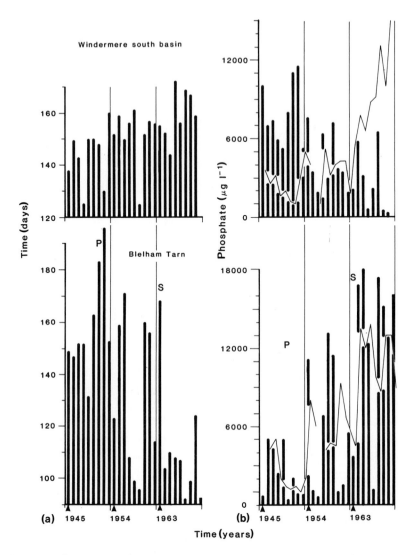

Fig. 12.4 Changes in the timing and the magnitude of the Spring *Asterionella* peak in Windermere and Blelham Tarn between 1945 and 1971. P: the introduction of piped water to the main village of Blelham. S: the introduction of a sewage system. (a) the time taken after Dec 31 to reduce the silica to a minimum value. (b) (bars) the maximum abundance of *Asterionella* (0–18 000 cells ml^{-1}) and (continuous line) the concentration of phosphate in the preceeding nine months (0–18 μg l^{-1}) (from Lund, 1972).

Parasitism is a further complicating factor. Parasitic fungal infections may seriously reduce or delay the *Asterionella* blooms so that the maximum cell number in any year is reduced. Lund (1972) gave the mean weekly number of cells of *Asterionella* in Windermere Southern Basin and Blelham for a number of years. Windermere showed a declining trend in *Asterionella* crops whereas Blelham showed the opposite. Irregular peaks occurred throughout both trends (Fig. 12.4). In the case of Blelham Tarn a clear sequence of warm springs interspersed with periods of parasitic attacks resulted in strong year to year fluctuations. Low crops occurred in 1950–2 and 1960–1 in warm Junes and parasitism reduced the crops in 1956 and 1966 (A. Irish, personal communication).

The long-term trends in *Asterionella* in the two lakes do not appear to be due to nutrient loadings as the trends in N and P are similar (Figs 12.1 and 12.2). More recent information indicates that *Asterionella* in Blelham reached a peak in 1965 and subsequently declined (Lund, 1979) whereas the same species reached a minimum in the Southern Basin of Windermere in 1969 and subsequently recovered. The long-term fluctuations in *Asterionella* must be seen as a result of the interaction of basin morphology, critical depth and the growth of the organism. The discrepancy in the mean depths of the two lakes will mean that *Asterionella* may grow rapidly in Blelham in the year when growth in Windermere will be delayed. Resuspension and regrowth of the *Asterionella* may occur if a partial or complete turnover follows the spring stratification. Turnover will serve to replenish the stocks of silica in surface waters and reduce the pH. As long as some vertical turbulence continues, growth may continue all summer. This was well illustrated by Bethge (1953). The effectiveness of wind mixing depends on basin size and the orientation to the prevailing wind. Large differences are to be expected in Windermere and Blelham as the two basins differ greatly in size and wind exposure. The growth patterns of *Asterionella* will therefore differ greatly from year to year depending on the particular sequence of events in spring and summer each year.

(c) *Fluctuations in zooplankton populations*

Stratification is followed each year in these lakes by the growth of a variety of nanoplankton. Lund (1961) documented the seasonal cycles of these organisms between 1956 and 1959 and showed considerable variability between basins and years. The nanoplankton blooms in 1959 were earlier than in previous years. Warm Junes are characterized by the growth of large populations of species such as *Cryptomonas* which form a large part of the diet of zooplankton. The growth of these 'r' strategist species is followed by a zooplankton outburst and a temporary clear water phase when the grazing pressure is briefly sufficient to clear the water. The abundance of zooplankton in the Northern Basin of Windermere shows strong long-term

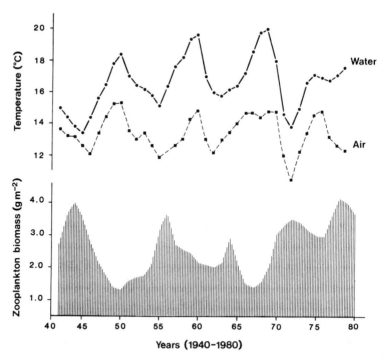

Fig. 12.5 The time series of mean June air and water temperatures in Windermere compared to the mean summer biomass of zooplankton at the North Basin buoy.

fluctuations (Fig. 12.5). In good years the early summer outburst of zoo-plankton, characteristically dominated by species of *Bosmina*, *Daphnia* and *Diaptomus*, is followed by a second pulse of *Diaptomus* in August. Good 'zooplankton years' may therefore be characterized initially by the early onset of warming and stratification followed by the growth of phytoplankton in suitable size classes. If the summer continues to be warm, however, then strong stratification ensues and no further entrainment of hypolimnetic nutrients occurs. In the English lakes this leads to a rapid switch from r to K phytoplankton and the growth of large, inedible species. Thus the zooplankton do poorly if the summer continues to be warm as the supply of suitably sized species runs out.

In cooler years there are frequent periods of wind and rain in summer (particularly July and August) which entrain hypolimnetic water into the epilimnion and stimulate outbursts of new production and the growth of edible species. These outbursts result in subsequent increases in zooplankton abundance as the zooplankton track the availability of food. There is a clear tendency for the zooplankton in Windermere to do better in the periods between the warm years (around the mid-years of each decade, Fig.

12.3) but there are poor years at times other than the known warm years. Years in which the perch and pike in the lake showed high larval survival (see below) also tend to be poor zooplankton years, so there is a suggestion of 'top down' control by larval fish predation as well as extensive evidence of 'bottom up' control by the fluctuating availability of food. Evidence for a climatological influence on the 'zooplankton years' may be seen in the correspondence between the periodicities in N and P in the water and the zooplankton abundance; particularly in the years between 1965 and 1975. The overall trophic state of the Northern Basin of Windermere has changed little in the last 40 years but the long-term, climatologically induced periodicities were nevertheless present.

(d) *Summer phytoplankton populations*

The year-to-year fluctuations in summer populations are also a function of nutrient loadings and of climate. Sedimentation of the spring diatom pulse removes silica and other nutrients from surface waters. The conversion of the remaining dissolved nutrient pools into animal biomass also depletes these pools further but rapid phytoplankton growth may continue as long as nutrients are being turned over by grazing. The domination of the late summer phytoplankton by K-strategists is usually most successful in well-stratified water. Under these circumstances dissolved nutrient pools may drop to very low levels, as turnover slows and nutrients are tied up in the algal biomass. The low levels of DIN in the water during the summer *Ceratium* blooms in Esthwaite are a good example of this (Harris *et al.*, 1979). Year-to-year changes in summer biomass, K species and nutrient

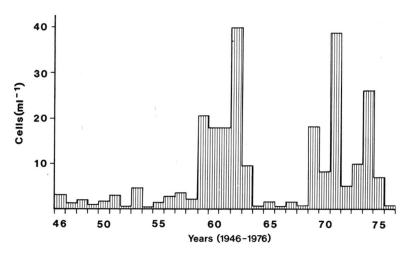

Fig. 12.6 The long term record of the abundance of *Ceratium hirundinella* in Blelham Tarn (from Lund, 1979).

depletion are therefore a function of the occurrence of warm stable epilimnia. This is well illustrated by Lund's (1979) records of Blelham where *Ceratium hirundinella* (Fig. 12.6) occurred abundantly in the years around 1960 and 1970. These were years of warm Junes and warm summers. 1955 and 1976 were also exceptionally warm in July and August. Lund (1972b) recorded a declining trend in blue-green algae in Esthwaite from 1946 to 1969 but superimposed upon the trend were year-to-year fluctuations in abundance which were also related to warm summers. Good blue-green years occurred around 1950, in 1955, in 1960–1, in 1966 and again in 1969.

There was a tendency for the good *Ceratium* years in Blelham Tarn to lag the periods of warm summers by a year or two. This effect would be brought about by the perennation of cysts and the consequent effect of one good year leading to heavy encystment and a good following year (Heaney *et al.*, 1983). Blue-green algae also rely on perennation so similar effects are to be expected. Summer phytoplankton populations of this type should be expected to show persistence in time from year to year. It has been argued up to now that the phytoplankton are essentially opportunist and that growth is usually exponential. Only the late summer K species appear to show density dependent effects and a reduction in relative growth rates brought about by nutrient limitation (Chapter 7). As the seasonal succession of species is ordered by the sedimentation of populations and the depletion of soluble nutrient pools in surface waters, there is a way in which phytoplankton can interact with one another by virtue of their persistence and through their effects on the cycling of nutrients at a scale of years. This may be the reason for the long-term cyclic behaviour of Esthwaite Water which exhibits periods of dominance by *Ceratium* which alternate with periods of dominance by *Microcystis*. The sediments also act as a long-term integrator of algal production and have a long-term effect on the cycling of nutrients with lake basins.

(e) *Long term changes in biomass*

The complete record of weekly data from Blelham Tarn is summarized in Fig. 12.7 where total phytoplankton volume is plotted for the last 40-plus years. Increased DIP loadings have led to increased biomass and have altered the seasonal succession of species. The early pattern is one of a spring pulse of *Asterionella* followed by little further biomass accumulation: the typical oligotrophic lake pattern. The periodic nature of the spring *Asterionella* bloom is clearly visible. As the nutrient loadings to the Tarn increased summer pulses appeared, and it is possible to see clearly the effects of changing land and water use in the basin. The changes which occurred in 1953–5 and 1962 are clearly visible. The introduction of piped water supplies and sewers had a marked effect on the seasonal biomass cycles of the Tarn. The summer pulses of algal biomass became more pronounced as the P-loadings increased and the summer blooms became more dependent on external nutrient inputs. The effect of warm summers on these populations is also visible. The large summer blooms of recent years are associated

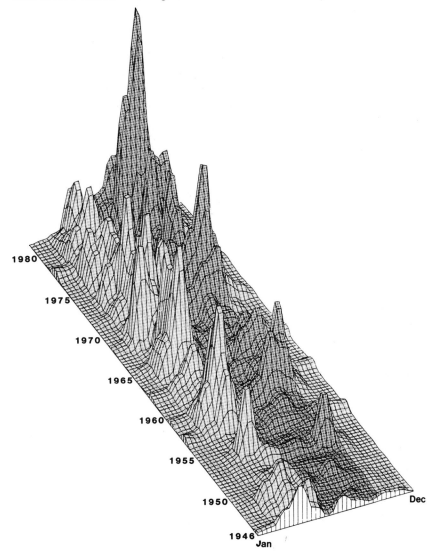

Fig. 12.7 The changes in total algal biomass observed in Blelham Tarn since 1946. I am indebted to John Lund for permission to reproduce this figure. Data from weekly counts. In 1951 a piped water system was installed and in 1962 a sewage system. Spring and summer peaks are shaded for clarity.

with a shift towards blue-greens which resulted from the altered N:P ratios in the nutrient loadings.

In the early years the spring *Asterionella* peak increased in size in response to increased nutrient loadings. The summer biomass pulse remained small until the nutrient loadings exceeded the capacity of the spring bloom to remove nutrients from surface waters. Once this point was passed the summer bloom appeared abruptly. Over time the summer peak broadened as the supply of nutrients in summer increased. The long-term sequence was one of initial biomass limitation by the TP pool followed, as the TP increased, first by light limitation, and then by N limitation. The changing fluxes of energy and nutrients determined the course of the year-to-year changes in the species succession. The blooms of blue-green algae in later years show increased total volume and a partial evasion of light limitation. Because of the reduced self-shading by large cells or colonies the growth habit of those species which grow as large clumps or flakes enables them to pack more biomass into the photic zone. Figure 12.7 beautifully illustrates the effects of climate and nutrient loadings on the seasonal cycles of species from year to year. The data from the English Lakes provides a unique record of these events.

(f) *Recruitment to the perch and pike populations in Windermere*

Perhaps the most striking evidence of climatologically induced events in Windermere comes from data on the recruitment of pike and perch. Recruitment into successive year classes has been estimated since the 1940s and the data have been published recently (Le Cren *et al.*, 1977; Craig *et al.*, 1979; Kipling and Frost, 1970; Kipling, 1983a). These data are an excellent example of Cushing's 'match–mismatch' hypothesis as applied to freshwater fish populations. Perch spawn in Windermere in May and the fish larvae feed planktonically for the first two weeks of life. Thorpe (1977a) reviewed the biology of perch and showed that there is a strong latitudinal effect on the timing of spawning. Perch spawn at temperatures between 8 °C and 11 °C and the spawning cue appears to be rising water temperatures; or as Thorpe (1977b) puts it 'during a period of accelerating temperature increase'. This ensures that spawning coincides with stratification and the spring outburst of 'new' production. Le Cren *et al.* (1977) and Craig *et al.* (1979) noted a strong effect of summer temperatures on perch recruitment and there is indeed an effect of temperature on larval growth and survivorship. With the exception of the 1955 year class in the Southern Basin of Windermere (1955 was exceptionally warm late in the summer) the pattern of good and bad years for perch matches the periodicity of warm Junes exceptionally well (Fig. 12.8).

There is also a remarkable correspondence between years of high perch

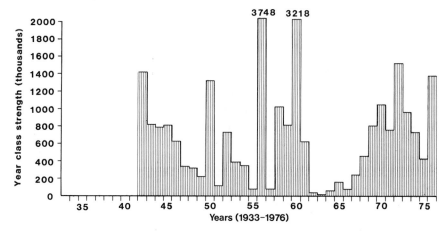

Fig. 12.8 Recruitment to the Windermere perch populations since 1942. Note the strong similarity between this figure and the temperature data in Fig. 12.3.

recruitment and the fluctuations in the zooplankton populations. Strong perch recruitment corresponded in time with poor zooplankton years. The mechanism appears to be a classic case of 'match–mismatch'. However, the good perch years appear to be correlated with the timing of the spring zooplankton outburst. Warm springs tend to be early springs and vice versa, so that even though the zooplankton biomass is reduced in warm years there is sufficient biomass early in the year to support good perch recruitment. Timing is all important. There is also good evidence for a climatolgoical

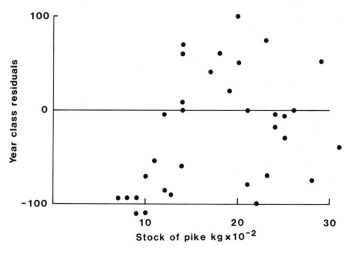

Fig. 12.9 After removal of a strong temperature effect the pike data in Windermere reveals a clear stock/recruitment curve which indicates the usual density dependence.

effect in that strong perch year classes occurred simultaneously in a number of the English Lakes (Le Cren *et al.*, 1977); notably in 1949 and 1955. These good year classes occurred in lakes with or without the presence of pike, the major predator. The good pike years also corresponded to good perch years (Kipling, 1983b; Craig, 1982). Figure 12.9 shows the changes in pike recruitment since the 1940s and illustrates the fact that the climatological changes in the English Lake District in the 1960s and 1970s resulted in fluctuations in all chemical and biological components in the water, from alkalinity (Sutcliffe *et al.*, 1982) to the predators.

Cushing (1982) extended the 'match–mismatch' hypothesis to study the effects of climate on commercial fisheries. He argued that the major influence on fish stocks was not density dependent interactions between adult stocks but was a climatological influence of larval survivorship during a brief period of planktonic feeding. This is not to say that density dependent interactions do not occur, it is simply that such interactions appear as residuals about a strong climatological trend.

The link between primary production and larval fish survivorship is clear when the feeding of larval fish is examined. Smyly (1952) and Guma'a (1978) examined the food and feeding habits of larval perch in Windermere and showed that the larvae changed their food as they grew. The gape height of the mouth determined what could be eaten. The first sources of food were algae, diatoms and rotifers changing to cyclopoid nauplii, *Bosmina*, *Daphnia* and copepods as the larvae grew. Smyly (1952) examined the food of larval perch in 1947, 1948 and 1949 (years of declining zooplankton abundance, Fig. 12.5) and showed a strong interaction between food availability and the material in the guts of the larvae. In 1947 there was a strong June–July pulse of *Bosmina*, *Daphnia* and *Diaptomus* followed by a second *Diaptomus* pulse in August. Only rarely did less than 80% of the larvae have zooplankton in their guts. In 1948 and 1949 the June–July pulse of *Daphnia* failed totally and the percentage of larvae with zooplankton in their guts fell to less than 20% in both years. Smyly (1952) recorded cannabalism during these periods. The zooplankton data indicates that in 1949 the zooplankton outburst was very late with no great biomass accumulation until the end of June. It is difficult to tell whether the poor zooplankton years are entirely a function of poor food supply or are partly due to a combination of poor food availability and strong predation by fish larvae. The possible effect of predation by larval fish has not been studied but it seems unlikely that, at the densities of natural fish populations, the predation can have much effect (Cushing, 1983). Clearly, more information is required to fully detail the links between climate, primary production, zooplankton and fish but the data from the English Lakes is highly suggestive and points to the importance of events in June at the time of the onset of summer stratification. In this respect the parallels between freshwater and marine systems are clear (Cushing, 1982).

12.1.2 THE LAURENTIAN GREAT LAKES

The Laurentian Great Lakes of Canada and the United States of America represent about one third of the world's freshwater and, by limnological standards, are all very large bodies of water. The surface areas are such that the horizontal mixing times of the basins are of the order of a month and the depth of the mixed layer in summer is of the order of 15 m. Maximum water depths are at least 200 m in most lakes. As the horizontal mixing times of the basins are much longer than the growth rates of the phytoplankton, significant inshore/offshore differences in abundance and species composition are found. Furthermore, physical processes in the coastal zones such as internal waves and thermal bars (Chapter 3) lead to the entrapment of nutrient loadings in the coastal zones. The fact that the basins are, when stratified, neither horizontally nor vertically homogeneous means that any sampling strategy to detect long-term changes must take the basin scale patchiness into account.

(a) *Chemistry*

As might be expected, long-term trends in such conservative parameters as TDS are the easiest to detect with confidence. Beeton (1965, 1969) reviewed the long-term changes in TDS, calcium, sodium and potassium, chloride and sulphate in the five major lakes and showed that the changes were related to basin geochemistry, geomorphology and human settlement patterns. Lake Superior, the coldest, most northerly lake which is embedded in the rocks of the Canadian Shield, has shown no changes since 1890, partly because there is little settlement in the basin. Lakes Huron and Michigan show some increases in TDS, sulphate and chloride since the turn of the century, while lakes Erie and Ontario show considerable increases in all parameters. These last two lakes are the most southerly of the five, lie in softer rocks surrounded by fertile soils and have the highest population densities in their drainage basins. Increases in these conservative substances come from a number of sources including run-off, industrial and urban waste disposal and the use of de-icing salt on the roads in winter. Examination of the data in Beeton's (1965, 1969) papers shows that there is little good data on water chemistry prior to 1935 and the number of early samples is very limited. This, coupled with a number of political considerations including an overriding concern for the reversal of the effects of eutrophication, has led to an unwarranted emphasis on recent events in these lakes. It is only possible to fully interpret trends in time when the recent changes in the lakes are seen in a longer perspective.

There are real problems with the interpretation of data if there are inconsistencies in the sampling and analytical methods used over a period of years. There has apparently been a long-term decline in the silica levels in Lake Michigan since 1925. This long-term decline in silica was described by Schelske and Stoermer (1971). Schelske and Stoermer linked the increased loadings of TP to the lake with the reported silica decline and explained the links in terms of eutrophication. They warned that the silica limitation of diatom growth might ultimately occur and that the primary production may become dominated by blue-greens. There are two problems with this argument however. Firstly, Parker and Edgington (1976) measured the concentration of diatom frustules in the sediments of Lake Michigan and showed that silica regeneration from diatoms was rapid; both within the water column and from the sediment surface. By attempting to calculate the various terms of the mass balance of silica in the Lake, Parker and Edgington showed (as might be expected) that the decrease in silica from late winter to late spring was quantitatively similar to the silica flux associated with the sedimentation of the spring diatom crop. Regeneration within the water column and sediments returned all but a small fraction of the silica to the water column at autumn turnover. The flux of silica to the permanent sediments equalled the inputs to the lake and, as far as could be ascertained, the whole lake was at steady state at a scale of years. Secondly, Shapiro and Swain (1983) looked critically at the time series of data and showed that the major silica decrease was a function of a change of laboratories and a change of analytical method. Shapiro and Swain suggested that in the absence of cross calibrations, of appropriate standards, and of inter-laboratory comparisons such a long-term data set may be rendered almost useless. Unfortunately the apparent silica decline in Lake Michigan has appeared in a number of textbooks, has figured largely in litigation and has been widely quoted as a consequence of eutrophication in the lake.

The changes in TP over time in the Great Lakes are more difficult to reconstruct because of the highly non-conservative nature of their spatial and temporal distributions. Few good early analyses are available and few stations were sampled. Chapra (1977, 1980) has used a mass balance approach to model the TP concentrations in Great Lakes waters and has shown that two major events influenced the changes in concentration over time. The first major event in the lower lakes (Erie and Ontario) was the clearing of the watersheds for agriculture around 1850. This led to a major change in water chemistry and phytoplankton populations (Harris and Vollenweider, 1982; Frederick, 1981). The second major event was the introduction of phosphate detergents which, superimposed on the input of urban wastes from rising populations, led to a second marked rise in TP after 1950. Changes in TP in the upper lakes are much less marked as population densities and land use practices have changed less in these basins in the last 100 years.

(b) *Changes in phytoplankton*

There are a number of sources of long-term phytoplankton data from the Great Lakes but they vary greatly in quality. Perhaps the most complete sets in terms of frequency are those taken from the water intakes of Chicago (Lake Michigan) and Cleveland (Lake Erie). The Chicago data consists of almost daily counts on water samples collected since 1927 and the Cleveland water intake data began in 1919. As suggested above these data which begin in the early years of this century contain no information on the precolonization period. Only palaeolimnological information can be used to reconstruct events from that period.

Lake Erie
Early (pre-1900) events in Lake Erie have been reconstructed by Harris and Vollenweider (1982) who showed that significant changes in phytoplankton populations occurred in the late 1800s. Frederick's (1981) analysis of another Lake Erie core showed that a big increase in diatom abundance occurred just above the horizon in which ragweed pollen first became abundant. The presence of ragweed is a good indicator of cleared ground. Harris and Vollenweider (1982) showed that oligotrophic conditions persisted in Lake Erie until about 1850.

By the turn of the century the transition to mesotrophy was well under way and this was nicely illustrated by the transition in *Melosira* species. *M. distans* and *M. italica* which are indicators of oligotrophic conditions were dominant before 1850. *M. italica* showed a transitory maximum around the turn of the century as *M. distans* became less common. At the same time *M. islandica* and *M. granulata* appeared. *M. granulata* is a good indicator of eutrophic conditions. Species of *Fragilaria* showed a similar replacement after the effects of phosphate detergents were felt. *F. crotonensis*, which was increasingly abundant from 1850–1950, was then replaced by *F. capucina*, the most abundant diatom in the surface layers of the core. Two other indicators of increased P loadings and increased TDS concentrations also appeared when the use of P detergents became common. *Coscinodiscus* appeared in 1950 and *Stephanodiscus binderanus* appeared around 1960. The agreement between the anecdotal information in the literature about the first dates of appearance of species in the lake, and the dates of first appearance of species in the core was remarkable (Harris and Vollenweider, 1982). The dates of first appearance of species could, in themselves, be used as markers for dating the core. Stoermer (1978) recorded the same sequence of species across the Great Lakes in waters of differing water quality.

Lake Erie showed a trend of rising phytoplankton abundance (Davis, 1964) consistent with the rising *TP* concentrations in the water and the replacement of species as the trophic status rose. The seasonal cycle of phytoplankton in the lake has also shown the effects of the increased *TP* loadings. In 1927 the record of phytoplankton abundance during the year at the Cleveland water intake showed spring and autumn peaks typical of oligotrophic conditions. The spring peak of *Asterionella* typically occurred early in April. As the P loading increased over the years, the spring and autumn pulses became larger (Fig. 12.10) until, by the end of 1960, the

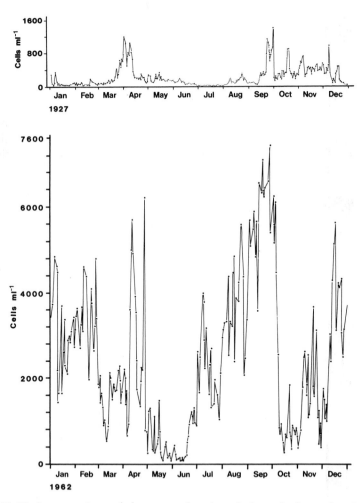

Fig. 12.10 A comparison of the seasonal cycles of phytoplankton abundance in Lake Erie in 1927 and 1962 (from Davis, 1964).

seasonal cycle exhibited typical eutrophic standing crops throughout the year. There was then only a brief clear water phase in May–June (Chapter 9). The changes in the seasonal cycles of phytoplankton abundance brought about by increased P loadings in Lake Erie were very similar to those noted in Blelham Tarn (Fig. 12.7) despite the much greater basin dimensions.

It is important to note that Fig. 12.10 illustrates not only an increase in the phytoplankton biomass and the eventual dominance of the summer crop over the spring crop (as in Fig. 12.7), but also a change in the timing of the spring peak. Up to 1946 the spring peak occurred in March and April. After that date the spring peak became earlier (January–February) and became dominated first by *Melosira* and finally by *Stephanodiscus* (Davis, 1964). This all but eliminated the mid winter minimum. It should be remembered that *S. binderanus* became common in 1960. In this case an increase in P loadings altered the timing of the spring peak of production drastically and the change was abrupt. This may have dramatic effects on other food chain components. *Stephanodiscus* appears to be capable of rapid growth under low light conditions if given sufficient nutrients (Heaney and Sommer, 1984). The earlier spring peak in phytoplankton biomass will result in a more rapid decline in silica and other nutrients and will also alter the timing of the transition to *r*-strategists and summer populations. The interaction of physical and chemical factors during this critical phase is very important.

Lake Michigan

The equally long series of phytoplankton records from the Chicago water intake have been summarized by Damann (1945, 1960) and by Makarewicz and Baybutt (1981). There are problems with this data set as taxonomic inconsistencies render interpretation rather difficult. It should be remembered that the changes in the P-loadings to Lake Michigan have been rather less than the changes in Lake Erie as the effluent from Chicago was diverted south, down the Mississippi river system many years ago. TP concentrations off Chicago have changed little in the last twenty years but there has perhaps been some decrease since 1975 as the effects of controls on P-loadings have taken effect. The long-term changes in the net phytoplankton biomass off Chicago (Fig. 12.11) show large cyclic oscillations. There is a fairly regular 15 year fluctuation in all species if the three dominant diatoms are left out. Superimposed on this fluctuation is a high amplitude fluctuation in *Tabellaria*, *Stephanodiscus tenuis* and *S. binderanus* with a similar periodicity.

Makarewicz *et al.* (1979) showed that *Stephanodiscus* species dominated the spring (March–April) flora of Lake Michigan in 1975 but were rare in 1937. In 1937 the dominant spring diatoms were *Fragilaria* and *Asterionella* with the *Asterionella* reaching peak abundance in May and June. Thus increases in the nutrient loading have, as in Lake Erie, stimulated the growth of *Stephanodiscus* in the coastal zone and have brought the spring blooms forward from May–June to February–March. The change in the seasonal cycle occurred between the early and late 1950s; precisely the time when the large outburst of *Tabellaria* and *Stephanodiscus* occurred (Fig. 12.11).

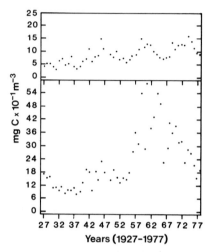

Fig. 12.11 Changes in the abundance of phytoplankton off Chicago from 1927 to 1977. Lower panel: total net phytoplankton biomass, upper panel: total biomass minus *Tabellaria* and *Stephanodiscus*. (From Makarewicz and Baybutt, 1981.)

The problem with these long term phytoplankton data is the fact that they represent only coastal zone populations. There is a marked thermal bar in spring in Lake Michigan (Fig. 3.5) which traps both nutrients and phytoplankton close inshore. Even in summer the internal waves which are propagated in the coastal zone serve to concentrate the longshore velocity components and trap nutrients and phytoplankton within a few kilometres of the coast for days at a time. Inshore and offshore phytoplankton populations differ markedly in the Great Lakes (Nalewajko, 1967; Holland and Beeton, 1972). Water intake data is therefore not representative of the lake as a whole and the phytoplankton records will be heavily influenced by the physical processes in the coastal zone (Harris, 1983). Data from large lakes is therefore not as easy to interpret as that from smaller bodies of water. As in the English Lakes, there is a strong effect of both climate and nutrient loadings on the long-term records from the Great Lakes. It is therefore dangerous to assume that the decrease in inshore populations in Lake Michigan since 1960 is solely due to a reduction in nutrient loadings (Danforth and Ginsberg, 1980). There are a number of components that are due to the long-term fluctuations of the abundance of Great Lakes phytoplankton and even the early nineteenth century observers noted great year-to-year fluctuations (Harris and Vollenweider, 1982). These fluctuations were, even then, sufficient to make the interpretation of long-term trends very difficult.

(c) *Anoxia in Lake Erie*

The interpretation of trends in time when the signal to noise ratios is high is not just an academic question. Almost every recent trend in Great Lakes water quality has been interpreted in terms of eutrophication and the reversal thereof. There are good political and economic reasons for this bias but the result does a disservice to science. The decline in silica in Lake Michigan was explained in terms of eutrophication and the stimulation of diatom growth (Schelske and Stoermer, 1977) but the evidence does not support this contention. Clearly there has been a massive increase in loadings of nutrients and other substances to the Great Lakes since the end of the last century. The clearest trends are in conservative substances. The trends in non-conservative substances are more complex and Lake Erie provides a good example of the problems of interpretation.

The TP loadings to Lake Erie increased sharply in the 1950s and by 1960 severe water quality degradation was observed inshore, in the central basin (particularly near Cleveland), and in the shallow western basin. The Detroit River was, and is, a major source of nutrient loadings to the lake. In the 1960s the disturbance of the seasonal production cycle (Fig. 12.10) led to large amounts of algal material sedimenting out in the central basin. This 'displaced metabolism' led to severe anoxia in bottom waters as, by a quirk of fate, the central basin of the lake not only has very high nutrient loadings but also has a very shallow hypolimnion. The depth of the hypolimnion may be as little as one or two metres depending on meteorological factors, and the hypolimnetic stocks of oxygen are consequently small.

An examination of the apparent relationships between TP loadings and hypolimnial anoxia by Dobson and Gilbertson (1971) revealed a (remarkably) linear relationship between the two. The functional link between the loadings and hypolimnial anoxia was, as explained above, the increase in algal biomass brought about by the increased TP loadings and the subsequent sedimentation and decomposition of that biomass. The excellent linear relationship between loadings and anoxia was subsequently used by those concerned with the management of water quality to determine a critical loading of TP which would achieve a stated minimum level of dissolved oxygen by the end of the summer stratification. Considerable sums of money have been spent in controlling the P loads to the Lake to achieve this end. This management goal was bolstered by the type of empirical model discussed in the previous chapter.

A recent reappraisal of these relationships by Charlton (1980b) has exposed a number of problems and reopened the debate about the fate of the Lake. In retrospect it is surprising that the Dobson and Gilbertson (1971) relationship was so good. The *TP* loadings do not enter the central basin directly but flow into the coastal zone where much biological activity is centred. Therefore the relationship between coastal zone loadings and anoxia offshore is complicated by residence times, algal growth rates and transport processes. Vollenweider's models (Chapter 11) assume complete horizontal mixing and are therefore not truly applicable to a system with large-scale, offshore transport at scales slower than the algal growth rates. Furthermore these models assume reasonable (and constant) sedimentation rates. It is clear that in the Great Lakes system there is long-term variability in sedimentation rates which, through an interaction between physical and biological processes, can greatly influence *TP* concentrations in the water over time periods much shorter than the water residence times in the basins (Fraser, 1980).

Dobson and Gilbertson's (1971) trend in hypolimnial anoxia since 1929 was obtained through an analysis of oxygen data from profiles taken in the central basin of the Lake. Two immediate problems are apparent. First there was little early sampling so any trend in time is heavily dependent on a few, early data points. Secondly, the stations sampled differed from year to year as did the areas of hypolimnetic anoxia. 'Representative' stations had to be chosen to arrive at an oxygen depletion rate for each particular year. This problem of altering the positions of sampling stations has bedevilled much of the monitoring and surveillance work in the Great Lakes over the years. The spatial and temporal fluctuations in all non-conservative parameters require that inshore and offshore waters be defined and stations blocked up for trend analysis (Kwiatkowski, 1982). As the trends differ markedly from station to station then it is possible to produce quite different trends depending on the choice of blocks and stations. Offshore areas are, of course, more stable over time than inshore areas and less prone to high frequency 'noise' components. Unfortunately it is the inshore areas which receive the loadings, which are physically perturbed and which produce most of the perceived water quality problems.

Charlton's (1980b) reappraisal of the Lake Erie anoxia problem used a more general relationship between algal biomass, hypolimnial temperature and hypolimnial thickness (Charlton, 1980a). When examined this way (and when the historical data was recalculated) not only did Dobson and Gilbertson's (1971) trend become much weaker (Fig. 12.12) but it became clear that physical processes had a massive effect on the anoxia on any given year. The data from 1929, upon which much depends, was obtained in a year of high outflows and above average rainfall. This led to higher lake levels; and a rise of only 1 m was very significant when the average hypolimnion depth was only 4 m. The early estimates of the oxygen depletion rate were therefore biased by high water levels, thicker hypolimnia and

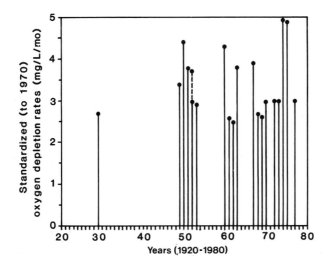

Fig. 12.12 The standardized oxygen depletion rate in the hypolimnion of Lake Erie according to Charlton (1980b). The clear trend discerned by earlier workers is no longer apparent.

were consequently low. When corrected for year-to-year changes in hypolimnion thickness and temperature the trend of anoxia with time almost vanished. Charlton (1980a) calculated that a 50% reduction in algal biomass brought about by a reduction in P loadings would still result in anoxia in certain years: a conclusion in accord with the historical evidence brought together by Harris and Vollenweider (1982). Once again biological processes are strongly influenced by year to year differences in climate.

Lam *et al.* (1983) have shown that it is possible to model the processes of hypolimnetic anoxia in Lake Erie by applying a segmented box model to the problem. The basic model is a one dimensional physical model of the Lake which can predict epilimnion and hypolimnion temperatures and depths. Using reasonable estimates of vertical entrainment and mixing between the layers, and horizontal advection and diffusion, it is possible to achieve good agreement between the observed and predicted distributions of temperature and phosphorus in the Lake. All that was required then was to link the biological processes of production and decomposition to the basic physico–chemical model. Production of algal biomass was modelled by the simple empirical relationship between chlorophyll and DIP of the type discussed in Chapter 11. Simple empirical relationships between DIP and TP and TP and oxygen consumption rates were also employed. The fit between the observed and predicted oxygen distributions was remarkable (Fig. 12.13) and the model led directly to some significant conclusions about the relationships between P-loadings and anoxia in the central basin.

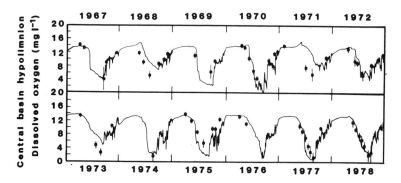

Fig. 12.13 The excellent fit between Lam *et al*'s (1983) model and the observed oxygen concentration in the hypolimnion of Lake Erie.

Over the 12 year period of the model there was a general reduction in *TP* concentrations which resulted from the reduction in *TP* loadings to the Lake since 1972. The general reduction was, however, strongly modified by occasional loading pulses, resuspension from the bottom and regeneration from anoxic sediments. The effect of *TP* removal on the rate of oxygen depletion in the hypolimnion was masked by the large variations in oxygen concentration brought about by physical processes such as variations in hypolimnial thickness and temperature, interbasin transport and vertical diffusion. Even when these physical processes were accounted for the effects of *TP* reduction were further masked by the inaccuracies of the loading estimates themselves and by the long-term integrative effect of the sediment oxygen demand. Variations in sediment oxygen demand extend over periods (> 50 years) much longer than the year-to-year variations in climate and hypolimnial thickness. In short, the results of the model eloquently support the assertion of Charlton (1980a) that even a 50% reduction in *TP* loadings to Lake Erie would result in occasional anoxia in years when a warm, thin hypolimnion developed. This is also in accord with the review of the long-term state of the Lake by Harris and Vollenweider (1982). They showed that the most radical change in the trophic state of the Lake occurred in the middle of the last century when the watersheds were cleared and that, in years with thin hypolimnia, anoxia could have occurred before that.

The results of the massive clean up campaign in Lake Erie are therefore by no means obvious. The signal to noise ratio is low (Kwiatkowski, 1982) and it is difficult to detect any long-term trends in data which is strongly seasonal. The year-to-year changes in the physical structure of the Lake preclude any simple solution to the anoxia problem. The greatest changes in water quality are clearly to be expected in inshore areas when the balance between inputs, algal growth and offshore transport are changed. Lake Erie has undergone long-term changes in lake level (Charlton, 1980b) and temperature (Beeton, 1961, see Fig. 12.14) in the last century and these have had a major effect on the growth of phytoplankton and the history of anoxia.

(d) Changes in the food chain in Lake Erie

The long-term prognosis for the state of the Lake is further complicated by the massive changes in the food chain within the Lake. Overfishing, degradation of spawning grounds, eutrophication and the introduction of exotic species have totally altered the fish populations in all the Great Lakes (Beeton, 1969; Regier and Hartmann, 1973). These changes have been most noticeable in the lower lakes and an historical survey by Whillans (1977) showed clearly that major changes in the fish populations in Lakes Ontario and Erie had taken place before the turn of the century. Again there are questions about the effects of such changes on the rest of the food chain. Are there any 'top down' effects or are the changes in the rest of the food chain components due to climate, lake physics and loadings? Wells (1970) argued that the introduction of the planktivorous Alewife (*Alosa pseudoharengus*) to Lake Michigan had reduced the populations of the larger zooplankton in much the same fashion as suggested by Brooks and Dodson (1965). In the late 1950s the Lake Herring (*Coregonus* sp.) populations were replaced by the Alewife and this, it was suggested, increased the grazing pressure on the larger zooplankton species.

Whether or not the apparent change in zooplankton populations was due to the effects of predation is not clear as Gannon (1972) put forward an alternative explanation. Changes in the size structure of zooplankton populations can be brought about by changes in the growth of suitable prey species of phytoplankton. There is little or no historical data for nanoplankton populations in the Great Lakes; the sampling and counting techniques used to collect the long-term data sets did not allow the smaller phytoplankton to be counted (Makarewicz and Baybutt, 1981). In Lake Michigan it appears that the major peaks and troughs in the phytoplankton record (Fig. 12.11) correspond in a rough way with the good and bad years for planktivorous fish (Brooks *et al.*, 1984). The correspondence between the fish and phytoplankton records is most likely due to a form of Cushing's 'match–mismatch' hypothesis than to a direct 'top down' control of the phytoplankton by the fish. There are a number of reasons for this. Firstly, the phytoplankton records are dominated by large, chain-forming diatoms that occur early in the year and are not grazed significantly because of their large size. There are only small populations of zooplankton early in the year. Secondly, zooplankton predation in Lake Michigan was less than in smaller lakes (Wells, 1970) as the prey are more likely to evade predation in a larger lake. Thirdly, an examination of the fish catch records for all the

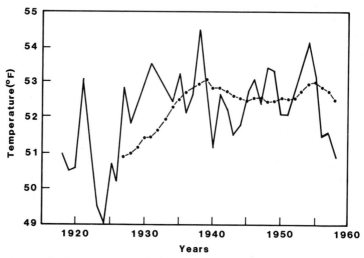

Fig. 12.14 The long term record of temperature in Lake Erie (from Beeton, 1961). The dotted line is a 10y moving average.

Great Lakes (Beeton, 1969) shows that good year classes occurred simultaneously in all the Lakes. This is a good indication of a strong climatological effect as the trophic conditions of the various Lakes varied greatly. In short it appears that, just as in the English Lakes, the phytoplankton populations in the Great Lakes have undergone long-term changes due to alterations in the nutrient loadings and long-term climatological fluctuations. These environmental changes may be seen in all the components of the food chain. The sequence of events in spring which control the size and timing of the spring bloom are once again seen to be important.

12.1.3 THE NORTH SEA AND THE ENGLISH CHANNEL

The mechanisms which operate in lakes are equally important in the oceans. Cushing (1975, 1982) has elegantly demonstrated the links between climate, nutrients, primary production and fish production in the oceans and was the originator of the 'match–mismatch' hypothesis (Cushing, 1962, 1966, 1969). The North Sea is one of the most intensively studied areas of the world's oceans. Physical, chemical and biological measurements have been made in North Sea waters for many decades but, as with the English Lakes, the most complete data sets begin after the war. The long term North Sea data comes from the Continuous Plankton Recorder (CPR) surveys. In Chapter 9 I showed how such surveys were used to examine the seasonal patterns of phytoplankton and copepod abundance. The long-term, year-to-year comparisons will be discussed in this chapter.

The year-to-year variations in phytoplankton biomass in the oceans are different from those in lakes for two reasons which are linked to the much increased horizontal scales of the oceanic basins. Firstly, large scale horizontal redistribution of species can occur through the influence of ocean currents. This can produce apparent long-term fluctuations in abundance due to long-term changes in advective components (Cushing, 1982). Secondly, ocean basins are less dependent on external nutrient loadings and more dependent on internal recycling (Chapters 4 and 7) and, therefore, any long-term changes are more likely to be due to the effects of climate than to the effects of altered nutrient loadings. Two components of variability are visible in the long-term North Sea data. There has been a long-term increase in colour in some areas as well as a noticeable decline in diatom abundance since 1948. Combined with this there has been an increase in dinoflagellates in recent years, particularly since 1970. These long-term changes in phytoplankton have been noted by a number of authors (Reid, 1975, 1977; Colebrook *et al.*, 1978; Dickson and Reid, 1983; Robinson, 1983). Some have concentrated on the decline in average abundance of individual major species (Robinson, 1983), but the overall pattern is clear from the seasonal plots. If anything the changing patterns of seasonal abundance are more revealing than the changes in mean annual abundance.

Superimposed upon this long-term change in phytoplankton composition is a quasi cyclic pattern very similar to the periodicity noted in the English Lakes data (Figs 12.3 and 12.5). Indeed the similarity between the two data sets is quite striking in some respects, and strongly suggests a common source of variability. Climatological variability in the area of the North Sea and the English Channel has been analysed extensively in recent years. Events in the English Channel in particular have indicated strong coupling between physical and biological processes. Long-term changes in the biota off Plymouth have been recorded since the 1920s and the cyclic changes that have occurred have been termed the Russell cycle after one of the principal investigators. (Strictly speaking the term cycle may be something of a misnomer as the end point of the long-term fluctuation is not the same as the starting point in this case.)

(a) *Changes in the climate*

Cushing (1966), Russell *et al.* (1971) and Southward (1974, 1980) have reviewed the changes in the English Channel and have demonstrated that the observed changes in winter phosphorus, zooplankton and young fish may be linked to long- and short-term fluctuations in water temperatures. Since the 1920s there has been a period of warming very similar to that observed in Lake Erie (Fig. 12.14). This long-term trend in sea surface temperatures peaked around 1940–50 and has subsequently shown a decline (Fig. 12.15). The biological changes off Plymouth are linked to the changes in the relative influence of southerly and northerly species with the balance shifting from cool water, northerly species to warm water, southerly species and back again. This raises another problem which is not encountered in lakes. Over periods of 10–20 years, the geographical distribution of species in the oceans may fluctuate in response to long-term changes in climate. Cushing (1966) and Southward (1980) reviewed these changes and showed that they extended to rocky shore invertebrates as well as plankton. During the warm years from 1930–70 there was a northerly extension of boreal species into the Arctic and of warm temperate species into boreal regions. The south coast of England is very close to the southern limit of a number of northern species and thus such changes are easily seen. The switch from herring or mackerel to pilchards and back was brought about in this way. Southward (1980) saw the origins of the Russell cycle in these climatically induced shifts in distribution together with changes in the biotic interactions within the food chain (for example the change from herring in the 1920s to mackerel in the 1970s). Southward (1974) noted that there was historical evidence for long-term fluctuations in the herring and pilchard fisheries off Cornwall. Fluctuations, of undetermined periodicity, have occurred for at least 400 years.

The climatological basis of these fluctuations and of the ten-year cycle seen in both the North Sea plankton and in the English Lakes has been thoroughly investigated (Southward et al., 1975; Dickson et al., 1975; Colebrook and Taylor, 1979). Since 1940 there has been a strong correlation between sea surface temperature and sun spot activity (Southward et al., 1975). The close link between solar activity and sea surface temperatures led Southward et al. to suggest that these changes were a direct result of local air/sea interactions rather than an advective pattern depending on large-scale ocean currents such as the Gulf Stream. This conclusion was borne out by Dickson et al. (1975) and Colebrook and Taylor (1979) who showed that there were systematic changes in the strength of the westerly winds over the British Isles brought about by pressure differences over the North Atlantic and Europe. Colebrook and Taylor (1979) showed that, while sea surface temperatures over the North Atlantic were greatly influenced by the Gulf Stream, those in the North Sea and the Channel were of local origin. Throughout the year northerly winds caused cooling of North Sea waters while southerly winds caused warming.

Thus the clear ten-year cycle in June water temperatures (Fig. 12.3) in the English Lakes and the fluctuations in sea surface temperature off Plymouth (Fig. 12.15) have a similar origin. Years with reduced westerly and northerly air flows, warmer summers and more southerly winds produce higher surface water temperatures. Dickson and Reid (1983) noted that the

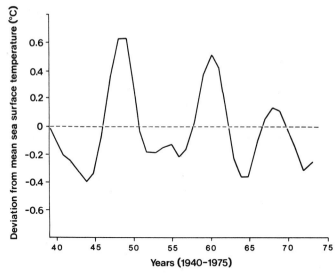

Fig. 12.15 The time series of water temperatures from Marsden Square 145D (45–50°N 10–50°W) during the period 1940 to 1973. Data filtered to display the long-term trends (after Colebrook and Taylor, 1979).

greatest reduction in northerly and westerly air flows occurred in the first quarter of the year, so that the warm periods of higher sea and lake surface temperatures have their origins early in each year. The biological effects of such warm periods are also most significant early in the year.

(b) *Changes in the plankton*

This insight into the climatological and physical mechanisms in the North Sea may now be used to interpret the long-term CPR records. During the period from 1948 to the late 1970s there was a marked decline in diatom abundance in all areas (Robinson, 1983) accompanied, most noticeably, by a reduction in the length of the diatom growth period. The warm periods around 1959–60 and 1968–9 led to a reduction in diatom biomass late in the summer. This is precisely what would be expected if the warm years led to reduced vertical turbulence and early sedimentation of the diatom crop. Cooler years with more westerly winds led to late stratification, a larger spring diatom pulse and the resuspension and subsequent regrowth of diatoms later in the year. The decline in the annual average abundance of diatoms may be as much the result of a shortening of the season as an absolute reduction in numbers.

There is also the expected interaction between water depth, the critical depth and nutrient status. As was seen in the discussion on seasonal cycles

(Chapter 9), the deep waters off the continental shelf produce smaller, later phytoplankton crops than inshore areas. This leads to the situation in oceanic areas such as B1 (Fig. 9.1) where the trend in time is reversed as compared to shelf waters and light limitation in cool windy years leads to reduced diatom growth. There is some indication that the diatom peaks in areas such as B1, B2, C1, C2 and D2 have occurred later with each passing year. From 1958–80 the peak diatom abundance has apparently shifted from March–April to April–May. The ten-year cycle of warm years is superimposed on a longer term cooling trend which has tended to delay the spring pulse.

The development of phytoplankton 'colour' (presumed to reflect the abundance of small r-strategist flagellates) also depends on hydrographic conditions. In areas outside the North Sea (B1, B2, D3, D4) the warm years led to increased colour as small species replaced the diatoms. In such areas the seasonal cycle of phytoplankton is one of low overall biomass and brief duration (Chapter 9) with another small late summer or autumn pulse. In the shallower, more nutrient rich waters of the North Sea the colour pattern is more similar to the diatom and dinoflagellate patterns as the biomass is higher and the growth period is longer. In the more eutrophic North Sea waters, the larger species are a more important component of the total phytoplankton biomass. Thus in these areas (C1, C2, D1, D2) the colour patterns (Fig. 12.16) tend to show minima in spring in warm years rather than maxima. In the plots of areas C1 and C2 the distinction between spring diatoms and late summer dinoflagellates is clear whereas in D1 and D2 the patterns are more complicated.

A long-term trend is apparent in the plots of phytoplankton colour from the North Sea in that, from about 1965 onwards, there was a tendency for the peak in colour to occur later in the year (D1, D2) and for the spring and late summer peaks to merge (C2). This was coincident with the decline in diatoms in these areas. The shift of the peak of colour was apparently due to both a relative reduction in the diatom (W) peak and a relative rise in the importance of the r species. This is analogous to the shift in the spring pulse in the Great Lakes where the peak became earlier as the relative proportions of *Stephanodiscus* and *Asterionella* changed; only in this case the shift was the result of climatological events rather than nutrient loadings. Glover *et al.* (1972, 1974) showed that there was a trend towards a later spring pulse from 1948 to the 1970s and this was used by Cushing (1982) to explain a number of events concerning changes in the abundance of zooplankton and fish.

According to Southward *et al.* (1975) and Southward (1980) the peak of the long-term temperature cycle off Plymouth occurred around 1945 so that throughout the period for which CPR records are available the cyclic

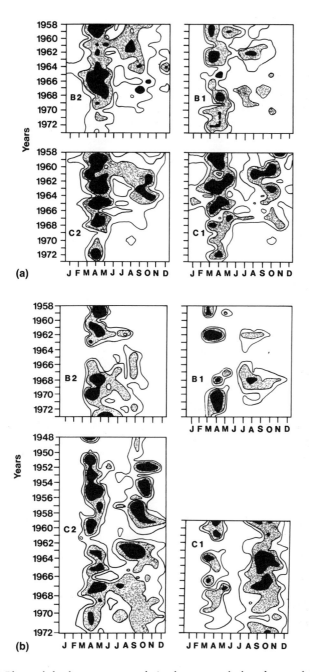

Fig. 12.16 Plots of the long term trends in the seasonal abundance of (a) diatoms and (b) 'colour' as revealed by the CPR network. CPR areas as in Fig. 9.1.

ten-year temperature fluctuations were superimposed on a downward trend. Even though there were peaks in surface temperatures at the end of each decade, the 1969 peak was lower than those of the 1930s and 1940s. This may account for the trend towards later phytoplankton pulses and a tendency for the colour peak to extend later in the year. A cooling trend would be expected to produce later stratification and weaker stratification thus delaying the growth of the r-strategist species. Algal production would also be stimulated by the entrainment of nutrients across a weaker thermocline in summer. In addition, Glover *et al.* (1972) documented a trend towards a reduction in total solar input between 1949 and 1969 which would also have the effect of delaying the early summer phytoplankton peak.

The dinoflagellates (Fig. 12.17) show some complex patterns of distribution

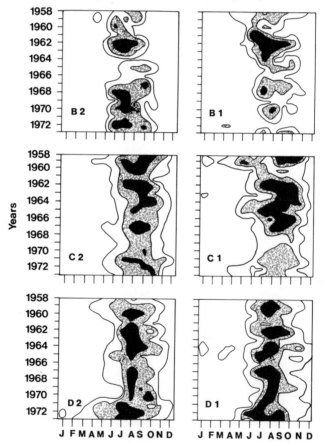

Fig. 12.17 Plots of the long-term trends in the seasonal cycles of dinoflagellates as revealed by the CPR network. CPR areas as in Fig. 9.1.

in space and time but tend to occur (as expected) in late summer. There is good evidence that some species respond to stable water columns in warmer years. Overall, however, the average yearly abundance of dinoflagellates has not declined in recent years (Robinson, 1983). This may well be because the greatest temperature anomaly occurred early in the year (Dickson and Reid, 1983) before the growth of these species. Dickson and Reid (1983) showed that a marked increase in the abundance of three species of *Ceratium* since 1969 was coupled with an abrupt increase in the southerly wind component and, hence, a less strongly mixed water column in late summer. Overall the complex patterns of phytoplankton growth since 1948 may be interpreted in terms of the same model as that developed in lakes (Fig. 9.8). Late winter and spring diatom pulses (*W*) give way to early summer growths of smaller species (*r*) which are, in turn, followed by the slower growing (*K*) species. As in lakes, the relative importance of the three groups in the seasonal cycle depends on the interplay of climate, hydrography and nutrients.

(c) *Changes in the rest of the food chain*

Cushing (1982) has reviewed the simultaneous changes in the other food chain components of the North Sea at some length. Briefly, from 1948–75 there were marked decreases in total zooplankton biomass, total copepods and the length of the zooplankton growing season (Glover *et al.*, 1972, 1974; Colebrook, 1978a,b; Bainbridge *et al.*, 1978). Not all zooplankton declined in abundance; *Calanus* sp. and *Temora* sp. changed little, *Pleuromamma* sp., *Euchaeta* sp. and *Acartia* sp. increased, while *Pseudocalanus* sp. and *Paracalanus* sp. declined dramatically. Colebrook (1978a,b) showed that the decline in total zooplankton abundance contained both 8–10 year and 3 year cyclic components which were associated with the temperature cycles and westerly wind indices discussed above. Cushing (1982) explained the fluctuations in the numbers of *Calanus* from year to year, in terms of advection of Norwegian water into the North Sea. *Calanus finmarchius* is a northern species and the changes in *Calanus* abundance reflect the changing influence of northern waters. Changes in the abundance of the smaller species (*Pseudo-* and *Paracalanus*) appeared to be dependent on two factors; an inverse relationship with each year's westerly wind component and a correlation with large scale water mass advection two years before. If *Pesudo-* and *Paracalanus* are regarded as southern species then the decline in abundance becomes clear. The overall cooling trend since 1945 reflected a more northerly influence but peaks in abundance still occurred in warm years (around 1950, 1960 and 1970, Fig. 12.18). The populations of *Temora* sp. have changed little since 1948 (Glover *et al.*, 1972) but the peak in abundance has become significantly later with the passing years.

Cushing (1982) has reviewed the evidence for the effects of climate on

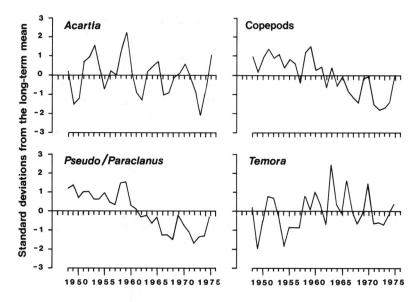

Fig. 12.18 Long-term changes in the abundance of the major groups of zooplankton in CPR area D1. (From Glover *et al.*, 1972, 1974; Colebrook, 1978a,b).

fisheries in the North Sea at some length. Using the 'match–mismatch' hypothesis he examined the effects of the climatological and biological changes since the 1940s on the commercial fish stocks. More recently Cushing (1984) has examined the statistical relationships between climatological and biological events and the recruitment to the North Sea haddock and cod stocks since 1933. There was a dramatic increase in the stocks of gadoid fish after 1962; the 1962 year class of haddock was 25 times larger than the average of 1918–61 and after that date the catches of prime gadoids rose by a factor of 3.15. Cushing (1984) showed that the recruitment to the cod and haddock stocks was related to the changing phase relationships between spawning and the peak of *Calanus* production. Using CPR data Cushing showed that the average monthly peak of *Calanus* production was delayed from April–May before 1961 to May–June after that year. A detailed year by year correlation of cod and haddock recruitment showed statistically significant relationships between cod recruitment and the delay in *Calanus* production. There was also a significant negative correlation between cod recruitment and sea surface temperature in March. Haddock recruitment peaked when the *Calanus* production peaked in May.

These correlations provide the first statistically reliable evidence for the 'match–mismatch' hypothesis and bolster the case for considerable 'bottom up' control in the food chain. As with the perch and pike in Windermere there was evidence for an interaction between cod and haddock as the cod is the top predator in the North Sea but otherwise the predominant mechanism appeared to be physical. Koslow (1983) attempted to address the 'bottom up' or 'top down' question by modelling the interactions between food chain components in the North Sea. He found that the food chain model of the North Sea system could not explain the shifts between species that had been observed and concluded that the zooplankton were food-limited rather than predator-limited. According to the model, fish predation could only limit the biomass of large herbivores at unrealistically high levels of fish biomass. Accordingly 'bottom up' control was the favoured mechanism.

The CPR data therefore documents the changing physical and biological conditions in the North Sea remarkably well. The climatological parameters which produce year-to-year changes in the planktonic ecosystems of lakes and the oceans are essentially the same and the biological responses are the same in both systems also. The oceanographic explanations of long-term trends are more complex than limnological explanations because of the large scale biogeographic and advective changes which occur. Fundamentally, however, the biology of the oceans and of lakes is very similar and both systems indicate that 'bottom up' control of the food chain is very important. The sequence of events leading to spring stratification in temperate waters is a key determinant of subsequent biological activity at all levels in the food chain.

References

Abele, L.G. (1976) Comparative species richness in fluctuating and constant environments: coral associated decapod crustaceans, *Science*, **192**, 461–3.

Abrams, P. (1976) Limiting similarity and the form of the competition coefficient, *Theor. Pop. Biol.*, **8**, 356–75.

Allan, J.D. and Goulden, C.E. (1980) Some aspects of reproductive variation among freshwater zooplankton, in *Evolution and Ecology of Zooplankton Communities* (ed. W. C. Kerfoot), University Press of New England, Hanover, New Hampshire, pp. 388–410.

Allen, T.F.H. (1977) Scale in microscopic algal ecology, *Phycologia*, **16**, 253–7.

Allen, H.E. and Kramer, J.R. (1972) *Nutrients in Natural Waters*, Wiley–Interscience, New York.

Allen, T.F.H. and Starr, T.B. (1982) *Hierarchy: Perspectives for Ecological Complexity*, University of Chicago Press, Chicago.

American Chemical Society (1971) Non equilibrium systems in natural water chemistry, *Advances in Chemistry*, Ser 106, American Chemical Society, Washington.

Amezaga E.De., Goldman, C.R. and Stull, E.A. (1973) Primary productivity and rate of change of biomass of various species of phytoplankton in Castle Lake California, *Verh. Int. Verein. Limnol.*, **18**, 1768–75.

Andersson, G., Berggren, H., Cronberg, G. and Gelin, C. (1978) Effects of planktivorous and benthivorous fish on organisms and water chemistry in eutrophic Lakes, *Hydrobiologia*, **59**, 9–15.

Andrews, J.C. and Scully-Power, P. (1976) The structure of an East Australian Current anticyclonic eddy, *J. Phys., Oceanogr.*, **6**, 756–65.

Antia, N.J., McAllister, C.D., Parsons, T.R. *et al.* (1963) Further measurements of primary production using a large-volume plastic sphere, *Limnol. Oceanogr.*, **8**, 166–83.

Armesto, J.J. and Contreras, L.C. (1981) Saxicolous lichen communities: non-equilibrium systems? *Am. Nat.*, **118**, 597–604.

Atkins, W.R.G. (1924) On the vertical mixing of sea water and its importance for the algal plankton, *J. Mar. Biol. Assn UK*, **13**, 319–24.

Atkins, W.R.G. (1930) Seasonal variations in the phosphate and silicate content of sea water in relation to the phytoplankton crop. V. November 1927 to April 1929, compared to earlier years from 1923, *J. Mar. Biol. Assn UK*, **16**, 821–52.

Auclair, A.N. and Goff, F.G. (1971) Diversity relations of upland forests in the Western Great Lakes area, *Am. Nat.*, **105**, 499–508.

Azam, F., Fenchel, T., Field, J.G., *et al.* (1983) The ecological role of water column microbes in the sea, *Mar. Ecol. Prog. Ser.*, **10**, 257–63.

Bainbridge, V., Forsyth, D.C.T. and Canning, D.W. (1978) The plankton in the northwestern North Sea 1948 to 1974, *Rapp. P-v. Réun. Cons. Int. Explor. Mer*, **172**, 297–404.

Bannister, T.T. (1974a) Production equations in terms of chlorophyll concentration, quantum yield, and upper limit to production, *Limnol. Oceanogr.*, **19**, 1–12.

Bannister, T.T. (1974b) A general theory to steady state phytoplankton growth in a nutrient saturated mixed layer, *Limnol. Oceanogr.*, **19**, 13–30.

Bannister, T.T. (1979) Quantitative description of steady-state, nutrient saturated algal growth, including adaptation, *Limnol. Oceanogr.*, **24**, 76–96.

Bannister, T.T. and Laws, E.A. (1980) Modelling phytoplankton carbon metabolism, in *Primary Productivity in the Sea* (ed. P. Falkowski), Plenum, New York, pp. 241–58.

Banse, K. (1974) The nitrogen to phosphorus ratio in the photic zone of the sea and the elemental composition of the plankton, *Deep Sea Res.*, **21**, 767–71.

Banse, K. (1976) Rates of growth, respiration and photosynthesis of unicellular algae as related to cell size – a review, *J. Phycol.*, **12**, 135–40.

Barber, R.T., Zuta, S., Kogelschatz, J. and Chavez, F. (1983) Temperature and nutrient conditions in the E. equatorial Pacific, October 1982, *Trop. Oc. Atmos. News.* **16**, 15–17.

Barnett, T.P. (1977) An attempt to verify some theories of El Nino, *J. Phys. Oceanogr.*, **7**, 633–47.

Beadle, L.C. (1981) *The Inland Waters of Tropical Africa*, 2nd edn, Longmans, London.

Beeton, A.M. (1961) Environmental changes in Lake Erie, *Trans. Am. Fish. Soc.*, **90**, 153–9.

Beeton, A.M. (1965) Eutrophication of the St. Lawrence Great Lakes, *Limnol. Oceanogr.*, **10**, 240–54.

Beeton, A.M. (1969) Changes in the environment and biota of the Great Lakes, in *Eutrophication* (ed. G. A. Rohlich), National Academy of Science, Washington, pp. 150–87.

Begon, M. and Mortimer, A.M. (1981) *Population Ecology*, Blackwell, Oxford.

Berman, T. (1972) Profiles of chlorophyll concentrations by *in vivo* fluorescence: some limnological implications, *Limnol. Oceanogr.*, **17**, 616–18.

Berman, T. and Eppley, R.W. (1974) The measurement of phytoplankton parameters in nature, *Sci. Prog. (Oxf.)*, **61**, 219–39.

Bethge, H. (1953) Beiträge zur Kenntnis des Teichplanktons II, *Ber. Deutsch. Bot. Ges.*, **66**, 93–101.

Bidwell, R.G.S. (1977) Photosynthesis and light and dark respiration in freshwater algae, *Can. J. Bot.*, **55**, 809–18.

Birch, L.C. and Ehrlich, P.R. (1967) Evolutionary history and population biology, *Nature*, **214**, 349–52.

Birchfield, G.E. and Davidson, D.R. (1967) A case study of coastal currents in Lake Michigan, *Proc. 10th Conf. Gt Lakes Res.*, University, Michigan. Gt Lakes Res. Div., pp. 264–73.

Bishop, J.K.B., Edmond, J.M. and Ketten, D.R. (1977) The chemistry, biology and vertical flux of particulate matter from the upper 400 m of the equatorial Atlantic Ocean, *Deep Sea Res.*, 24, 511–48.

Bishop, J.K.B., Collier,. R.W., Ketten, D.R. and Edmond, J.M. (1980) The chemistry, biology and vertical flux of particulate matter from the upper 1500 m of the Panama Basin, *Deep Sea Res.*, 27, 615–40.

Blackadar, A.K. and Tennekes, H. (1968) Asymptotic similarity in neutral barotropic atmospheric layers, *J. Atmos. Sci.*, 25, 1015–20.

Blanton, J.O. (1974) Some characteristics of nearshore currents along the north shore of Lake Ontario, *J. Phys. Oceanogr.*, 4, 415–24.

Blasco, D., Estrada, M. and Jones, B. (1980) Relationship between the phytoplankton distribution and composition and the hydrography in the Northwest African upwelling region near Cabo Corbeiro, *Deep Sea Res.*, 27, 799–822.

Boersma, P.D. (1978) Breeding patterns of Galapagos penguins as an indicator of oceanographic conditions. *Science*, 200, 1481–3.

Bogorov, B.G. (1958) Perspectives in the study of seasonal changes of plankton and of the number of generations at different latitudes, in *Perspectives in Marine Biology* (ed. A. A. Buzzati–Traverso), University of California Press, Berkeley, pp. 145–58.

Bold, H.C. and Wynne, M.J. (1978) *Introduction to the Algae*, Prentice Hall, Englewood Cliffs, New Jersey.

Borgmann, U. (1982) Particle-size conversion efficiency and total animal production in pelagic ecosystems, *Can. J. Fish. Aq. Sci.*, 39, 668–74.

Bormann, F.H. and Likens, G.E. (1979) Catastrophic disturbance and the steady state in northern hardwood forests, *Am. Sci.*, 67, 660–9.

Bougis, P. (1974) *Ecologie du plancton marin, I, Le Phytoplancton* Masson et Cie, Paris.

Bowden, K.F. (1970) Turbulence II, *Oceanogr. Mar. Biol. Ann. Rev.*, 8, 11–32.

Boyce, F.M. (1974) Some aspects of Great Lakes physics of importance to biological and chemical processes, *J. Fish. Res. Bd. Can.*, 31, 689–730.

Boyce, F.M. (1977) Response of a coastal boundary layer on the north shore of Lake Ontario to a fall storm, *J. Phys. Oceanogr.*, 7, 719–32.

Boyd, C.E. and Lawrence, J.M. (1967) The mineral composition of several freshwater algae, *Proc. 20th Ann. Conf. SE Assoc. Game and Fish Commissioners*, USA.

Brezonik, P.L. (1972) Nitrogen: sources and transformations in natural waters, in *Nutrients in Natural Waters*, (eds H. E. Allen and J. R. Kramer), Wiley Interscience, New York, pp. 1–50.

Briand, F. and McCauley, E. (1978) Cybernetic mechanisms in lake plankton systems: how to control undesirable algae. *Nature*, 273, 228–30.

Brock, T.D. (1973) Lower pH limit for the existence of blue-green algae: evolutionary and ecological implications, *Science*, 179, 480–3.

Brockmann, C., Fahrbach, E., Huyer, A. and Smith, R.L. (1980) The poleward undercurrent along the Peru coast: 5 to 15°S, *Deep Sea Res.*, 27A, 847–6.

Broecker, W.S. (1973) Factors controlling CO_2 content in the oceans and atmosphere, in *Carbon and the Biosphere* (eds G. M. Woodwell and E. V. Pecan), *Proc.*

24th. Brookhaven Symp., Tech. Inf. Centre, US Atomic Energy Committee, (NTIS CONF-720510), pp. 32–50.

Broecker, W.S. (1974) *Chemical Oceanography*, Harcourt Brace Jovanovich Inc., New York.

Brooks, A.S., Warren, G.J., Boraas, M.E., Sale, D.B. and Edgington, D.N. (1984) Long term phytoplankton changes in Lake Michigan: cultural eutrophication or biotic shifts? *Verh. Int. Verein. Limnol.*, **22**, 452–59.

Brooks, J.L. and Dodson, S.I. (1965) Predation, body size and composition of plankton, *Science*, **150**, 28–35.

Brown, E.J., Harris, R.F. and Koonce, J.F. (1978) Kinetics of phosphate uptake by aquatic micro-organisms: Deviations from a simple Michaelis–Menten equation, *Limnol. Oceanogr.*, **23**, 26–34.

Brugam, R.B. and Patterson, C. (1983) The A/C (Araphidinae/Centrales) ratio in high and low alkalinity lakes in eastern Minnesota, *Freshwat. Biol.*, **13**, 47–56.

Brylinsky, M. and Mann, K.H. (1973) An analysis of factors governing productivity in lakes and reservoirs, *Limnol. Oceanogr.*, **18**, 1–14.

Buckingham, S., Walters, C.J. and Kleiber, P. (1975) A procedure for estimating gross production, net production and algal carbon content using ^{14}C, *Verh. Int. Verein. Limnol.*, **19**, 32–38.

Burmaster, D.E. (1979a) The continuous culture of phytoplankton: mathematical equivalence among three unsteady state models, *Am. Nat.*, **113**, 123–34.

Burmaster, D.E. (1979b) The unsteady continuous culture of phosphate limited *Monochrysis lutheri* Droop: experimental and theoretical analysis, *J. Exp. Mar. Biol. Ecol.*, **39**, 167–86.

Canfield, D.E. and Bachmann, R.W. (1981) Prediction of total phosphorus concentrations, chlorophyll a, and secchi depths in natural and artificial lakes, *Can. J. Fish. Aq. Sci.*, **38**, 414–23.

Cannon, D., Lund, J.W.G. and Sieminska, J. (1961) The growth of *Tabellaria flocculosa* (Roth), Kütz var *flocculosa* (Roth) Knuds, under natural conditions of light and temperature, *J. Ecol.*, **49**, 277–87.

Canter, H.M. (1979) Fungal and protozoan parasites and their importance to the ecology of the phytoplankton, *Ann. Rep. Freshwat. Biol. Assn*, **47**, 43–50.

Caperon, J. (1969) Time lag in population growth response of *Isochrysis galbana* to a variable nitrate environment, *Ecology*, **50**, 188–92.

Capone, D.G. and Carpenter, E.J. (1982) Nitrogen fixation in the marine environment, *Science*, **217**, 1140–2.

Cassie, R.M. (1959) Micro distribution of plankton, *NZ Jl. Sci.*, **2**, 398–409.

Caswell, H. (1976) Community structure: a neutral model analysis, *Ecol. Monogr.*, **46**, 327–54.

Caswell, H. (1978) Predator mediated coexistence – a non-equilibrium model, *Am. Nat.*, **112**, 127–54.

Caswell, H. (1982) Life history theory and the equilibrium status of populations, *Am. Nat.*, **120**, 317–39.

Chapra, S.C. (1977) Total phosphorus model for the Great Lakes, *J. Env. Eng. Div. ASCE*, **103**, 147–61.

Chapra, S.C. (1980) Simulation of recent and projected total phosphorus trends in Lake Ontario, *J. Great Lakes Res.*, **6**, 101–12.

Charlton, M.N. (1980a) Oxygen depletion in Lake Erie: has there been any change? *Can. J. Fish. Aq. Sci.*, **37**, 72–81.

Charlton, M.N. (1980b) Hypolimnion oxygen consumption in lakes: discussion of productivity and morphometric effects, *Can. J. Fish Aq. Sci.*, **37**, 1531–9.

Checkley, D.M. Jr (1980) Food limitation of egg production by a marine, planktonic copepod in the sea off S. California, *Limnol. Oceanogr.*, **25**, 991–5.

Cheney, R.E. (1977) Synoptic observations of the oceanic frontal system east of Japan, *J. Geophys. Res.*, **82**, 5459–68.

Cody, M.L. (1974a) Optimization in ecology, *Science*, **183**, 1156–64.

Cody, M.L. (1974b) *Competition and the structure of bird communities*, Princeton University Press, Princeton, New Jersey.

Colebrook, J.M. (1978a) Changes in the zooplankton of the North Sea, 1948 to 1973, *Rapp. P-v. Réun. Cons. Int. Explor. Mer*, **172**, 390–6.

Colebrook, J.M. (1978b) Continuous plankton records: zooplankton and environment, North-East Atlantic and North Sea, 1948–1975, *Oceanologica Acta*, **1**, 9–23.

Colebrook, J.M. (1979) Continuous plankton records: seasonal cycles of phytoplankton and copepods in the North Atlantic Ocean and the North Sea, *Mar. Biol.*, **51**, 23–32.

Colebrook, J.M. (1982) Continuous plankton records: seasonal variations in the distribution and abundance of plankton in the North Atlantic Ocean and the North Sea, *J. Plankton Res.*, **4**, 435–62.

Colebrook, J.M. and Robinson, G.A. (1961) The seasonal cycle of plankton in the North Sea and the North-Eastern Atlantic Ocean, *J. Cons. perm. int. Explor. Mer.*, **26**, 156–65.

Colebrook, J.M. and Robinson, G.A. (1964) Continuous plankton records: annual variations of abundance of plankton, 1948–1960, *Bull. Mar. Ecol.*, **6**, 52–69.

Colebrook, J.M. and Robinson G.A. (1965) Continuous plankton records: seasonal cycles of phytoplankton and copepods in the north-eastern Atlantic and the North Sea, *Bull. Mar. Ecol.*, **6**, 123–39.

Colebrook, J.M. and Taylor, A.H. (1979) Year to year changes in sea-surface temperature, North Atlantic and North Sea, 1948 to 1974, *Deep Sea Res.*, **26A**, 825–50.

Colebrook, J.M., Reid, P.C. and Coombs, S.H. (1978) Continuous plankton records: a change in the plankton of the southern North Sea between 1970 and 1972, *Mar. Biol.*, **45**, 209–13.

Colinvaux, P. (1973) *Introduction to Ecology*, Wiley, New York.

Colinvaux, P. and Steinitz, M. (1980) Species richness and area in Galapagos and Andean Lakes. Equilibrium phytoplankton communities and the paradox of the zooplankton, in *Evolution and Ecology of Zooplankton Communities* (ed. W. C. Kerfoot), University Press of New England, Hanover, New Hampshire, pp. 697–711.

Collos, Y. and Slawyk, G. (1980) Nitrogen uptake and assimilation by marine phytoplankton, in *Primary Production in the Sea* (ed. P. Falkowski), Plenum, New York, pp. 195–212.

Connell, J.H. (1975) Some mechanisms producing structure in natural communities: a model and evidence from field experiments, in *Ecology and Evolution of*

Communities (eds M. L. Cody and J. M. Diamond) Belknap Press, Harvard University, Cambridge, Mass., pp. 460–90.

Connell, J.H. (1978) Diversity in tropical rain forests and coral reefs, *Science*, **199**, 1302–10.

Connell, J.H. (1980) Diversity and the coevolution of competitors, or the ghost of competition past, *Oikos*, **35**, 131–8.

Connor, E.F. and McCoy, E.D. (1979) The statistics and biology of the species area relationship, *Am. Nat.*, **113**, 791–833.

Connor, E.F. and Simberloff, D. (1979) The assembly of species communities: chance or competition, *Ecology*, **60**, 1132–40.

Coombs, J. and Greenwood, A.D. (1976) Compartmentation of the photosynthetic apparatus in *The Intact Chloroplast* (ed. J. Barber), Elsevier/N. Holland, Amsterdam, pp. 1–51.

Cooper, L.H.N. (1933a) Chemical constituents of biological importance in the English Channel I, November 1930 to January 1932; phosphate, silicate, nitrate, nitrite and ammonia, *J. Mar. Biol. Assn UK*, **18**, 677–728.

Cooper, L.H.N. (1933b) Chemical constituents of biological importance in the English Channel III, June to December 1932; phosphate, silicate, nitrate, hydrogen ion concentration with a comparison with wind records, *J. Mar. Biol. Assn UK*, **19**, 55–62.

Cooper, L.H.N. (1938) Phosphate in the English channel, 1933–38 with a comparison with earlier years, 1916 and 1923–32, *J. Mar. Biol. Assn UK*, **23**, 181–95.

Cornett, R.J. and Rigler, F.H. (1979) Hypolimnetic oxygen deficits: their prediction and interpretation, *Science*, **205**, 580–1.

Coté, B. and Platt, T. (1983) Day-to-day variations in the spring–summer photosynthetic parameters of coastal marine phytoplankton, *Limnol. Oceanogr.*, **28**, 320–44.

Craig, J.F. (1982) Population dynamics of Windermere perch, *50th Ann. Rep. Freshwater Biol. Assn*, Windermere, UK, 49–59.

Craig, J.F., Kipling, C., LeCren, E.D. and McCormack, J. (1979) Estimates of the numbers, biomass and year class strengths of perch (*Perca fluviatilis* L.) in Windermere from 1967 to 1977 and some comparisons with earlier years, *J. Anim. Ecol.*, **48**, 315–25.

Csanady, G.T. (1971) On the equilibrium shape of the thermocline in a shore zone, *J. Phys. Oceanogr.*, **1**, 263–70.

Csanady, G.T. (1972) Response of large stratified lakes to wind, *J. Phys. Oceanogr.*, **2**, 3–13.

Csanady, G.T. (1974) Spring thermocline behaviour in Lake Ontario during IFYGL, *J. Phys. Oceanogr.*, **4**, 425–45.

Csanady, G.T. (1975) Hydrodynamics of large lakes, *Ann. Rev. Fluid Mech.*, **7**, 357–86.

Csanady, G.T. (1978) Water circulation and dispersal mechanisms, in *Lakes, Chemistry, Geology, Physics* (ed. A. Lerman), Springer, Berlin, pp. 21–64.

Cunningham, A. and Maas, P. (1978) Time lag and nutrient storage effects in the transient growth response of *Chlamydomonas reinhardii* in nitrogen limited batch and continuous culture, *J. Gen. Microbiol.*, **104**, 227–31.

Cushing, D.H. (1959) The seasonal variation in oceanic production as a problem in population dynamics, *J. Cons. perm. int. Explor. Mer*, **24**, 455–64.

Cushing, D.H. (1962) An alternative method of estimating critical depth, *J. Cons. perm. int. Explor. Mer*, **27**, 131–40.

Cushing, D.H. (1966) Biological and hydrographic changes in British Seas during the last thirty years, *Biol. Rev.*, **41**, 221–58.

Cushing, D.H. (1969) The regularity of the spawning season of some fishes, *J. Cons. perm. int. Explor. Mer.*, **33**, 81–92.

Cushing, D.H. (1975) *Marine Ecology and Fisheries*, Cambridge University Press, Cambridge.

Cushing, D.H. (1978) Biological effects of climate change, *Rapp. Proc. Vérb. Réun. Cons. Int. Expl. Mer*, **173**, 107–16.

Cushing, D.H. (1982) *Climate and Fisheries*, Academic Press, London.

Cushing, D.H. (1983) Are fish larvae too dilute to affect the density of their food organisms? *J. Plankton Res.*, **5**, 847–54.

Cushing, D.H. (1984) The gadoid outburst in the North Sea, *J. Cons. perm. int. Expl. Mer*, **41**, 159–66.

Cushing, D.H. and Dickson, R.R. (1976) The biological response in the sea to climate changes, *Adv. Mar. Biol.*, **14**, 1–22.

Dake, J.M.K. (1972) Evaporative cooling of a body of water, *Water Resources Res.*, **8**, 1087–1091.

Damann, K.E. (1945) Plankton studies of Lake Michigan I. Seventeen years of plankton data collected at Chicago, Illinois, *Am. Midl. Nat.*, **34**, 769–96.

Damann, K.E. (1960) Plankton studies of Lake Michigan II. Thirty-three years of continuous plankton and coliform bacteria data collected from Lake Michigan at Chicago, *Trans. Am. Micr. Soc.*, **79**, 397–404.

Danforth, W.F. and Ginsberg, W. (1980) Recent changes in the phytoplankton of Lake Michigan near Chicago, *J. Great Lakes Res.*, **6**, 307–14.

Darwin, C. (1857) Abstract from a letter of C. Darwin Esq. to Professor Asa Gray, Boston, US, dated Down, September 5th, 1857 in *The Collected Papers of Charles Darwin* (1977) (ed. P. H. Barrett), Vol. 2, University of Chicago Press, Chicago, pp. 8–10.

Darwin, C. (1859) *The Origin of Species*, John Murray, London.

Darwin, C. and Wallace, A.R. (1858) On the tendency of species to form varieties; and on the perpetuation of varieties and species by natural means of selection, in *The Collected Papers of Charles Darwin* (1977) (ed. P. H. Barrett), Vol. 2, University of Chicago Press, Chicago, pp. 3–8, 10–19.

Davis, C.C. (1964) Evidence for the eutrophication of Lake Erie from phytoplankton records, *Limnol. Oceanogr.*, **9**, 275–83.

Davis, S.N. and DeWiest, R.I.M. (1966) *Hydrogeology*, Wiley, New York.

Deacon, E.L., Sheppard, P.A. and Webb, E.K. (1956) Wind profiles over the sea and the drag at the sea surface, *Aust. J. Phys.*, **9**, 511–41.

De Jong, T.M. (1975) A comparison of three diversity indices based on their components of richness and evenness, *Oikos*, **26**, 222–7.

Delwiche, C.C. (1970) The nitrogen cycle, in *The Biosphere*, Freeman, San Francisco, pp. 69–80.

Denman, K.L. (1973) A time-dependent model of the upper ocean, *J. Phys. Ocea-nogr.*, **3**, 173–84.

Denman, K.L. (1976) Covariability of chlorophyll and temperature in the sea, *Deep Sea. Res.*, **23**, 539–50.

Denman, K.L. and Gargett, A.E. (1983) Vertical mixing and advection of phytoplankton in the upper ocean, *Limnol. Oceanogr.*, **28**, 801–15.

Denman, K.L. and Platt, T. (1976) The variance spectrum of phytoplankton in a turbulent ocean, *J. Mar. Res.*, **34**, 593–601.

Denman, K.L., Okubo, A. and Platt, T. (1977) The chlorophyll fluctuation spectrum in the sea, *Limnol. Oceanogr.*, **22**, 1033–8.

Derenbach, J.B., Astheimer, H., Hansen, H.P. and Leach, H. (1979) Vertical microscale distribution of phytoplankton in relation to the thermocline, *Mar. Ecol. Prog. Ser.*, **1**, 187–93.

Diamond, J.M. (1973) Distributional ecology of New Guinea Birds, *Science*, **179**, 759–69.

Diamond, J.M. (1975) Assembly of species communities, in *Ecology and Evolution of Communities* (eds M. L. Cody and J. M. Diamond), Belknap Press, Harvard University, Cambridge, Mass. pp. 342–444.

Diamond, J.M. (1978) Niche shifts and the rediscovery of interspecific competition, *Am. Sci.* **66**, 322–31.

Dickson, R.R. and Reid, P.C. (1983) Local effects of wind speed and direction on the phytoplankton of the Southern Bight, *J. Plankton Res.*, **5**, 441–55.

Dickson, R.R., Lamb, H.H., Malmberg, S.A. and Colebrook, J.M. (1975) Climatic reversals in northern North Atlantic, *Nature*, **256**, 479–82.

Dillon, P.J. and Rigler, F.H. (1974a) The phosphorus: chlorophyll relationships in lakes, *Limnol. Oceanogr.*, **19**, 767–73.

Dillon, P.J. and Rigler, F.H. (1974b) A test of a simple nutrient budget model predicting the phosphorus concentration in lakewater, *J. Fish. Res. Bd Can.*, **31**, 1771–8.

Dillon, T.M., Richman, J.G., Hansen, C.G. and Pearson, M.D. (1981) Near-surface turbulence measurements in a lake, *Nature*, **290**, 390–2.

DiToro, D.M. (1980) Applicability of cellular equilibrium and monod theory of phytoplankton growth kinetics, *Ecol. Modelling*, **8**, 201–18.

Dobson, H.H. and Gilbertson, M. (1971) Oxygen depletion in the central basin of Lake Erie: 1929–1970, *Proc. Conf. Gt Lakes Res.*, **14**, 743–48.

Donaghay, P.L., DeManche, J.M. and Small, L.F. (1978) On predicting phytoplankton growth rates from carbon: nitrogen ratios, *Limnol. Oceanogr.*, **23**, 359–62.

Dougherty, J.P. (1961) The anisotropy of turbulance at the meter level, *J. Atmos. Terr. Phys.*, **21**, 210–213.

Drever, J.I. (1982) *The Geochemistry of Natural Waters*, Prentice Hall, Englewood Cliffs, N.J.

Dring, M.J. and Jewson, D.H. (1982) What does ^{14}C uptake by phytokplankton really measure. A theoretical modelling approch. *Proc. Roy. Soc. Lond. B*, **214**, 351–68.

Droop, M.R. (1974) The nutrient status of algal cells in continuous culture, *J. Mar. Biol. Assn UK*, **54**, 825–55.

Drury, W.H. and Nisbet, I.C.T. (1973) Succession, *J. Arnold Arboretum*, **54**, 331–68.

Dugdale, R.C. (1976) Nutrient cycles, in *The Ecology of the Seas* (eds D. H. Cushing and J. J. Walsh), Saunders, Philadelphia, pp. 141–72.

Durbin, E.G., Durbin, A.G., Smayda, T.J. and Verity, P.G. (1983) Food limitation of production by adult *Acartia tonsa* in Narrangansett Bay, Rhode Island, *Limnol. Oceangr.*, **28**, 1199–213.

Durbin, E.G., Krawiec, R.W. and Smayda, T.J. (1975) Seasonal studies on the relative importance of different size fractions of phytoplankton in Narrangansett Bay (USA), *Mar. Biol.*, **32**, 271–87.

Edmond, J.M. and Von Damm, K. (1983) Hot springs on the ocean floor, *Sci. Am.*, **248**, 78–93.

Edmond, J.M., Measures, C. McDuff, R.E. *et al.* (1979) Ridge crest hydrothermal activity and the balance of the major and minor elements in the ocean: the Galapagos data, *Earth Planet. Sci. Lett.*, **46**, 1–18.

Edmondson, W.T. (1972) The present condition of Lake Washington, *Verh. Int. Verein. Limnol.*, **18**, 284–91.

Edmondson, W.T. and Lehman, J.T. (1981) The effect of changes in the nutrient income on the condition of Lake Washington, *Limnol. Oceanogr.*, **26**, 1–29.

Ehrlich, P.R. and Birch, L.C. (1967) The 'balance of nature' and 'population control', *Am. Nat.*, **101**, 97–107.

Ehrlich, P.R. and Holm, R.W. (1962) Patterns and populations, *Science*, **137**, 652–7.

Elbrächter, M. (1973) Population dynamics of *Ceratium* in coastal waters of the Kiel Bay, *Oikos. Suppl.*, **15**, 43–8.

Eldredge, N. and Gould, S.J. (1972) Punctuated equilibria: an alternative to phyletic gradualism, in *Models in Palaeobiology* (ed. T. J. M. Schopf) Freeman, Cooper and Co., San Francisco, 82–115.

Emery, K.O. and Csanady, G.T. (1973) Surface circulation of lakes and nearly land-locked seas, *Proc. Natl. Acad. Sci. USA*, **70**, 93–7.

Enfield, D.B. and Allen, J.S. (1980) On the structure and dynamics of monthly mean sea level anomalies along the Pacific coast of North and South America, *J. Phys. Oceanogr.*, **10**, 557–8.

Eppley, R.W. (1972) Temperature and phytoplankton growth in the sea, *Fishery Bull.*, **70**, 1063–85.

Eppley, R.W. (1980) Estimating phytoplankton growth rates in the central oligotrophic oceans. *Primary Productivity in the Sea* (ed. P. Falkowski), *Env. Sci. Res. 19*, (*Brookhaven Symposium Biology 31*) Plenum, New York, pp. 231–42.

Eppley, R.W. (1981) Relations between nutrient assimilation and growth in phytoplankton with a brief review of estimates of growth rate in the ocean, in *Physiological Bases of Phytoplankton Ecology* (ed. T. Platt), *Can. Bull. Fish. Aq. Sci.*, **210**, 251–63.

Eppley, R.W. and Peterson, B.J. (1979) Particulate organic matter flux and planktonic new production in the deep ocean, *Nature*, **282**, 677–80.

Eppley, R.W. and Sharp, J.H. (1975) Photosynthetic measurements in the Central North Pacific: the dark loss of carbon in 24h incubations, *Limnol. Oceanogr.*, **20**, 981–7.

Eppley, R.W., Venrick, E.L. and Mullin, M.M. (1973) A study of plankton dynamics and nutrient cycling in the central gyre of the N. Pacific ocean, *Limnol. Oceanogr.*, **18**, 534–51.

Falkowski, P. (ed.) (1980) Primary productivity in the sea, *Env. Sci. Res. 19*, (*Brookhaven Symposium Biology 31*) Plenum, New York.

Fasham, M.J.R. (1978) The application of some stochastic processes to the study of plankton patchiness, in *Spatial Pattern in Plankton Communities*, (ed. J. H. Steele), Plenum, New York, pp. 131–56.

Fay, P., Stewart, W.D.P., Walsby, A.E. and Fogg, G.E. (1968) Is the heterocyst the site of nitrogen fixation in blue-green algae? *Nature*, **220**, 810–12.

Fenchel, T. (1974) Intrinsic rate of natural increase: the relationship with body size, *Oecologia*, **14**, 317–26.

Fenchel, T. (1982a) Ecology of heterotrophic microflagellates. I, Some important forms and their functional morphology, *Mar. Ecol. Prog. Ser.*, **8**, 211–23.

Fenchel, T. (1982b) Ecology of heterotrophic microflagellates. II, Bioenergetics of growth, *Mar. Ecol. Prog. Ser.*, **8**, 225–31.

Ferguson, A.J.D., Thompson, J.M. and Reynolds, C.S. (1982) Structure and dynamics of zooplankton communities maintained in closed systems, with special reference to the algal food supply, *J. Plankton Res.*, **4**, 523–43.

Flower, R.J. and Battarbee, R. (1983) Diatom evidence for recent acidification of two Scottish Lochs. *Nature*, **305**, 130–3.

Fogg, G.E. (1982) Nitrogen cycling in sea waters, *Phil. Trans. Roy. Soc. Lond. B.*, **296**, 511–20.

Ford, D.E. and Stefan, H. (1980) Stratification variability in three morphometrically different lakes under identical meterological forcing, *Water. Res. Bull.*, **16**, 243–7.

Fortier, L. and Legendre, L. (1979) Le contrôl de la variabilité à court terme du phytoplancton estuarien: stabilité verticale et profondeur critique, *J. Fish Res. Bd Can.*, **36**, 1325–35.

Fournier, R.O. (1978) Biological aspects of the Nova Scotian shelfbreak fronts, in *Oceanic Fronts in Coastal Processes* (eds M. J. Bowman and W. E. Esaias), Springer, Berlin, pp. 69–77.

Fox, J.F. (1979) Intermediate disturbance hypothesis, *Science*, **204**, 1344–5.

Fox, J. F. (1981) Intermediate levels of soil disturbance maximise alpine plant diversity, *Nature*, **293**, 564–5.

Fraser, A.S. (1980) Changes in Lake Ontario total phosphorus concentrations, *J. Great Lakes Res.*, **65**, 83–7.

Frederick, V.R. (1981) Preliminary investigations of the algal flora in the sediments of Lake Erie, *J. Great Lakes Res.*, **7**, 404–8.

French, D.P., Furnas, M.J. and Smayda, T.J. (1983) Diel changes in nitrite concentration in the chlorophyll maximum in the Gulf of Mexico, *Deep Sea Res.*, **30**, 707–22.

Friebele, E.S., Correll, D.L. and Faust, M.A. (1978) Relationship between phytoplankton cell size and the rate of orthophosphate uptake: *in situ* observations of an estuarine population, *Mar. Biol.*, **45**, 39–52.

Fritsch, F.E. (1935) *The Structure and Reproduction of the Algae*, Vols I and II, Cambridge University Press, Cambridge.

Frost, B.W. (1980a) Grazing, in *The Physiological Ecology of Phytoplankton* (ed. I. Morris), Blackwell, Oxford, pp. 465–92.

Frost, B.W. (1980b) The inadequacy of body size as an indicator of niches in the zooplankton in, *Spec. Symp. Vol. 3 ASLO, Evolution and Ecology of Zooplankton Communities* (ed. W. C. Kerfoot), University Press of New England, Hanover, New Hampshire, pp. 742–53.

Gaarder, T. and Gran, H.H. (1927) Investigations of the production of plankton in the Oslo fjord, *Rapp. et Proc-Verb. Cons. Int. Explor. Mer*, 42, 3–48.

Gallegos, C.L., Church, M.R., Kelly, M.G. and Hornberger, G.M. (1983) Asynchrony between rates of oxygen production and inorganic carbon uptake in a mixed culture of phytoplankton. *Arch. Hyrdrobiol.*, 96, 164–75.

Ganf, G.G. (1969) Physiological and ecological aspects of the phytoplankton of Lake George, Uganda, PhD Thesis, University of Lancaster, Lancaster, Cumbria.

Ganf, G.G. (1974) Phytoplankton biomass and distribution in a shallow eutrophic lake (Lake George, Uganda), *Oecologia (Berl)*, 16, 9–29.

Ganf, G.G., Shiel, R.J. and Merrick, C.J. (1983) Parasitism: the possible cause of the collapse of a *Volvox* population in Mount Bold Reservoir, South Australia, *Aust. J. Mar. Freshwat. Res.*, 34, 489–94.

Gannon, J.E. (1972) Effects of eutrophication and fish predation on recent changes in the zooplankton crustacea species composition in Lake Michigan, *Trans. Am. Microsc. Soc.*, 91, 82–4.

Gause, G.F. (1964) *The Struggle for Existence*, Williams and Wilkins, Baltimore.

Garrels, K.M., Mackenzie, F.T. and Hunt, C. (1975) *Chemical cycles and the global environment*, W. Kauffmann, Los Altos, CA.

Gascon, D. and Leggett, W.C. (1977) Distribution, abundance, and resource utilization of littoral zone fishes in a response to a nutrient/production gradient in Lake Memphremagog, *J. Fish. Res. Bd Can.*, 34, 1105–17.

Gasse, F. and Tekaia, F. (1983) Transfer functions for estimating palaeoecological conditions (pH) from East African diatoms, *Hydrobiologia*, 103, 85–90.

Gasse, F., Talling, J.F. and Kilham, P. (1983) Diatom assemblages in East Africa: classification, distribution and ecology, *Rev. Hydrobiol. Trop.*, 16, 3–34.

George, D.G. and Heaney, S.I. (1978) Factors influencing the spatial distribution of phytoplankton in a small productive lake, *J. Ecol.*, 66, 133–55.

Gibbs, R.J. (1967) The geochemistry of the Amazon River System I, *Bull. Geol. Soc. Am.*, 78, 1203–32.

Gibson, C.E., Wood, R.B., Dickson, E.L. and Jewson, D.H. (1971) The succession of phytoplankton in Lake Neagh 1968–70, *Mitt. Internat. Verein. Limnol.*, 19, 146–60.

Gieskes, W.W.C. and Kraay, G.W. (1977) Continuous plankton records: changes in the plankton of the North Sea and its eutrophic southern bight from 1948 to 1975, *Neth. J. Sea Res.*, 11, 334–64.

Gieskes, W.W.C., Kraay, G.W. and Baars, M.A. (1979) Current ^{14}C methods for measuring primary production: gross underestimates in oceanic waters, *Neth. J. Sea Res.*, 13, 58–78.

Gilbert, P.M. and Goldman, J.C. (1981) Rapid ammonium uptake by marine phytoplankton, *Mar. Biol. Lett.*, 2, 25–31.

Gleason, H.A. (1917) The structure and development of the plant association, *Bull. Torrey Bot. Club*, 43, 463–81.

Gleason, H.A. (1926) The individualistic concept of the plant association, *Bull. Torrey Bot. Club*, 53, 7–26.

Gliwicz, Z.M. (1969) The share of algae, bacteria and tripton in the food of the pelagic zooplankton of lakes of various trophic characteristics. *Bull. Acad. Pol. Sci. Cl. II Sér. Sci. Biol.*, 17, 159–65.

Gliwicz, Z.M. (1977) Food size selection and seasonal succession of filter feeding zooplankton in an eutrophic lake, *Ekol. Polska*, 25, 179–255.

Gliwicz, Z.M. (1980) Filtering rates, food size selection and feeding rates in clado-cerans – another aspect of interspecific competition in filter feeding zooplankton, in *Evolution and Ecology of Zooplankton Communities* (ed. W. C. Kerfoot), University Press of New England, Hanover, New Hampshire, pp. 282–91.

Gliwicz, Z.M. and Hillbricht-Ilkowska, A. (1975) Ecosystem of the Mikolajskie lake. Elimination of phytoplankton biomass and its subsequent fate in lake through the year, *Pol. Arch. Hydrobiol.*, 22, 39–52.

Gliwicz, Z.M. and Prejs, A. (1977) Can planktivorous fish keep in check planktonic crustacean populations? A test of the size efficiency hypothesis in typical Polish lakes, *Ekologia Polska*, 25, 567–91.

Glover, R.S., Robinson, G.A. and Colebrook, J.M. (1972) Plankton in the North Atlantic – an example of the problems of analysing variability in the environment, in *Marine pollution and Sea Life* Fishing News (Books), West Byfleet, England pp. 439–45.

Glover, R.S., Robinson, G.A. and Colebrook, J.M. (1974) Marine biological surveil-lance, *Environment and Change*, 2, 395–402.

Goering, J.J. (1972) The role of nitrogen in eutrophic processes, in *Water Pollution Microbiology* (ed. R. Mitchell), Wiley Interscience, New York, pp. 43–68.

Goldman, J.C. (1973) Carbon dioxide and pH effect on species succession of algae, (comment on Shapiro (1973) *Science*, 179, 382–4) *Science*, 182, 306–7.

Goldman, J.C. (1977a) Steady state growth of phytoplankton in continuous culture: comparison of internal and external nutrient equations, *J. Phycol.*, 13, 251–8.

Goldman, J.C. (1977b) Temperature effects on phytoplankton growth in continuous culture, *Limnol. Oceanogr.*, 22, 932–6.

Goldman, J.C. (1979) Outdoor algal moss culture I, *Appl. Wat. Res.*, 13, 1–19; Photosyntehtic yield limitations II, *Appl. Wat. Res.*, 13, 119–36.

Goldman, J.C. (1980) Physiological processes, nutrient availability and the concept of relative growth rate in marine phytoplankton ecology, in *Primary Production in the Sea* (ed. P. Falkowski), Plenum, New York, pp. 179–94.

Goldman, J.C. (1984) Oceanic nutrient cycles, in *Flow of Energy and Materials in Marine Ecosystems: Theory and Practice* (ed. M. J. Fasham), Plenum, New York, pp. 137–70.

Goldman, J.C. and Carpenter, E.J., (1974) A kinetic approach to the effect of temperature and algal growth, *Limnol. Oceanogr.*, 19, 756–66.

Goldman, J.C. and Dennett, M.R. (1983) Effect of nitrogen source on short-term light and dark CO_2 uptake by a marine diatom, *Mar. Biol.*, 76, 7–16.

Goldman, J.C. and Gilbert, P.M. (1983) Kinetics of inorganic nitrogen uptake by phytoplankton, in *Nitrogen in the Marine Environment*, (eds E. J. Carpenter and D. G. Capone), Academic Press, New York, pp. 233–274.

Goldman, J.C. and Mann, R. (1980) Temperature influenced variations in speciation and chemical composition of marine phytoplankton in outdoor mass cultures, *J. Exp. Mar. Biol. Ecol.*, **46**, 29–40.

Goldman, J.C. and Ryther, J.H. (1976) Temperature influenced species competition in mass cultures of marine phytoplankton, *Biotechnology and Bioengineering*, **18**, 1125–44.

Goldman, J.C., McCarthy, J.J. and Peavey, D.G. (1979) Growth rate influence on the chemical composition of phytoplankton in oceanic waters, *Nature*, **279**, 210–5.

Goldman, J.C., Oswald, W.J. and Jenkins, D. (1974) The kinetics of inorganic carbon limited algal growth, *J. Water Poll. Cont. Fed.*, **46**, 554–74.

Goldman, J.C., Porcella, D.B., Middlebrooks, E.J. and Toerien, D.F. (1972) The effects of carbon on algal growth, *Wat. Res.*, **6**, 637–79.

Goldman, J.C., Taylor, C.D. and Gilbert, P.M. (1981) Non-linear time course uptake of carbon and ammonium by marine phytoplankton, *Mar. Ecol. Prog. Ser.*, **6**, 137–48.

Goldschmidt, R. (1940) *The Material Basis of Evolution*, Yale University Press, New Haven.

Golterman, H.L. (1973) Natural phosphate sources in relations to phosphate budgets: a contribution to the understanding of eutrophication, *Water Res.*, **7**, 3–17.

Gran, H.H. and Braarud, T. (1935) A quantitative study of the phytoplankton in the Bay of Fundy and the Gulf of Maine, *J. Biol. Bd Canada*, **1**, 210–27.

Granhall, U. and Lundgren, A. (1971) Nitrogen fixation in Lake Erken, *Limnol. Oceanogr.*, **16**, 711–9.

Grant, P.R. (1981) Speciation and the adaptive radiation of Darwin's finches, *Am. Sci.*, **69**, 653–63.

Grant, P.R. and Abbott, I. (1980) Interspecific competition, island biogeography and null hypotheses, *Evolution*, **34**, 332–41.

Grenney, W.J., Bella, D.A. and Curl, H.C. (1973) A theoretical approach to interspecific competition in phytoplankton communities, *Am. Nat.*, **107**, 405–25.

Grime, J.P. (1973) Competitive exclusion in herbaceous vegetation, *Nature*, **242**, 344–7.

Grubb, P.J. (1977) The maintenance of species richness in plant communities: the importance of the regeneration niche, *Biol. Rev.*, **52**, 107–45.

Guillen, O.G. and Calienes, R.Z. (1981) Biological productivity and El Nino, in *Resource Management and Environmental Uncertainty* (eds M. Glantz and D. Thompson), Wiley Interscience, New York, pp. 255–82.

Guma'a, S.A. (1978) The food and feeding habits of young perch, *Perca fluviatilis*, in Windermere, *Freshwat. Biol.*, **8**, 177–87.

Haffner, G.D., Harris, G.P. and Jarai, M.K. (1980) Physical variability and phytoplankton communities III. Vertical structure in phytoplankton populations, *Arch. Hydrobiol.*, **89**, 363–81.

Hairston, N.G., Smith, F.E. and Slobodkin, L.B. (1960) Community structure, population control, and competition, *Am. Nat.*, **94**, 421–5.

Haliwell, G.R. and Mooers, C.N.K. (1979) The space-time structure of the shelf water-slope water and Gulf Stream surface temperature fronts on associated warm-core eddies, *J. Geophys. Res.*, **84**, 7707–25.

Hall, D.J., Threlkeld, S.T., Burns, C.W. and Crowley, P.H (1976) The size efficiency hypothesis and the size structure of zooplankton communities, *Ann. Rev. Ecol. Syst.*, 7, 177–208.

Hammer, V.T., Walker, K.F. and Williams, W.D. (1973) Derivation of daily phytoplankton production estimates from short-term experiments in some shallow, eutrophic Australian saline lakes, *Aust. J. Mar. Freshwater Res.*, 24, 259–66.

Hanson, J.M. and Leggett, W.C. (1982) Empirical prediction of fish biomass and yield, *Can. J. Fish. Aq. Sci.*, 39, 257–63.

Hardy, A.C. (1935a) A further example of the patchiness of plankton distributions, *Deep Sea Res. (Suppl., Papers Mar. Biol. Oceanogr.)*, 3, 7–11.

Hardy, A.C. (1935b) The continuous plankton recorder: a new method of survey, *Rapp, et Proc. Verb. Cons. Int. Expl. Mer*, 94, 36–47.

Hardy, A.C. (1936) Observations on the uneven distribution of oceanic plankton, *Discovery Rep.*, 11, 513–38.

Harris, E. (1959) The nitrogen cycle in Long Island Sound, *Bull. Bingham Oceanogr. Coll.*, 17, 31–65.

Harris, G.P. (1973a) Diel and annual cycles of net plankton photosynthesis in Lake Ontario, *J. Fish. Res. Bd Can.*, 30, 1779–87.

Harris, G.P. (1973b) Vertical mixing mechanisms and their effects on primary production of phytoplankton, *Sci. Ser. No 33*, Inland Waters Directorate Canada Center for Inland Waters, Burlington, Ontario.

Harris, G.P. (1978) Photosynthesis, productivity and growth: the physiological ecology of phytoplankton, *Arch. Hydrobiol. Beih. Ergeb. Limnol.*, 10, 1–171.

Harris, G.P. (1980a) Spatial and temporal scales in phytoplankton ecology. Mechanisms, methods, models and management, *Can. J. Fish. Aq. Sci.*, 37, 877–900.

Harris, G.P. (1980b) The measurement of photosynthesis in natural populations of phytoplankton, in *The Physiological Ecology of Phytoplankton* (ed. I. Morris). Blackwell, Oxford, pp. 129–87.

Harris, G.P. (1980c) The relationship between chlorophyll a fluorescence, diffuse attenuation changes and photosynthesis in natural phytoplankton populations, *J. Plankton Res.*, 2, 109–27.

Harris, G.P. (1982) Spatial and temporal scales of physical and biological processes in the Laurention Great Lakes and the importance of non-equilibrium phytoplankton dynamics, *Atti. del 4th Congresso della AIOL*.

Harris, G.P. (1983) Mixed layer physics and phytoplankton populations; studies in equilibrium and non-equilibrium ecology, *Prog. Phyc. Res.*, 2, (eds F. E. Round and D. Champman), Elsevier, Amsterdam, pp. 1–52.

Harris, G.P. (1984) Phytoplankton productivity and growth measurements: past, present and future, *J. Plankton Res.*, 6, 219–37.

Harris, G.P. and Lott, J.N.A. (1973a) Light intensity and photosynthetic rates in phytoplankton, *J. Fish Res. Bd Can.*, 30, 1771–8.

Harris, G.P. and Lott, J.N.A. (1973b) Observations of Langmuir circulations in Lake Ontario, *Limnol. Oceanogr.*, 18, 584–9.

Harris, G.P. and Piccinin, B.B. (1977) Photosynthesis by natural phytoplankton populations, *Arch. Hydrobiol.*, 80, 405–57.

Harris, G.P. and Piccinin, B.B. (1980) Physical variability and phytoplankton communities IV. Temporal changes in the phytoplankton community of a physically variable lake, *Arch. Hydrobiol.*, 89, 447–73.

Harris, G.P. and Piccinin, B.B. (1983) Phosphorus limitation and carbon metabolism in a unicellular alga: interaction between growth rate and the measurement of net and gross photosynthesis, *J. Phycol.*, **19**, 185–92.

Harris, G.P. and Smith, R.E.H. (1977) Observations of small scale spatial patterns in phyoplankton populations, *Limnol. Oceanogr.*, **22**, 887–99.

Harris, G.P. and Vollenweider, R.A. (1982) Paleolimnological evidence of early eutrophication in Lake Erie. *Can. J. Fish. Aq. Sci.*, **39**, 618–26.

Harris, G.P., Heaney, S.I. and Talling, J.F. (1979) Physiological and environmental constraints in the ecology of the planktonic dinoflagellate *Ceratium hirundinella*, *Freshwat. Biol.*, **9**, 413–28.

Harris, G.P. Piccinin, B.B., Haffner, G.D. *et al.* (1980a) Physical variability and phytoplankton communities I. The descriptive limnology of Hamilton Harbour, *Arch. Hydrobiol.*, **88**, 303–27.

Harris, G.P., Haffner, G.D. and Piccinin, B.B. (1980b) Physical variability and phytoplankton communities II. Primary productivity by phytoplankton in a physically variable environment, *Arch. Hydrobiol.*, **88**, 393–425.

Harris, G.P., Piccinin, B.B. and van Ryn, J. (1983) Physical variability and phytoplankton communities V. Cell size, niche diversification and the role of competition, *Arch. Hydrobiol.*, **98**, 215–39.

Harrison, P.J. and Turpin, D.H. (1982) The manipulation of physical, chemical and biological factors to select species from natural phytoplankton communities, in *Marine Mesocosms* (eds G. D. Grice and M. R. Reeve), Springer, New York, pp. 275–89.

Harrison, W.G. (1980) Nutrient regeneration and primary production in the sea, in *Primary Productivity in the Sea* (ed. P. Falkowski), *Env. Sci. Res. 19*, (*Brookhaven Symposium Biology 31*), Plenum, New York, pp. 433–60.

Heaney, S.I. (1976) Temporal and spatial distribution of the dinoflagellate *Ceratium hirundinella* (O. F. Müller) within a small productive lake, *Freshwat. Biol.*, **6**, 531–42.

Heaney, S.I. and Sommer, U. (1984) Changes of algal biomass as carbon, cell number and volume, in bottles suspended in Lake Constance, *J. Plankton Res.*, **6**, 239–47.

Heaney, S.I. and Talling, J.F. (1980) Dynamic aspects of dinoflagellate distribution patterns in a small productive lake, *J. Ecol.*, **68**, 75–94.

Heaney, S.I., Chapman, D.V. and Morison, H.R. (1983) The role of the cyst stage in the seasonal growth of the dinoflagellate *Ceratium hirundinella* within a small productive lake, *Br. Phyc. J.*, **18**, 47–59.

Heckel, D.G. and Roughgarden, J. (1980) A species near its equilibrium size in a fluctuating environment can evolve a lower intrinsic rate of increase, *Proc. Natl. Acad. Sci. USA*, **77**, 7497–500.

Hecky, R.E. and Fee, E.J. (1981) Primary production and rates of algal growth in Lake Tanganyika, *Limnol. Oceanogr.*, **26**, 532–47.

Hecky, R.E. and Kilham, P. (1973) Diatoms in alkaline, saline lakes: ecology and geochemical complications, *Limnol. Oceanogr.*, **18**, 53–71.

Heinrich, A.K. (1962) The life histories of plankton animals and seasonal cycles of plankton communities in the oceans, *J. Cons. perm. int. Explor. Mer*, **27**, 15–24.

Hellebust, J.A. and Craigie, J.S. (eds) (1978) *Handbook of Phycological Methods. Physiological and Biochemical Methods*, Cambridge University Press, Cambridge.

Heller, M.D. (1977) The phased division of the freshwater dinoflagellate *Ceratium hirundinella* and its use as a method of assessing growth in natural populations, *Freshwat. Biol.*, 7, 527–33.

Hendrickson, J.A. (1981) Community-wide character displacement re-examined, *Evolution*, 35, 794–810.

Hentschel, E. and Wattenberg, H. (1931) Plankton und Phosphat in der Oberflächenschicht des Südatlantischen Ozeans, *Ann. Hydrogr. Berl.*, 58, 273–7.

Herbland, A. and LeBoutellier, A. (1981) The size distribution of phytoplankton and particulate organic matter in the Equatorial Atlantic Ocean: importance of ultraseston and consequences, *J. Plankton Res.*, 3, 659–73.

Heron, A.C. (1972) Population ecology of a colonizing species: the pelagic tunicate *Thalia democratica* II. Population growth rate, *Oecologia (Berl.)*, 10, 294–312.

Hesselein, R. and Quay, P. (1973) Vertical eddy diffusion rates in the thermocline of a small stratified lake, *J. Fish. Res. Bd Can.*, 30, 1491–500.

Hobbie, J.E. and Likens, G.E. (1973) Output of phosphorus, dissolved organic carbon and fine particulate carbon from Hubbard Brook watersheds, *Limnol. Oceanogr.*, 18, 734–42.

Hobson, C.A., Morris, W.J. and Pirquet, K.T. (1976) Theoretical and experimental analysis of the ^{14}C technique and its use in studies of primary production, *J. Fish. Res. Bd. Can.*, 33, 1715–21.

Holland, H.D. (1978) *The Chemistry of the Atmosphere and Oceans*, Wiley Interscience, New York.

Holland, R.E. and Beeton, A.M. (1972) Significance to eutrophication of spatial differences in nutrients and diatoms in Lake Michigan, *Limnol. Oceanogr.*, 17, 88–96.

Holligan, P.M. and Harbour, D.S. (1977) The vertical distribution and succession of phytoplankton in the Western English Channel, *J. Mar. Biol. Assn UK*, 57, 1075–93.

Hoogenhout, H. and Amesz, J. (1965) Growth rates of photosynthetic microorganisms in laboratory cultures, *Arch. Mikrobiol.*, 50, 10–25.

Horne, A.J. and Fogg, G.E. (1970) Nitrogen fixation in some English Lakes, *Proc. Roy. Soc. Lond. B*, 175, 351–66.

Horwood, J.W. (1978) Observations on spatial heterogeneity of surface chlorophyll in one and two dimensions, *J. Mar. Biol. Assn. UK*, 58, 487–502.

Hrbacek, J., Dvorakova, M., Korinek, V. and Prochazkova, L. (1961) Demonstration of the effect of the fish stock on the species composition of zooplankton and the intensity of metabolism of the whole plankton association, *Verh. Int. Verein. Limnol.*, 14, 192–5.

Huber, W.C., Harleman, D.R.F. and Ryan, P.J. (1972) Temperature prediction in stratified reservoirs. *J. Hydraulics Div. ASCE*, 98: HY4, 645–66.

Hurlbert, S.H. (1971) The non-concept of species diversity. A critique and alternative parameters, *Ecology*, 52, 577–86.

Hurlburt, H.E., Kindle, J.C. and O'Brien, J.J. (1976) A numerical simulation of the onset of El Nino. *J Phys. Oceanogr.*, 621–31.

Huston, M. (1979) A general hypothesis of species diversity, *Am. Nat.*, 113, 81–101.

Hutchinson, G.E. (1941) Ecological aspects of succession in natural populations, *Am. Nat.*, **75**, 406–18.

Hutchinson, G.E. (1953) The concept of pattern in ecology, *Proc. Acad. Nat. Sci. Philadelphia*, **105**, 1–11.

Hutchinson, G.E. (1957a) *A Treatise on Limnology Vol. 1, Geography, Physics and Chemistry*, Wiley, New York.

Hutchinson, G.E. (1957b) Concluding remarks. *Cold Spring Harbor Symp. Quant. Biol.*, **22**, 415–57.

Hutchinson, G.E. (1961) The paradox of plankton, *Am. Nat.*, **95**, 137–45.

Hutchinson, G.E. (1965) *The Ecological Theatre and the Evolutionary Play*, Yale University Press, New Haven.

Hutchinson, G.E. (1967) *A Treatise on Limnology*, Vol. II, Wiley, New York.

Hutchinson, G.E. (1973) Eutrophication, *Am. Sci.*, **61**, 269–79.

Hutchinson, G.E. (1974) *De rebus planctonibus, Limnol. Oceanogr.*, **19**, 360–1.

Hutchinson, G.E. (1978) *An Introduction to Population Ecology*, Yale University Press, New Haven.

Huttunen, P. and Meriläinen, J. (1983) Interpretation of lake quality from contemporary diatom assemblages, *Hydrobiologia*, **103**, 91–7.

Idyll, C.P. (1973) The anchovy crisis, *Sci. Am.*, **228**, 22–9.

Imboden, D.M. (1974) Phosphorus model of lake eutrophication, *Limnol. Oceanogr.*, **19**, 297–304.

Imboden, D.M. and Lerman, A. (1978) Chemical models of Lakes, in *Lakes: Chemistry, Geology, Physics* (ed. A. Lerman), Springer, New York, pp. 341–56.

Jackson, G.A. (1980) Phytoplankton growth and zooplankton grazing in oligotrophic oceans, *Nature*, **284**, 439–41.

Jackson, G.A. (1983) Zooplankton grazing effects on ^{14}C based phytoplankton production measurements: a theoretical study, *J. Plankton Res.*, **5**, 83–94.

Jacques, G., Fiala, M., Neveux, J. and Panouse, M. (1976) Fertilization de communantes phytoplanctoniques II. Cas d'un milieu eutrophe: upwelling des cotes du Sahara espagnol, *J. Exp. Mar. Biol. Ecol.*, **24**, 165–75.

Jacques, G., Descolas-Gros, C., Grall, J.R., and Sournia, A. (1979) Distribution du phytoplancton dans le partie antarctique de l'Ocean Indienne en fin d'ete, *Int. Rev. Res. Hydrobiol.*, **64**, 609–628.

Janus, L.L. and Vollenweider, R.A. (1982) The OECD Cooperative programme on Eutrophication. Canadian contribution. Summary report (Science Series 131), National Water Resources Institute, IWD Canada Centre for Inland Waters, Burlington, Ontario.

Janzen, D.H. (1980) When is it coevolution? *Evolution*, **34**, 611–2.

Janzen, D.H. (1981) Evolutionary ecology of personal defence, in *Physiological Ecology: an Evolutionary Approach to Resource Use* (eds C. R. Townsend and P. Calow), Blackwell, Oxford, pp. 145–64.

Jarnefelt, H. (1956) Zur limnologie einizer Gewasser Finlands XVI mit besonderer Berucksichtigung des Planktons, *Ann. Zool. Soc. Zool. Bot. Fenn Vanamo*, **17**(1), 1–201.

Jassby, A.D. and Goldman, C.R. (1974) Loss rates from a lake phytoplankton community, *Limnol. Oceanogr.*, **19**, 618–27.

Jaworski, G.H.M., Talling, J.F. and Heaney, S.I. (1981) The influence of carbon dioxide depletion on growth and sinking rate of two planktonic diatoms in culture, *Br. Phyc. J.*, **16**, 395–410.

Jenkin, P.M. (1937) Oxygen production by the diatom *Coscinodiscus excentricus* (Ehr.) in relation to submarine illumination in the English Channel, *J. Mar. Biol. Assn UK*, **22**, 301–43.

Jenkins, R.M. (1982) The morphoedaphic index and reservoir fish production, *Trans. Am. Fish. Soc.*, **111**, 135–40.

Jenkins, W.J. (1980) Tritium and ^{3}He in the Sargasso Sea, *J. Mar. Res.*, **38**, 533–69.

Jenkins, W.J. (1982) Oxygen utilization rates in North Atlantic subtropical gyre and primary production in oligotrophic systems, *Nature*, **300**, 246–8.

Jewson, D.H. (1976) The interaction of components controlling net phytoplankton photosynthesis in a well mixed lake (Lough Neagh, Northern Ireland) *Freshwat. Biol.*, **6**, 551–76.

Jewson, D.H., Rippey, B.H. and Gilmore, W.K. (1981) Loss rates from sedimentation, parasitism and grazing during the growth, nutrient limitation, and dormancy of a diatom crop, *Limnol. Oceanogr.*, **26**, 1045–56.

Johnson, L. (1981) The thermodynamic origin of ecosystems, *Can. J. Fish. Aq. Sci.*, **38**, 571–90.

Johnson, P.J. and Sieburth, J. McN. (1979) Chroococcoid cyanobacteria in the sea: a ubiquitous and diverse phototrophic biomass, *Limnol. Oceanogr.*, **24**, 928–35.

Jones, I.S.F. and Kenney, B.C. (1977) The scaling of velocity fluctuations I the surface mixed layer, *J. Geophys. Res.*, **82**, 1392–6.

Jones, J.G. (1972) Studies on freshwater micro-organisms: phosphatase activity in lakes of differing degrees of eutrophication, *J. Ecol.*, **60**, 777–91.

Jones, J.G. (1976) The microbiology and decomposition of seston in open water and experimental enclosures in a productive lake, *J. Ecol.*, **64**, 241–78.

Jones, J.R. and Hoyer, M.V. (1982) Sportfish harvest predicted by summer Chlorophyll a concentration in midwestern lakes and reservoirs, *Trans. Am. Fish. Soc.*, **111**, 176–9.

Jones, R.A. and Lee, G.F. (1982) Recent advances in assessing impact of phosphorus loads on eutrophication-related water quality, *Water Res.*, **16**, 503–15.

Jones, R.I. (1977) The importance of temperature conditioning to the respiration of natural phytoplankton communities, *Br. Phyc. J.*, **12**, 277–85.

Jones, R.I. (1978) Adaptations to fluctuating irradiance by natural phytoplankton communities, *Limnol. Oceanogr.*, **23**, 920–6.

Kalff, J. (1970) Arctic lake ecosystems, in *Antarctic Ecology* (ed. M. W. Holdgate), Vol. 2, Academic Press, London, pp. 651–63.

Kalff, J. and Knoechel, R. (1978) Phytoplankton and their dynamics in oligotrophic and eutrophic lakes, *Ann. Rev. Ecol. Syst.*, **9**, 475–95.

Kerfoot, W.C. (ed.) (1980) *Evolution and Ecology of Zooplankton Communities*, University Press New Hampshire, Hanover, New Hampshire.

Kerfoot, W.C. and DeMott, W.R. (1980) Foundations for evaluating community interactions: The use of enclosures to investigate coexistence of *Daphnia* and *Bosmina*, in *Evolution and Ecology of Zooplankton Communities* (ed. W. C. Kerfoot) (Special Symposium Vol. 3 ASLO), University Press of New England, Hanover, New Hampshire, pp. 725–41.

Kerr, S.R. (1974) Theory of size distribution in ecological communities, *J. Fish. Res. Bd Can.*, **31**, 1859–62.

Kerr, S.R. (1982) The role of external analysis in fisheries science, *Trans. Am. Fish. Soc.*, **111**, 165–70.

Kerr, S.R. and Martin, N.V. (1970) Trophic-dynamics of lake trout production systems, in *Marine Food Chains* (ed. J. H. Steele) Oliver and Boyd, Edinburgh, pp. 365–76.

Kester, D.R. and Pytkowicz, R.M. (1977) Natural and anthropogenic changes in the global carbondioxide system, in *Global Chemical Cycles and their Alterations by Man* (ed. W. Stumm), Dahlem Konferenzen, pp. 99–120. Bernhardt Berlin.

Ketchum, B. H. and Corwin, N. (1965) The cycle of phosphorus in a plankton bloom in the Gulf of Maine, *Limnol. Oceanogr.*, **10**, R148–R161.

Ketchum, B.H., Ryther, J.H., Yentsch, C.S. and Corwin, N. (1958a) Productivity in relation to nutrients, *Rapp. Proc. Verb Réun. Cons. Int. Perm. Expl. Mer*, **144**, 132–40.

Ketchum, B.H., Vaccaro, R.F. and Corwin, N. (1958b) The annual cycle of phosphorus and nitrogen in New England coastal waters, *J. Mar. Res.*, **17**, 282–301.

Kierstead, H. and Slobodkin, L.B. (1953) The size of water masses containing plankton blooms, *J. Mar. Res.*, **12**, 141–7.

Kilham, P. and Kilham, S.S. (1980) The evolutionary ecology of phytoplankton, in *The Physiological Ecology of Phytoplankton* (ed. I. Morris), Blackwell, Oxford, pp. 571–98.

Kimmel, B.L. (1983) Size distribution of planktonic autotrophy and microheterotrophy: implications for organic carbon flow in reservoir food webs, *Arch. Hydrobiol.*, **97**, 303–19.

King, D.L. (1970) The role of carbon in eutrophication, *J. Wat. Poll. Control Fed.*, **42**, 2035–51.

Kipling, C. (1983a) Changes in the growth of pike (*Esox lucius*) in Windermere, *J. Anim. Ecol.*, **52**, 647–57.

Kipling, C. (1983b) Changes in the population of pike (*Esox lucius*) L.) in Windermere from 1944 to 1981 *J. Anim. Ecol*, **52**, 989–99.

Kipling, C. and Frost, W.E.(1970) A study of the mortality, population numbers, year class strengths, production and food consumption of pike, *Esox lucius* L. in Windermere from 1944 to 1962, *J. Anim. Ecol.*, **39**, 115–57.

Kipling, C. and Roscoe, M.E. (1977), Surface water temperature of Windermere, *Occas. Publ. No. 2*, Freshwater Biological Association, Windermere.

Klein, P. and Coantic, M. (1981) A numerical study of turbulent processes in the marine upper layers, *J. Phys. Oceanogr.*, **11**, 849–63.

Knauer, G.A., Martin, J.H. and Bruland, K.W. (1979) Fluxes of particulate carbon, nitrogen and phosphorus in the upper water column of the Northeast Pacific, *Deep Sea Res.*, **26**, 97–108.

Knauss, J.A. (1978) *Introduction to Physical Oceanography*, Prentice Hall, Englewood Cliffs, New Jersey.

Knoechel, R. and Kalff, J. (1978) An *in situ* study of the productivity and population dynamics of five freshwater plankton diatom species, *Limnol. Oceanogr.*, **23**, 195–218.

Knowlton, N. Lang, J.C., Rooney, M.C. and Clifford, P. (1981) Evidence for delayed mortality in hurricane-damaged Jamaican staghorn corals, *Nature*, **294**, 251–2.

Koblentz-Mishke, O.J., Vedernikov, V.I. and Shirshov, P.P. (1976) A tentative comparison of primary production and phytoplankton quantities at the ocean surface, *Marine Sci. Comm.*, **2**, 357–74.

Koestler, A. (1967) *The Ghost in the Machine*, Macmillan, London and New York.

Koslow, J.A. (1983) Zooplankton community structure in the North Sea and Northeast Atlantic: development and test of a biological model, *Can. J. Fish Aq. Sci.*, **40**, 1912–24.

Kraus, E.B. (ed.) (1977) Modelling and prediction of the upper layers of the ocean, *Proc. NATO Adv. Study Inst.*, Pergamon, Oxford.

Kraus, E.B. and Turner, J.S. (1967) A one-dimensional model of the seasonal thermocline 2. The general theory and its consequences, *Tellus*, **19**, 98–106.

Krebs, C.J. (1978) *Ecology, the Experimental Analysis of Distribution and Abundance*, 2nd edn, Harper and Row, New York.

Kuentzel, L.E. (1969) Bacteria, carbon dioxide and algal blooms, *J. Wat. Poll. Cont. Fed.* **41**, 1737–47.

Kuhn, T.S. (1970) The structure of scientific revolutions, in *Foundations in the Unity of Science* (eds O. Neurath, R. Carnap and C. Morris), 2nd edn, University of Chicago Press, Chicago, pp. 53–272.

Kullenberg, G.E.B. (1971) Vertical diffusion in shallow waters, *Tellus*, **23**, 129–35.

Kullenberg, G.E.B. (1978) Vertical processes and the vertical–horizontal coupling, in *Spatial Pattern in Plankton Communities* (ed. J. H. Steele), Plenum, New York, pp. 43–710.

Kwiatkowski, R.E. (1982) Trends in Lake Ontario surveillance parameters, 1974–1980, *J. Great Lakes Res.*, **8**, 648–59.

Lam, D.C.L., Schertzer, W.M. and Fraser, A.S. (1983) Simulation of Lake Erie water quality responses to loading and weather variations, *Sci. Ser. No. 134 National Water Res. Inst.*, Inland Waters Directorate, Canada Centre for Inland Waters Burlington, Ontario.

Lamb, H.H. (1972) *Climates: Present, Past and Future*, Vol. 1, *Fundamentals and Climate Now*, Methuen, London.

Lamb, H.H. (1977) *Climates, Past, Present, Future*, Vol. II, *Climatic History and the Future*, Methuen, London.

Lampert W. and Schober, V. (1980) The importance of threshold food concentrations, *Evolution and Ecology of Zooplankton Communities* (ed. W. C. Kerfoot) University Press, New England, Hanover, New Hampshire, pp. 264–7.

Lane, P. and Levins, R. (1977) The dynamics of aquatic systems, 2, The effects of nutrient enrichment on model plankton communities, *Limnol. Oceanogr.*, **22**, 454–71.

Large, W.G. and Pond, S. (1981) Open ocean momentum flux measurements in moderate to strong winds, *J. Phys. Oceanogr.*, **11**, 324–36.

Larkin, P.A. and Northcote, T.G. (1969) Fish as indices of eutrophication, in *Eutrophication* (ed. G. A. Rohlich), National Academy of Science, Washington, pp. 256–73.

Larsen, D.P. and Mercier, H.T., (1976) Phosphorus retention capacity of lakes, *J. Fish Res. Bd Can.*, **33**, 1742–50.

Laws, E.A. (1975) The importance of respiration losses in controlling the size distribution of marine phytoplankton, *Ecology*, **56**, 419–26.

Laws, E.P. and Bannister, T.T. (1980) Nutrient and light-limited growth of *Thallassiosira fluviatilis* in continuous culture with implications for phytoplankton growth in the ocean, *Limnol. Oceanogr.*, **25**, 457–73.

Leah, R.T., Moss, B. and Forrest, D.E. (1980) The role of predation in causing major changes in the limnology of a hypereutrophic lake, *Int. Rev. Ges. Hydrobiol.*, **65**, 223–47.

Lean, D.R.S. (1973a) Phosphorus dynamics in lake water, *Science*, **179**, 678–80.

Lean, D.R.S. (1973b) Movements of phosphorus between its biologically important forms in lake water, *J. Fish. Res. Bd Can.*, **30**, 1525–36.

Lean, D.R.S. and Pick, F.R. (1981) Photosynthetic response of lake plankton to nutrient enrichment: a test for nutrient limitation, *Limnol. Oceanogr.*, **26**, 1001–19.

Lean, D.R.S. and White, E. (1983) Chemical and radiotracer measurements of phosphorus uptake by lake plankton, *Can. J. Fish. Aq. Sci.*, **40**, 147–55.

Le Borgne, R. (1982) Zooplankton production in the eastern tropical Atlantic Ocean: Net growth efficiency and P:B in terms of carbon, nitrogen and phosphorus, *Limnol. Oceanogr.*, **27**, 681–98.

LeCren, E.D., Kipling, C. and McCormack, J. (1977) A study of the numbers, biomass and year class strengths of perch (*Perca fluviatilis* L.) in Windermere from 1941 to 1966, *J. Anim. Ecol*, **46**, 281–307.

Legendre, L. and Demers, S. (1984) Towards dynamic biological oceanography and limnology, *Can. J. Fish. Aq. Sci.*, **41**, 2–19.

Lehman, J.T. (1980) Nutrient recycling as an interface between algae and grazers in freshwater communities, in *Evolution and Ecology of Zooplankton Communities* (ed. W. C. Kerfoot) University Press of New England, Hanover, New Hampshire, pp. 251–63.

Lerman, A. (1971) Time to chemical steady-states in lakes and ocean, in *Nonequilibrium Systems in Natural Water Chemistry* (ed. R. F. Gould), Chemical Series 106, American Chemical Society Washington, pp. 30–76.

Levins, R. (1968) *Evolution in Changing Environments*, Princeton University Press, Princeton, 1–120.

Levins, R. (1969) The effect of random variations of different types on population growth, *Proc. Natl. Acad. Sci.*, **62**, 1061–5.

Levins, R. (1979) Coexistence in a variable environment, *Am. Nat.*, **114**, 765–83.

Lewin, R.A. (1962) *Physiology and Biochemistry of Algae*, Academic Press, New York.

Lewis, W.M. Jr (1978a) Analysis of succession in a tropical phytoplankton community and a new measure of succession, *Am. Nat.*, **112**, 401–14.

Lewis, W.M. Jr (1978b) A compositional, phytogeographical and elementary structural analysis of the phytoplankton in a tropical lake: Lake Lanao, Philippines, *J. Ecol.*, **66**, 213–26.

Li, W.K.W. and Goldman, J.C. (1981) Problems in estimating growth rates of marine phytoplankton from short term ^{14}C assays, *Microb. Ecol.*, **7**, 113–21.

Li, W.K.W. and Harrison, W.G. (1982) Carbon flow into the end products of photosynthesis in short and long incubations of a natural phytoplankton population, *Mar. Biol.*, **72**, 175–82.

Livingstone, D.A. (1963) Chemical composition of lakes and rivers, *US Geol. Surv. Professional Paper*, **440-G**, 1–64.

Lotka, A.J. (1925) *Elements of Physical Biology*, Williams and Wilkins, Baltimore, (reprinted in 1956 as *Elements of Mathematical Biology*, Dover, New York.)

Lumley, J.L. (1964) The spectrum of nearly inertial turbulence in a stably stratified fluid, *J. Atmos. Sci.*, **21**, 99–102.

Lund, J.W.G. (1949) Studies on *Asterionella*, I. The origin and nature of the cells producing seasonal maxima, *J. Ecol.*, **37**, 389–419.

Lund, J.W.G. (1950) Studies on *Asterionella formosa* (Hass.) II. Nutrient depletion and the spring maximum, Pt. I, Observations on Windermere, Esthwaite Water and Blelham Tarn, *J. Ecol.*, **38**, 1–14.

Lund, J.W.G. (1950) Studies on *Asterionella formosa* (Hass.) II. Nutrient depletion and the spring maximum, Pt. II, Discussion, *J. Ecol.*, **38**, 15–35.

Lund, J.W.G. (1954) The seasonal cycle of the plankton diatom *Melosira italica* (Ehr) Kutz subsp. *subarctica* (O. Mull), *J. Ecol.*, **42**, 151–79.

Lund, J.W.G. (1954–55) Chemical analysis in ecology illustrated from Lake District tarns and lakes II. Algal differences, *Proc. Linn. Soc. Lond., 167th session*, **167**, 165–71.

Lund, J.W.G. (1955) Further observations on the seasonal cycle of *Melosira italica* (Ehr) Kutz subsp. *subarctica* (O. Mull), *J. Ecol.*, **43**, 90–102.

Lund, J.W.G. (1961) The periodicity of algae in three English lakes, *Verh. Int. Verein Limnol.*, **14**, 147–54.

Lund, J.W.G. (1964) Primary production and periodicity of phytoplankton, *Verh. Int. Verein. Limnol.*, **15**, 37–56.

Lunds, J.W.G. (1969) Phytoplankton, in *Eutrophication* (ed. G. A. Rohlich) National Academy of Science, Washington, pp. 306–30.

Lund, J.W.G. (1971a) An artificial alteration of the seasonal cycle of the plankton diatom *Melosira italica* subsp. *subarctica* in an English lake, *J. Ecol.*, **59**, 521–33.

Lund, J.W.G. (1971b) The seasonal periodicity of three planktonic desmids in Windermere, *Mitt. Int. Verein. Limnol.*, **19**, 3–25.

Lund, J.W.G. (1972a) Eutrophication, *Proc. Roy. Soc. Lond. B*, **180**, 371–82.

Lund, J.W.G. (1972b) Changes in the biomass of blue-green and other algae in an English Lake from 1945–69, in *Proceedings of Symposium on Taxonomy and Biology of Blue-Green Algae*, University of Madras, Madras, pp. 305–27.

Lund, J.W.G. (1972c) Preliminary observations on the use of large experimental tubes in lakes, *Verh. Int. Verein. Limnol.*, **18**, 71–7.

Lund, J.W.G. (1975) The uses of large experimental tubes in lakes, in *Proceedings on Symposium on the Effects of Storage on Water Quality'*, Reading University (March 1975), Water Res. Centre, Medmenham, Bucks, pp. 291–312.

Lund, J.W.G. (1978) Changes in the phytoplankton of an English lake, 1945–1977, *Gidrobiol. Zh. Kiev.*, **14**, 10–27.

Lund, J.W.G. (1979) Changes in the phytoplankton of an English lake, 1945–1977, *Hydrobiol. J.* **14**, 6–21.

Lund, J.W.G. (1981) Investigations on phytoplankton with special reference to water usage, Occasional Publication No. 13, Freshwater Biological Association, Windermere, pp. 1–64.

Lund, J.W.G. and Reynolds, C.S. (1982) The development and operation of large limnetic enclosures in Blelham Tarn, English Lake District, and their contribution to phytoplankton ecology, *Prog. Phycol. Res.*, 1, 1–65.

Lund J.W.G., Kipling, C. and Le Cren, E.D. (1958) The inverted microscope method of estimating algal numbers and the statistical basis of estimations by counting, *Hydrobiologia*, 11, 143–70.

Lynch, M. (1979) Predation, competition, and zooplankton community structure: an experiemental study, *Limnol. Oceanogr.*, 24, 253–72.

Lynch, M. (1983) Estimation of size specific mortality rates in zooplankton populations by periodic sampling, *Limnol. Oceanogr.*, 28, 533–45.

Lynch, M. and Shapiro, J. (1981) Predation, enrichment, and phytoplankton community structure, *Limnol. Oceanogr.*, 26, 86–102.

Macan, T.T. (1970) *Biological Studies of the English Lakes*, Longman, London.

MacArthur, R.H. (1960) On the relative abundance of species, *Am. Nat.*, 94, 25–36.

MacArthur, R.H. (1962) Some generalised theorems of natural selection, *Proc. Natl. Acad. Sci., USA*, 48, 1893–1897.

MacArthur, R.H. (1972) *Geographical Ecology*, Harper and Row, New York.

MacArthur, R.H. and Levins, R. (1964) Competition, habitat selection and character displacement in a patchy environment, *Proc. Natl. Acad. Sci. US*, 51, 1207–10.

MacArthur, R.H. and Wilson, E.O. (1967) *The Theory of Island Biogeography*, Princeton University Press, Princeton, New Jersey.

Machta, L. (1973) Prediction of CO_2 in the atmosphere, in *Carbon and the Biosphere* (eds G. M. Woodwell and E. V. Pecon), Tech. Inf. Centre, USAEC, pp. 21–31.

Mackereth, F.J. (1953) Phosphorus utilization by *Asterionella formosa* (Hass), *J. Exp. Bot.*, 4, 296–313.

Maguire, B. Jr (1963) The passive dispersal of small aquatic organisms and their colonization of small bodies of water, *Ecol. Mongr.*, 33, 161–85.

Maguire, B. Jr (1971) Community structure of protozoans and algae with particular emphasis on recently colonized bodies of water, *Res. Div. Monogr. Va. Polytech. Inst. St. Univ.*, 3, 121–49 Blacksburg, Va; (reprinted 1977 in *Aquatic Microbial Communities*, (ed. J. Cairns), Garland Ref. Lib. Sci Technol. 15, 355–97, Garland, New York.)

Maguire, B. Jr and Niell, W.E. (1971) Species and individual productivity in phytoplankton communities, *Ecology*, 52, 903–7.

Makarewicz, J.C. and Baybutt, R.I. (1981) Long term (1927–1978) changes in the phytoplankton community of Lake Michigan at Chicago, *Bull. Torrey Bot. Club*, 108, 240–54.

Makarewicz, J.C., Baybutt, R.I. and Damann, K. (1979) Changes in the apparent temperature optima of the plankton of Lake Michigan at Chicago, Illinois, *J. Fish. Res. Bd Can.*, 361169–73.

Malone, F.D. (1968) An analysis of current measurements in Lake Michigan, *J. Geophys. Res.*, 73, 7065–81.

Malone, T.C. (1971a) The relative importance of nanoplankton and netplankton as primary producers in tropical oceanic and neritic phytoplankton communities, *Limnol. Oceanogr.*, **16**, 633–9.

Malone, T.C. (1971b) The relative importance of nanoplankton and net plankton as primary producers in the California current system, *Fish Bull.*, **69**, 799–820.

Malone, T.C. (1977) Light saturated photosynthesis by phytoplankton size fractions in the New York Bight (USA), *Mar. Biol.*, **42**, 281–92.

Malone, T.C. (1980a) Algal size, in *The Physiological Ecology of Phytoplankton* (ed. I. Morris), Blackwell, Oxford, pp. 433–63.

Malone, T.C. (1980b) Size fractionated primary productivity of marine phytoplankton in *Primary Productivity in the Sea* (ed. P. Falkowski), *Env. Sci. Res. 19, (Brookhaven Symposium Biology 31)*, Plenum, New York, pp. 301–19.

Manzi, J.J., Stofan, P.E. and Dupuy, J.L. (1977) Spatial heterogeneity of phytoplankton populations in estuarine surface microlayers, *Mar. Biol.*, **41**, 29–38.

Margalef, R. (1958) Temporal succession and spatial heterogeneity in phytoplankton, in *Perspectives in Marine Biology* (ed. A. A. Buzzati-Traverso), University California Press, Berkeley, pp. 323–49.

Margalef, R. (1961) Communication of structure in planktonic populations, *Limnol. Oceanogr.*, **6**, 124–8.

Margalef, T. (1963a) Succession in marine populations, in *Advancing Frontiers of Plant Sciences 2* (ed. R. Vira), Institute for Advanced Science Culture, New Delhi, India, pp. 137–88.

Margalef, R. (1963b) On certain unifying principles in ecology, *Am. Nat.*, **97**, 357–74.

Magalef, R. (1964) Correspondence between the classic types of lakes and the structural and dynamic properties of their populations, *Verh. Int. Verein. Limnol.*, **15**, 169–75.

Margalef, R. (1967) Some concepts relative to the organisation of plankton, *Oceanogr. Mar. Biol. Ann. Rev.*, **5**, 257–89.

Margalef, R. (1968) *Perspectives in Ecological Theory*, University Chicago Press, Chicago.

Margalef, R. (1975) External factors and ecosystem stability, *Schweiz. Z. Hydrol.*, **37**, 102–17.

Margalef, R. (1978a) Phytoplankton communities in upwelling areas. The example of NW Africa, *Oecol. Aquatica*, **3**, 97–132.

Margalef, R. (1978b) Life-forms of phytoplankton as survival alternatives in an unstable environment, *Oceanologica Acta*, **1**, 493–509.

Margalef, R. (1978c) General concepts of population dynamics and food links, in *Marine Ecology IV, Dynamics* (ed. O. Kinne), Wiley New York, pp. 617–704.

Margalef, R. (1980) *La Biosfera*, Ediciones Omega, Barcelona.

Margalef, R. and Mir, M. (1979) Phytoplankton of Spanish reservoirs as dependent from environmental factors and potential indicator of water properties, *Attie del Convegro sui Bacini Lacustri Artificiali, Sassari, Ottobre 1977* pp. 191–206.

Margalef, R., Mir, M. and Estrada, M. (1982) Phytoplankton composition and distribution as an expression of properties of reservoirs, *Canadian Water Res.*, **7**, 26–49.

Marra, J. (1978a) Effect of short term variations in light intensity on photosynthesis of a marine phytoplankter-laboratory simulation study, *Mar. Biol.,* 46, 191–202.

Marra, J. (1978b) Phytoplankton photosynthetic response to vertical movement in mixed layers, *Mar. Biol.,* 40, 203–8.

Marra, J. (1980) Vertical mixing and primary production in *Primary Productivity in the Sea* (ed. P. Falkowski), *Env. Sci. Res. 19, (Brookhaven Symposium Biology 31),* Plenum, New York, pp. 121–38.

Marra, J. and Heinemann, K. (1982) Photosynthesis response by phytoplankton to sunlight variability, *Limnol. Oceanogr.,* 27, 1141–53.

Marra, J., Landriau, G. and Ducklow, H.W. (1981) Tracer kinetics and plankton rate processes in oligotrophic oceans, *Mar. Biol. Lett,* 2, 215–23.

Martin, D.M. and Goff, D.R. (1973) The role of nitrogen in the aquatic environment, *Contr. Dep. Limnol. Acad. Nat. Sci. (Philadelphia),* 2, 1–46.

Matthews, S.W. (1981) New world of the ocean, *National Geograpic,* 160, 792–832.

Matuszek, J.E. (1978) Empirical predictions of fish yields of large N. American Lakes, *Trans. Am. Fish. Soc.,* 107, 385–94.

May, R.M. (1973) *Stability and Complexity in Model Ecosystems,* Princeton University Press, Princeton.

May, R.M. (1974) General Introduction in *Ecological Stability* (eds M.B. Usher and M. H. Williamson) Chapman and Hall, London, pp. 1–14.

May, R.M. (1975) Patterns of species abundance and diversity in *Ecology and Evolution of Communities,* (eds M. L. Cody and J. M. Diamond), Belknap Press, Harvard University, Cmabridge, Mass, pp. 81–120.

May, R.M. (1976) *Theoretical Ecology – Principles and Applications,* Blackwell, Oxford, pp. 142–62.

Mayr, E. (1958) Change of genetic environment and evolution, in *Evolution as a Process* (eds J. Huxley, A. C. Hardy and E. B. Ford), George Allen and Unwin, London, pp. 157–80.

Mayr, E. (1961) Cause and effect in biology, *Science,* 134, 1501–6.

Mayr, E. (1977) Darwin and Natural Selection, *Am. Sci.,* 65, 321–7.

McCarthy, J.J. (1980) Nitrogen, in *The Physiological Ecology of Phytoplankton* (ed. I. Morris), Blackwell, Oxford, pp. 191–234.

McCarthy, J.J. (1981) The kinetics of nutrient utilization in *The Physiological Bases of Phytoplankton Ecology* (ed. T. Platt), *Can. Bull. Fish Aq. Sci.,* 210, 211–33.

McCarthy, J.J. and Goldman, J.C. (1979) Nitrogenous nutrition of marine phytoplankton in nutrient-depleted waters, *Science,* 203, 670–2.

McCarthy, J.J., Taylor, W.R. and Loftus, M.E. (1974) Significance of nanoplankton in the Chesapeake Bay estuary and problems associated with the measurement of nanoplankton productivity, *Mar. Biol.,* 24, 7–16.

McCauly, E. and Briand, F. (1979) Zooplankton grazing and phytoplankton species richness: field tests of the predation hypothesis, *Limnol. Oceanogr.,* 24 243–52.

McCauley, E. and Kalff, J. (1981) Empirical relationships between phytoplankton and zooplankton biomass in lakes, *Can. J. Fish. Aq. Sci.,* 38, 458–63.

McNaughton, S.J. and Wolf, L.L. (1970) Dominance and the niche in ecological systems, *Science,* 167, 131–9.

Menzel, D.W. and Ryther, J.H. (1960) The annual cycle of primary production in the Sargasso Sea off Bermuda, *Deep Sea Res.*, **6**, 351–66.

Melack, J.M. (1976) Primary productivity and fish yields in tropical lakes, *Trans. Am. Fish. Soc.*, **105**, 575–80.

Melack, J.M. (1979) Temporal variability of phytoplankton in tropical lakes, *Oecologia (Berl)*, **44**, 1–7.

Melack, J.M. (1981) Photosynthetic activity of phytoplankton in tropical African soda lakes, *Hydrobiologia*, **81**, 71–85.

Mellor, G.L. and Durbin, P.A. (1975) The structure and dynamics of the ocean surface mixed layer, *J. Phys. Oceanogr.*, **5**, 718–28.

Monin, A.S. and Yaglom, A.M. (1971) *Statistical Fluid Mechanics*, MIT Press. Cambridge, Mass.

Monin, A.S., Kamenkovich, V.M. and Kort, V.G. (1977) *Variability of the Oceans*, (translated from Russian by J. J. Lumley) Wiley, New York.

Monod, J. (1942;) Recherches sur la croissance des cultures bactériennes, 2nd edn, *Herman at Cie*, Paris.

Monteith, J.L. (1973) *Principles of Environmental Physics*, Arnold, London.

Mooers, C.N.K., Flagg, C.N. and Boicourt, W.C. (1978) Prograde and retrograde fronts, in *Oceanic Fronts in Coastal Processes* (eds M. Bowman and W. E. Esias), Springer, Berlin, pp. 43–58.

Morel, F. and Morgan, J. (1972) A numerical method for computing equilibria in aqueous chemical systems, *Env. Sci. Technol.*, **6**, 58–67.

Morowitz, H.J. (1980) The dimensionality of niche space, *Theor. Biol.*, **86**, 259–63.

Morris, I. (1967) *An Introduction to the Algae*, Hutchinson, London.

Morris, I. (ed.) (1980a) *The Physiological Ecology of Phytoplankton*, (Studies in Ecology 7) Blackwell, Oxford.

Morris, I. (1980b) Paths of C assimilation in marine phytoplankton, in *Primary Productivity in the Sea* (ed. P. Falkowski), *Env. Sci. Res. 19 (Brookhaven Symposium Biology 31)* Plenum, New York, pp. 139–60.

Morris, I. (1981) Photosynthesis products, physiological state and phytoplankton growth in *The Physiological Bases of Phytoplankton Ecology* (ed. T. Platt), *Can. Bull. Fish. Aq. Sci.*, **210**, 83–102.

Morris, I. and Glover, H.E. (1974) Questions on the mechanism of temperature adaptation in marine phytoplankton, *Mar. Biol.*, **24**, 147–54.

Morris, I., Yentsch, C.M. and Yentsch, C.S. (1971a) Relationship between light carbon dioxide fixation and dark carbon dioxide fixation by marine algae, *Limnol. Oceanogr.*, **16**, 854–8.

Morris, I., Yentsch, C.M. and Yentsch, C.S. (1971b) The physiological state with respect to nitrogen of phytoplankton from low-nutrient subtropical water as measured by the effect of ammonium ion on dark carbon-dioxide fixation, *Limnol. Oceanogr.*, **16**, 859–68.

Mortimer, C.H. (1941–42) The exchange of dissolved substances between mud and water in lakes, *J. Ecol.*, **29**, 280–329; **30**, 147–201.

Mortimer, C.H. (1974) Lake hydrodynamics, *Mitt. Internat. Verein. Limnol.*, **20**, 124–97.

Mortimer, C.H. (1975) Substantive corrections to SIL communications (IVL Mitteilungen) Nos 6 and 20, *Verh. Internat. Verein. Limnol.*, **19**, 60–72.

Mortimer, C.H. (1979) strategies for coupling data collection and analysis with dynamic modelling of lake motions, in *Hydrodynamics of Lakes* (eds W. H. Graf and C. H. Mortimer), Elsevier, Amsterdam, pp. 183–222.

Mortonson, J.A. and Brooks, A.S. (1980) Occurrence of a deep nitrite maximum in Lake Michigan, *Can. J. Fish. Aq. Sci.,* 37, 1025–7.

Moss, B. (1972) The influence of environmental factors on the distribution of freshwater algae: and experimental study I. Introduction and the influence of calcium concentration, *J. Ecol.,* 60, 917–32.

Moss, B. (1973a) The influence of environmental factors on the distribution of freshwater algae: and experimental study II. The role of pH and the CO_2–HCO_3 system. *J. Ecol.,* 61, 157–77.

Moss, B. (1973b) The influence of environmental factors on the distribution of freshwater algae: an experimental study III. Effects of temperature, vitamin requirements and inorganic nitrogen compounds on growth, *J. Ecol.,* 61, 179–92.

Moss, B. (1973c) The influence of environmental factors on the distribution of freshwater algae: an experimental study IV. Growth of test species in natural lake waters, and conclusion, *J. Ecol.,* 61, 193–211.

Moss, B. (1973d) Diversity in freshwater phytoplankton, *Am. Midl. Nat,* 90, 341–55.

Munawar, M. and Nauwerk, A. (1971) The composition and horizontal distribution of phytoplankton in Lake Ontario during the year 1970, *Proc. 14th Conf. Gt Lakes Res. (Internat. Assoc. Gt Lakes Res.)* pp. 69–78.

Murdoch, W.W. (1966) Community structure, population control and competition – a critique, *Am. Nat.,* 100, 219–26.

Murphy, T.P. (1980) Ammonia and nitrate uptake in the lower Great Lakes, *Can J. Fish. Aq. Sci.,* 37, 1365–72.

Murthy, C.R. (1973) Horizontal diffusion in lake currents, in *Proc. Int. Symp. Hydrology of Lakes*, Helsinki. Int. Assoc. Hydrol. Sci. 109, pp. 327–334.

Myers, J. (1962) Laboratory cultures, in *Physiology and Biochemistry of the Algae* (ed. R. A. Lewin), Academic Press, New York, pp. 603–15.

Myers, J. and Graham, J.R. (1971) The photosynthetic unit in *Chlorella* measured by repetitive short flashes, *Pl, Physiol.,* 48, 282–86.

Nakashima, B.S. and Leggett, W.C. (1975) Yellow Perch *(Perca fluvescens)* biomass responses to different levels of phytoplankton and benthic biomass in Lake Memphremagog, Quebec-Vermont, *J. Fish. Res. Bd Can.,* 32, 1785–97.

Nalewajko, C. (1966) Photosynthesis and excretion in various plankton algae, *Limnol. Oceanogr.,* 11 1–10.

Nalewajko, C. (1967) Phytoplankton distribution in Lake Ontario, *Proc. 10th Conf. Gt Lakes Res.,* pp. 63–9.

Nalewajko, C. and Lean, D.R.S. (1980) Phosphorus, in *The Physiological Ecology of Phytoplankton* (ed. I. Morris), Blackwell, Oxford, pp. 235–58.

Nalewajko, C. and Lee, K. (1980) Light stimulation of phosphate uptake in marine phytoplankton, *Mar. Biol.* 74, 9–16.

Nalewajko, C., Lee, K. and Shear, H. (1981) Phosphorus kinetics in Lake Superior: light intensity and phosphorus uptake in algae, *Can. J. Fish. Aq. Sci.,* 38, 224–32.

Oakey, N.S. and Elliott, J.A. (1982) Dissipation within the surface mixed layer, *J. Phy. Oceanogr.,* 12, 171–85.

Oglesby, R.T. (1977a) Phytoplankton summer standing crop and annual productivity as functions of phosphorus loading and various physical factors, *J. Fish. Res. Bd Can.*, **34**, 2255–70.

Oglesby, R.T. (1977b) Relationships of fish yield to lake phytoplankton standing crop, production and morphoedaphic factors, *J. Fish. Res. Bd Can.*, **34**, 2271–79.

Oglesby, R.T. (1982) The MEI symposium – overview and observations, *Trans. Am. Fish. Soc.*, **111**, 171–75.

Ohle, W. (1934) Chemische und physikalische Untersuchungen norddeutscher Seen, *Arch. Hydrobiol.*, **26**, 386–464; 584–658.

Ohle, W. (1952) Die hypolimnische Kohlendioxyd – Akkumulation als produktionsbiologischer Indikator, *Arch. Hydrobiol*, **46**, 153–285.

Ohle, W. (1956) Bioactivity, production and energy utilization of lakes, *Limnol. Oceanogr.*, **1**, 139–49.

Ozmidov, R.V. (1965) Certain features of the oceanic turbulent energy spectrum, *Dokl. Acad. Nauk SSSR.* **161**, 828–32.

Palmer, M.D. (1973) Some kinetic energy spectra in a nearshore region of Lake Ontario, *J. Geophys. Res*, **78**, 3585–95.

Parker, J.I. and Edgington, D.N. (1976) Concentration of diatom frustules in Lake Michigan sediment cores, *Limnol. Oceanogr.*, **21**, 887–93.

Parsons, T.R. and Le Brasseur, R.J. (1970) The availability of food to different trophic levels in the marine food chain, in *Marine Food Chains* (ed. J. H. Steele), Oliver and Boyd, Edinburgh, pp. 325–43.

Parsons, T.R., Stephens, K. and Strickland, J.D.H. (1961) On the chemical composition of eleven species of marine phytoplankters, *J. Fish. Res. Bd Can.*, **18**, 1001–16.

Pavoni, M. (1963) Die Bedeutung des nanoplanktons im Vergleich zum Wetzplankton, *Schweiz, Z. Hydrol.*, **25**, 219–341.

Peet, R.K. (1974) The measurement of species diversity, *Ann. Rev. Ecol. Syst.*, **5**, 285–307.

Peet, R.K. (1975) Relative diversity indices, *Ecology*, **56**, 496–8.

Perry, M.J. and Eppley, R.W. (1981) Phosphate uptake by phytoplankton in the central North Pacific Ocean, *Deep Sea Res.*, **28**, 39–50.

Perry, M.J. (1976) Phosphate utilization by an oceanic diatom in phosphorus limited chemostat culture and in the oligotrophic waters of the central north Pacific, *Limnol. Oceanogr.*, **21**, 88–107.

Peters, R.H. (1976) Tautology in evolution and ecology, *Am. Nat.*, **110**, 1–12.

Peters, R.H. (1983) Size structure of the plankton community along the trophic gradient of Lake Memphremagog, *Can. J. Fish. Aq. Sci.*, **40**, 1770–8.

Peterson, B.J. (1978) Radiocarbon uptake: its relation to net particulate carbon production, *Limnol. Oceanogr.*, **23**, 179–84.

Peterson, B.J. (1980) Aquatic primary productivity and the ^{14}C-CO_2 method. *Ann. Rev. Ecol. Syst.*, **11**, 359–85.

Peterson, B.J., Barlow, J.P. and Savage, A.E. (1974) The physiological state with respect to phosphorus of Cayuga Lake phytoplankton, *Limnol. Oceanogr.*, **19**, 396–408.

Petersen, R. (1975) The paradox of the plankton: an equilibrium hypothesis, *Am. Nat.*, **109**, 35–49.

Phillips, D.W. (1978) Evaluation of evaporation from Lake Ontario during IFYGL by a modified mass transfer equation, *Water Resources Res.*, **14**, 197–205.

Phillips, O.M. (1966) *The Dynamics of the Upper Ocean*, Cambridge University Press, Cambridge.

Phillips, O.M. (1977) Entrainment, in *Modelling and Prediction of the Upper Ocean* (ed. E. B. Kraus), Pergamon, Oxford, pp. 92–101.

Pianka, E.R. (1978) *Evolutionary Ecology*, 2nd edn, Harper and Row, New York.

Pierrou, V. (1976) The global phosphorus cycle, in Ecological Bulletins NFR (Statens Naturvetenskupliga Forskiningsrad) No. 22. *Nitrogen, Phosphorus and Sulphur – Global Cycles, SCOPE* (Scientific Committee on Problems of the Environment) *report* 7, meeting at Orsundsbro, Sweden, Dec. 14–18, 1975, (eds B.H. Svensson, R. Soderlund), Swedish Natural Science Research Council, Stockholm, Sweden.

Plass, G.N. (1972) Relationship between atmospheric carbon dioxide amount and properties of the sea, *Env. Sci. Technol.*, **6**, 736–40.

Platt, J.R. (1964) Strong inference, *Science*, **146**, 347–53.

Platt, T. (1972) Local phytoplankton abundance and turbulence, *Deep Sea Res.*, **19**, 183–7.

Platt, T. (ed.) (1981) Physiological bases of phytoplankton ecology, *Can. Bull. Fish. Aq. Sci.*, **210**.

Platt, T., and Denman, K.L. (1975a) Spectral analysis in ecology, *Ann. Rev. Ecol. Syst.*, **6**, 189–210.

Platt, T. and Denman, K.L. (1975b) A general equation for the mesoscale distribution of phytoplankton in the sea, *Mem. Soc. R. Sci. Liege. 6 ser.*, **VII**, 31–42.

Platt, T. and Denman, K. (1977) Organisation in the pelagic ecosystem, *Helgoländer wiss. meeresunters*, **30**, 575–81.

Platt, T. and Denman, K. (1980) Patchiness in phytoplankton distributions, in *The Physiological Ecology of Phytoplankton* (ed. I. Morris) Blackwell, Oxford, pp. 413–32.

Platt, T. and Gallegos, C.L. (1981) Modelling primary production, in *Primary Productivity in the Sea* (ed. P. Falkowski), *Env. Sci. Res. 19, (Brookhaven Symposium Biology 31)*, Plenum, New York, pp. 339–62.

Platt, T., Dickie, L.M. and Trites, R.W. (1970) Spatial heterogeneity of phytoplankton in a nearshore environment, *J. Fish. Res. Bd Can.*, **27**, 1453–73.

Platt, T., Subba Rao, D.V. and Irwin, B. (1983) Photosynthesis of picoplankton in the oligotrophic ocean, *Nature*,, 702–4.

Pollingher, U. and Berman, T. (1977) Quantitative and qualitative changes in the phytoplankton of Lake Kinneret, Israel, 1972–1975, *Oikos*, **29**, 418–28.

Pomeroy, C.R. (1960) Residence time of dissolved phosphate in natural waters, *Science*, **131**, 1731–2.

Poole, R.W. (1976) Stochastic difference equation predictors of population fluctuations, *Theor. Pop. Biol.*, **9**, 25–45.

Porter, J.W.M Woodley, J.D., Smith, G.J. *et al.* (1981) Population trends among Jamaican reef corals, *Nature*, **294**, 249–50.

Porter, K.G. (1973) Selective grazing and differential digestion of algae by zooplankton, *Nature*, **244**, 179–80.

Porter, K.G. (1977) The plant-animal interface in freshwater ecosystems, *Am. Sci.,* **65**, 159–70.

Porter, K.G., Orcutt Jr, J.D., and Gerristen, J. (1983) Functional response and fitness in a generalist filter feeder *Daphnia magna* (Cladocera, Crustacea), *Ecology,* **64**, 735–42.

Postma, H. (1971) Distribution of nutrients in the sea and the oceanic nutrient cycle, in *Fertility of the Sea* (ed. J. D. Costlow), Gordon and Breach, New York, pp. 337–49.

Poulet, S.A. (1978) Comparison between five coexisting species of marine copepods feeding on naturally occurring particulate matter, *Limnol. Oceanogr.,* **23**, 1126–43.

Powell, T. and Richerson, P.J. (1985) Temporal variation, spatial heterogeneity and competition for resources in plankton systems: a theoretical model, *Am. Nat.,* **125**, 431–464.

Powell, T.M., Richerson, P.J., Dillon, T.M. *et al.* (1975) Spatial scales of current speed and phytoplankton biomass fluctuations in Lake Tahoe, *Science,* **189**, 1088–90.

Prepas, E.E. (1983a) Orthophosphate turnover time in shallow productive lakes, *Can J. Fish. Aq. Sci.,* **40**, 1412–8.

Prepas, E.E. (1983b) Total dissolved solids as a predictor of lake biomass and productivity, *Can. J. Fish. Aq. Sci.,* **40**, 92–5.

Prescott, G.W. (1968) *The Algae: a Review* Houghton Miflin Co., Boston.

Preston, F.W. (1984) The commonness and rarity of species, *Ecology,* **29**, 254–83.

Preston, F.W. (1962) The canonical distribution of commonness and rarity, *Ecology,* **43**, 185–215; 410–32.

Prigogine, I. (1978) Time, structure and fluctuations, *Science,* **201**, 777–85.

Prowse, G.A. and Talling, J.F. (1958) The seasonal growth and succession of plankton algae in the White Nile, *Limnol. Oceanogr.,* **3**, 222–38.

Quay, P.D., Broecker, W.S., hesslein, R.H. and Schindler, D.W. (1980) Vertical diffusion rates determined by tritium tracer tracer experiments in the thermocline and hypolimnion of two lakes, *Limnol. Oceanogr.,* **25**, 201–18.

Quarmby, C.M., Turpin, D.H. and Harrison, P.J. (1982), Physiological responses of two marine diatoms to pulsed additions of ammonium, *J. Exp. Mar. Biol. Ecol.,* **63**, 173–82.

Ragotzkie, R.A. (1978) Heat budgets of lakes, in *Lakes: Chemistry, Geology, Physics,* (ed. A. Lerman), Springer, Berlin, pp. 1–19.

Rai, H. (1982) Primary production of various size fractions of natural phytoplankton communities in a North German lake, *Arch. Hydrobiol.,* **95**, 395–412.

Rast, W., Jones, R.A. and Lee, G.F., (1983) Predictive capability of US OECD phosphorus loading-eutrophication response models, *J. Wat. Poll. Control Fed.,* **55**, 990–1003.

Raven, J.A. and Beardall, J. (1981) Respiration and photorespiration, *Can. Bull. Fish. Aq. Sci.,* **210**, 55–82.

Raymont, J.E.G. (1963) *Plankton and Productivity in the Oceans,* Pergamon Press, Oxford.

Raymont, J.E.G. (1980) *Plankton and Productivity in the Oceans, Vol. I, Phytoplankton,* 2nd edn, Pergamon Press, Oxford.

Redfield, A.C. (1958) The biological control of chemical factors in the environment, *Am. Sci.*, **46**, 205–22.

Redfield, A.C., Ketchum, B.H. and Richards, F.A. (1963) The influence of organisms on the composition of sea-water, in *The Sea* (ed. M. N. Hill), Vol. 2, Wiley Interscience, New York, pp. 26–79.

Regier, H.A. and Hartman, W.L. (1973) Lake Erie's fish community: 150 years of cultural stresses, *Science*, **180**, 1248–55.

Reid, P.C. (1975) Large scale changes in North Sea phytoplankton, *Nature*, **257**, 217–9.

Reid, P.C. (1977) Continuous plankton records: changes in the composition and abundance of the phytoplankton of the North-Eastern Atlantic ocean and the North Sea, 1958–1974, *Mar. Biol.*, **40**, 337–9.

Reid, P.C. (1978) Continuous plankton records: large-scale changes in the abundance of phytoplankton in the North Sea from 1958 to 1973, *Rapp. Proc. Verb Réun. Cons. Perm Int. Explor. Mer*, **172**, 384–9.

Reid, P.C., Hunt, H.G. and Jonas, T.D. (1983) Exceptional blooms of diatoms associated with anomalous hydrographic conditions in the Southern Bight in early 1977, *J. Plankton Res.* **5**, 755–65.

Renberg, I. and Hellberg, T. (1982) The pH history of lakes in SW Sweden, as caluculated from the subfossil diatom flora of the sediments, *Ambio.*, **11**, 30–3.

Reynolds, C.S. (1973a) The seasonal periodicity of planktonic diatoms in a shallow eutrophic lake, *Freshwat. Biol.*, **3**, 89–110.

Reynolds, C.S. (1973b) Phytoplankton periodicity of some north Shropshire meres, *Br. Phyc. J.*, **8**, 301–20.

Reynolds, C.S. (1976a) Sinking movements of phytoplankton indicated by a simple trapping method I.A. *Fragilaria* population, *Br. Phyc. J.*, **11**, 279–91.

Reynolds, C.S. (1976b) Sinking movements of phytoplankton indicated by a simple trapping method II. Vertical activity ranges in a stratified lake, *Br. Phyc. J.*, **11**, 293–303.

Reynolds, C.S. (1976c) Succession and vertical distribution of phytoplankton in response to thermal stratification in a lowland mere, with special reference to nutrient availability, *J. Ecol.*, **64**, 529–51.

Reynolds, C.S. (1978) Phosphorus and eutrophication of lakes – a personal view, in *Phosphorus in the Environment: its Chemistry and Biochemistry* (Ciba Foundation Symp. 57 (new series)) Elsevier, Amsterdam, pp. 201–28.

Reynolds, C.S. (1979) Seston sedimentation: experiments with *Lycopodium* spores in a closed system, *Freshwat. Biol.*, **9**, 55–76.

Reynolds, C.S. (1980a) Phytoplankton assemblages and their periodicity in stratifying lake systems, *Hol. Ecol.*, **3**, 141–59.

Reynolds, C.S. (1980b) Processes controlling the quantities of biogenic materials in lakes and reservoirs subject to cultural eutrophication. *Proc. of the Joint Anglo-Soviet Committee on Co-operation in the Field of Environmental Protection*, (ed. T. R. Graham), Department of the Environment, GOSGIMET Academy of Sciences of the USSR, pp. 45–62. *2nd Anglo-Soviet Seminar, Elaboration of the scientific bases for monitoring the quality of surface waters by hydrobiological indicators*, Windermere, UK, April 1979, pp. 24–6.

Reynolds, C.S. (1982) Phytoplankton periodicity: its motivation, mechanisms and manipulation, *50th Ann. Rep. Freshwat. Biol. Assn.*, Windermere, UK, pp. 60–75.

Reynolds, C.S. (1983a) Growth-rate responses of *Volvox aureus* (Ehrenb.) (Chlorophyta. Volvocales) to variability in the physical environment, *Br. Phyc. J.*, **18**, 433–442.

Reynolds, C.S. (1983b) A physiological interpretation of the dynamic responses of populations of a planktonic diatom to physical variability of the environment, *New Phytol.* **95**, 41–53.

Reynolds, C.S. (1984a) Phytoplankton periodicity: the interactions of form, function and environmental variability, *Freshwat. Biol.*, **14**, 111–42.

Reynolds, C.S. (1984b) *The Ecology of Freshwater Phytoplankton*, Cambridge University Press, Cambridge.

Reynolds, C.S. and Walsby, A.E. (1975) Water blooms, *Biol. Rev.*, **50**, 437–481.

Reynolds, C.S. and Wiseman, S.W. (1982) Sinking losses of phytoplankton in closed limnetic systems, *J. Plankton Res.*, **4**, 489–522.

Reynolds, C.S., Morison, H.R. and Butterwick, C. (1982) The sedimentary flux of phytoplankton in the south basin of Windermere, *Limnol. Oceanogr.*, **27**, 1162–75.

Reynolds, C.S., Wiseman, S.W. and Clarke, M.J.O. (1984) Growth- and loss-rate responses of phytoplankton to intermittent artificial mixing and their potential application to the control of planktonic algal biomass, *J. Appl. Ecol*, **21**, 11–39.

Reynolds, C.S., Harris, G.P. and Gouldney, D.N. (1985) Comparison of carbon-specific growth rates and rates of cellular increase of phytoplankton in large limnetic enclosures, *J. Plankton Res.*, **7** (in press).

Reynolds, C.S., Thompson, J.M., Ferguson, A.J.D. and Wiseman, S.W. (1982) Loss processes in the population dynamics of phytoplankton maintained in closed systems, *J. Plankton Res.*, **4**, 561–600.

Reynolds, C.S., Wiseman, S.W., Godfrey, B.M. and Butterwick, C. (1983) Some effects of artificial mixing on the dynamics of phytoplankton populations in large limnetic enclosures, *J. Plankton Res.*, **5**, 203–34.

Rhee, G.Y. (1979) Continuous culture in phytoplankton ecology, in *Advances in Aquatic Microbiology* (ed. M. R. Droop and H. W. Jannasch), Academic Press, London and New York, pp. 150–207.

Richards, F.A. (1958) Dissolved silicate and some related properties of some western N. Atlantic and Caribbean waters, *J. Mar. Res.*, **17**, 449–65.

Richards, F.A. (1965) Anoxic basins and fjords, in *Chemical Oceanography I* (eds J. P. Riley and G. Skirrow), Academic Press, London, pp. 611–43.

Richardson, P.L., Cheney, R.E. and Worthington, L.V. (1978) A census of Gulf Stream rings, spring 1975, *J. Geophys. Res.*, **83**, 6136–44.

Richerson, P.J., Armstrong, R. and Goldman, C.R. (1970) Contemporaneous disequilibrium: a new hypothesis to explain the 'paradox of the plankton', *Proc. Natl. Acad. Sci.*, **67**, 1710–14.

Richerson, P.J., Powell, T.M., Leigh-Abott, M.R. and Coil, J.A. (1978) Spatial heterogeneity in closed basins, in *Spatial Pattern in Plankton Communities* (ed. J. H. Steele), Plenum, New York, pp. 239–76.

Rigler, F.H. (1964) The phosphorus fractions and the turnover time of inorganic phosphorus in different types of lakes, *Limnol. Oceanogr.*, 9, 511–18.

Rigler, F.H. (1966) Radiobiological analysis of inorganic phosphorus in lake water, *Verh. Int. Verein. Limnol.*, 16, 465–70.

Rigler, F.H. (1975) The concept of energy flow and nutrient flow between trophic levels, in *Unifying Concepts in Ecology* (eds W. H. van Dobben and R. H. Lowe-McConnell), Junk, The Hague, pp. 15–26.

Rigler, F.H. (1982a) Recognition of the possible: an advantage of empiricism in ecology, *Can. J. Fish. Aq. Sci.*, 39, 1323–31.

Rigler, F.H. (1982b) The relation between fisheries management and limnology, *Trans. Am. Fish. Soc.*, 11, 121–31.

Riley, G.A. (1942) The relationship of vertical turbulence and spring distom flowerings, *J. Mar. Res.*, 5, 67–87.

Riley, G.A. (1946) Factors controlling phytoplankton populations on Georges Bank, *J. Mar. Res.*, 6, 54–73.

Riley, G.A. (1951) Oxygen, phosphate and nitrate in the Atlantic Ocean, *Bull. Bingham. Oceanogr. Coll.*, 13, 1–126.

Riley, G.A. (1953) Letters to the editor, *J. Cons. perm. int. Explor. Mer*, 19, 85–9.

Riley, G.A. (1956) Oceanography of Long Island Sound 1952–1954, IX. Production and utilization of organic matter, *Bull. Bingham. Oceanogr. Coll.*, 15, 324–44.

Riley, G.A. (1957) Phytoplankton of the north central Sargasso Sea, 1950–52, *Limnol. Oceanogr.*, 2, 252–70.

Riley, G.A. and Gorgy, S. (1948) Quantitative studies of the summer plankton populations of the Western North Atlantic, *J. Mar. Res.*, 7, 100–121.

Riley, G.A., Stommel, and Bumpus, D.F. (1949) Quantitative ecology of the plankton of the Western North Atlantic, *Bull. Bingham. Oceanogr. Coll.*, 12(3), 1–169.

Robinson, G.A. (1970) Continuous plankton records: variation in the seasonal cycle of phytoplankton in the North Atlantic, *Bull. Mar. Ecol.*, 6, 333–45.

Robinson, G.A. (1983) Continuous plankton records: phytoplankton in the North Sea, 1958–1980, with special reference to 1980, *Br. Phyc. J.*, 18, 131–9.

Robinson, G.A., Colebrook, J.M. and Cooper, G.A. (1975) The continuous plankton recorder survey: plankton in the ICNAF area, 1961–71, with special reference to 1971, *Res. Bull. Int. Comm. NW Atlantic Fish.*, 11, 61–71.

Rodhe, W. (1958) The primary production in lakes: some results and restrictions of the ^{14}C method, *Rapp. Proc. Verb Réun. Cons. Int. Perm Explor. Mer.*, 144, 122–128.

Rodhe, W. (1974) Plankton, planktic, planktonic, *Limnol. Oceanogr.*, 19, 360.

Rodhe, W., Vollenweider, R.A. and Nauwerck, A. (1958) The primary production and standing crop of phytoplankton, in *Perspectives in Marine Biology* (ed. A. A. Buzzati-Traverso) University of California Press, Berkeley, pp. 299–322.

Rogers, G.K. (1965) The thermal bar in the Laurentian Great Lakes, *Proc. 8th Conf. Great Lakes Res.*, University of Michigan Gt Lakes Res. Div., pp. 358–63.

Rogers, G.K. (1966) The thermal bar in Lake Ontario, spring 1965 and winter 1965, *Proc. 9th Conf. Gt Lakes Res.*, University of Michigan Gt Lakes Res. Div., pp. 369–74.

Rogers, G.K. (1968) Heat advection within Lake Ontario in spring and surface water transparency associated with the thermal bar, *Proc. 11th. Conf. Gt Lakes Res.*, Int. Assoc. Gt Lakes Res., pp. 480–6.

Ross, P.E. and Kalff, J. (1975) Phytoplankton production in L. Memphremagog Quebec (Canada) – Vermont (USA) *Verh. Int. Verein. Limnol.*, 19, 760–9.

Roughgarden, J. (1975a) Population dynamics in a stochastic environment: spectral theory for the linearized N-species Lotka–Volterra competition equations, *Theor. Pop. Biol.*, 7, 1–12.

Roughgarden, J. (1975b) A simple model for population dynamics in stochastic environments, *Am. Nat.*, 109, 713–36.

Round, F.E. (1971) The growth and succession of algal populations in freshwaters, *Mitt. Int. Verein. Limnol.*, 19, 70–99.

Round, F.E. (1973) *The Biology of the Algae*, 2nd edn, Edward Arnold, London.

Round, F.E. (1981) *The Ecology of Algae*, Cambridge, University Press, Cambridge.

Ruse, M. (1979) *The Darwinian Revolution*, University of Chicago Press, Chicago.

Russell, F.S. (1973) A summary of the observations on the occurrence of planktonic stages of fish off Plymouth 1924–1972, *J. Mar. Biol. Assn UK*, 53, 347–55.

Russell, F.S., Southward, A.J., Boalch, G.T. and Butler, E.I. (1971) Changes in biological conditions in the English Channel off Plymouth during the last half century, *Nature*, 234, 468–70.

Ruttner, F. (1952) Planktonstudien der Deutschen Limnologischen Sunda-Expedition, *Arch. Hydrobiol. Suppl.*, 21, 1–274.

Ryder, R.A. (1954) A method for estimating the potential fish production of North-temperate lakes, *Trans. Am. Fish. Soc.*, 94, 214–8.

Ryder, R.A. (1965) A method for estimating the potential fish production of North-temperate lakes, *Trans. Am. Fish. Soc.*, 94, 214–8.

Ryder, R.A. (1972) The limnology and fishes of oligotrophic glacial lakes in North America (about 1800 AD), *J. Fish. Res. Bd Can.*, 19, 617–28.

Ryder, R.A. (1982) The morphoedaphic index – use, abuse and fundamental concepts, *Trans. Am. Fish. Soc.*, 111, 154–64.

Ryder, R.A., Kerr, S.R., Loftus, K.H. and Regier, H.A. (1974) The morphoedaphic index, a fish yield estimator – review and evaluation, *J. Fish. Res. Bd Can.*, 31, 663–88.

Ryther, J.H. (1963) Geographic variations in productivity, in *The Sea* (ed. M. V. Hill), Vol. 3, Wiley Interscience, New York, pp. 347–80.

Ryther, J.H. (1969) Photosynthesis and fish production in the sea, *Science*, 166, 72–6.

Ryther, J.H. and Guillard, R.R.L. (1962) Studies of marine planktonic diatoms III. Some effects of temperature on respiration of five species, *Can. J. Micro Biol.*, 8, 447–53.

Ryther, J.H. and Yentsch, C.S. (1957) The estimation of phytoplankton production in the ocean from chlorophyll and light data, *Limnol. Oceanogr.*, 2, 281–6.

Sakshaug, E., Andersen, K., Myklestad, S. and Olsen, Y. (1983) Nutrient status of phytoplankton communities in Norwegian waters (marine, brackish and fresh) as revealed by their chemical composition, *J. Plankton Res.*, 5, 175–96.

Sale, P.F. (1977) The maintenance of high diversity in coral reef fish communities, *Am. Nat.*, 111, 337–59.

Saunders, P.M. (1971) Anticyclonic eddies formed from shoreward meanders of the Gulf Stream, *Deep Sea Res.*, **18**, 1207–19.

Schaffer, W.M. (1981) Ecological abstraction: the consequences of reduced dimensionality in ecological models, *Ecol. Monogr.*, **51**, 383–401.

Schelske, C.L. and Stoermer, E.F. (1971) Eutrophication, silica depletion and predicted changes in algal quality in Lake Michigan, *Science*, **173**, 423–4.

Schelske, C.L. and Stoermer, E.F. (1972) Phosphorus, silica and eutrophication of Lake Michigan, *Spec. Symp. Am. Soc. Limnol. Oceanogr.*, **1**, 157–71.

Schertzer, W.M. (1978) Energy budget and monthly evaporation estimates for Lake Superior, 1973, *J. Gt Lakes Res.*, **4**, 320–30.

Schindel, D.E. (1982) The gaps in the fossil record, *Nature*, **297**, 282–4.

Schindler, D.W. (1971) Carbon, nitrogen and phosphorus and the eutrophication of freshwater lakes, *J. Phycol.*, **7**, 321–9.

Schindler, D.W. (1974) Eutrophication and recovery in experimental lakes: implications for lake management, *Science*, **184**, 897–9.

Schindler, D.W. (1975) Whole-lake eutrophication experiments with phosphorus, nitrogen and carbon, *Verh. Int. Verein Limnol.*, **19**, 3221–31.

Schindler, D.W. (1977) Evolution of phosphorus limitation in lakes, *Science*, **195**, 260–2.

Schindler, D.W. (1978) Factors regulating phytoplankton production and standing crop in the world's freshwaters, *Limnol. Oceanogr.*, **23**, 478–86.

Schindler, D.W., Armstrong, F.A.J., Holmgren, S.K. and Brunskill, G.J. (1971) Eutrophication of lake 227, Experimental lakes area, Northwestern Ontario by addition of phosphate and nitrate, *J. Fish. Res. Bd Can.*, **28**, 1763–82.

Schindler, D.W., Brunskill, G.J., Emerson, S. *et al.* (1972) Atmospheroic carbon dioxide: its role in maintaining phytoplankton standing crops, *Science*, **177**, 1192–4.

Schindler, D.W., Kling, H., Schmidt, R.V. *et al.* (1973) Eutrophication of lake 223 by addition of phosphate and nitrate: the second, third and fourth years of enrichment, *J. Fish, Res. Bd Can.*, **30**, 1415–40.

Schlesinger, D.A. and Regier, H.A. (1982) Climatic and morphoedaphic indices of fish yields from natural lakes, *Trans. Am. Fish. Soc.*, **111**, 141–50.

Schlesinger, D.A. and Regier, H.A. (1983) Relationship between environmental temperature and yields of subarctic and temperate zone fish species, *Can. J. Fish. Aq. Sci.*, **40**, 1829–37.

Schlesinger, D.A., Molot, L.A. and Shuter, B.J. (1981) Specific growth rates of freshwater algae in relation to cell size and light intensity, *Can. J. Fish. Aq. Sci.*, **28**, 1052–8.

Schwinghamer, P. (1981) Characteristic size distributions of integral benthic communities, *Can. J. Fish. Aq. Sci.*, **38**, 1255–63.

Schwinghamer, P. (1983) Generating ecological hypotheses from biomass spectra using causal analysis: a benthic example, *Mar. Ecol. Prog. Ser.*, **13**, 151–66.

Seliger, H.H., Loftus, M.E. and Subba Rao, D.V. (1975) Dinoflagellate accumulations in Chesapeake Bay, in *Proc. 1st Int. Conf., Toxic dinoflagellate blooms*, Boston (ed. V. R. LoCicero), Massachusetts Science and Technology Foundation, Wakefield, Mass, pp. 181–205.

Semina, H.G. (1972) The size of phytoplankton cells in the Pacific Ocean, *Int. Revue. Ges. Hydrobiol.*, **57**, 177–205.

Sephton, D. (1980) Time series studies of phytoplankton in Hamilton Harbour, MSc thesis, McMaster University, Hamilton, Ontario.

Sephton, D.H. and Harris, G.P. (1984) Physical variability and phytoplankton communities VI. Day to day changes in primary productivity and species abundance, *Arch. Hydrobiol.*, **102**, 155–75.

Serruya, C. and Berman, T. (1975) Phosphorus, nitrogen and the growth of algae in Lake Kinneret, *J. Phycol.*, **11**, 155–62.

Shapiro, A.M. (1978) Letter to the Editor, *Am. Sci.*, **66**, 540–1.

Shapiro, J. (1973a) Blue-green algae: why they become dominant, *Science*, **179**, 382–4.

Shapiro, J. (1973b) Reply to a comment by J. C. Goldman (1973) *Science*, **182**, 306–7) *Science*, **182**, 307.

Shapiro, J. (1977) Biomanipulation – a neglected approach? MS of lecture presented at Plenary session of 40th Ann Meeting, American Society of limnology and Oceanography, Michigan State University June 20th, 1977.

Shapiro, J. (1978) The need for more biology in lake restoration, *Com. No. 183 Limnol. Res. Centre Univ. Minnesota*, Minn. Mimeo.

Shapiro, J. (1980) The importance of trophic level interactions to the abundance and species composition of algae in lakes, *Dev. Hydrobiol.*, **2**, 105–16.

Shapiro, J. and Swain, E.B. (1983) Lessons from the silica 'decline' in Lake Michigan, *Science*, **221**, 457–9.

Shapiro, J., Forsberg, B., Lamarra, V. *et al.* (1982) Experiments and experiences in biomanipulation: studies of biological ways to reduce algal and eliminate blue-greens, *Interim Rep. No. 19 Limnol. Res. Centre*, University of Minnesota Minneapolis, Minn. USA, EPA-600/3–82–096 Corvallis Env. Res. Lab. ORD US EPA Corvallis Oregon USA.

Shapiro, J., Lamarra, V. and Lynch, M. (1975) Biomanipulation: an ecosystem approach to lake restoration, *Proc. Symp. Water quality management through biological control* Florida (eds P. L. Brezonic and J. L. Fox), Dept of Environmental Engineering and Science and USEPA, University of Florida, Gainesville.

Sharp, J.H., Perry, M.J., Renger, E.H. and Eppley, R.W. (1980 Phytoplankton rate processes in the oligotrophic waters of the central North Pacific Ocean, *J. Plankton Res.*, **2**, 335–53.

Sheldon, R.W. and Parsons, T.W. (1967) A continuous size spectrum for particulate matter in the sea, *J. Fish. Res. Bd Can.*, **24**, 909–15.

Sheldon, R.W. and Sutcliffe Jr, W.H. (1978) Generation times of 3 h for Sargasso Sea microplankton determined by ATP analysis, *Limnol. Oceanogr.*, **23**, 1051–5.

Sheldon, R.W., Prakash, A. and Sutcliffe Jr, W.H. (1972) The size distribution of particles in the ocean, *Limnol. Oceanogr.*, **17**, 327–40.

Sheldon, R.W., Sutcliffe Jr, W.H. and Paranjape, M.A. (1977) Structure of pelagic food chain and relationship between plankton and fish production, *J. Fish. Res. Bd Can.*, **34**, 2344–53.

Sheldon, R.W., Sutcliffe, W.H. and Prakash, A. (1973) The production of particles in the surface waters of the ocean with particular reference to the Sargasso Sea, *Limnol. Oceanogr.*, **18**, 719–33.

Sherman, F.S., Imberger, J. and Corcos, G.M. (1978) Turbulence and mixing in stably stratified waters, *Ann. Rev. Fluid. Mech.*, **10**, 267–8.

Shoesmith, E.A. and Brook, A.J. (1983) Monovalent-divalent cation ratios and the occurrence of phytoplankton, with special reference to the desmids, *Freshwat. Biol.*, **13**, 151–5.

Shugart, H.H. and Hett, J.M. (1973) Succession: similarities of species turnover rates, *Science*, **180**, 1379–80.

Shugart, H.H. and West, D.C. (1981) Long-term dynamics of forest ecosystems, *Am. Sci.*, **69**, 647–52.

Shugart, H.H., Jr., Emanuel, W.R., West, D.C. and De Angelis, D.L. (1980) Environmental gradients in a simulation model of a beech-yellow poplar stand, *Math. Biosci.*, **50**, 163–170.

Shulenberger, E. and Reid, J.L. (1981) The Pacific shallow oxygen maximum, deep chlorophyll maximum, and primary productivity reconsidered, *Deep Sea Res.*, **28**, 901–20.

Shuter, B.J. (1978) Size dependence of phosphorus and nitrogen subsistence quotas in unicellular micro organisms, *Limnol. Oceanogr.*, **23**, 1248–55.

Silander, J.A. and Antonovics, J. (1982) Analysis of interspecific interactions in a coastal plant community – a perturbation approach, *Nature*, **298**, 557–60.

Simberloff, D. (1972) Models in biogeography, in *Models in Palaeobiology* (ed T. J. M. Schopf),Freeman, Cooper and Co., San Francisco, pp. 160–91.

Simberloff, D. (1978) Using island biogeographic distributions to determine if colonization is stochastic, *Am. Nat.*, **112**, 713–26.

Simberloff, D. (1983) Markov model, *Science*, **220**, 1275.

Simberloff, D.S. (1970) Taxonomic diversity of island biotas, *Evolution*, **24**, 23–47.

Simpson, E.H. (1949) Measurement of diversity, *Nature*, **163**, **688**.

Simpson, J.H. and Pingree, R.D. (1978) Shallow sea fronts produced by tidal stirring, in *Oceanic Fronts in Coastal Processes*, (eds H. J. Bowman and W. E. Esias) Springer-Verlag, Berlin, pp. 29–42.

Simpson, J.J. and Dickey, T.D. (1981) The relationship between downward irradiance and upper ocean structure, *J. Phys. Oceanogr.*, **11**, 309–323.

Skellam, J.G. (1951) Random dispersal in theoretical populations, *Biometrika*, **78**, 196–218.

Slobodkin, L.B. and Rapoport, A. (1974) An optimal strategy of evolution, *Quart. Rev. Biol.*, **49**, 181–200.

Smayda, T.J. (1970) The suspension and sinking of phytoplankton in the sea, *Oceanogr. Mar. Biol. Ann. Rev.*, **8**, 353–414.

Smayda, T. (1980) Phytoplankton species succession, in *The Physiological Ecology of Phytoplankton* (ed. I. Morris), Blackwell, Oxford, pp. 493–570.

Smith, D.F. and Horner, S.M.J. (1981) Tracer kinetic analysis applied to problems in marine biology, in *Physiological Bases of Phytoplankton Ecology* (ed. T. Platt), *Can. Bull. Fish. Aq. Sci.*, **210**, 113–29.

Smith, P.E. (1978) Biological effects of ocean variability: time and space scales of biological response, *Rapp. Proc. Vérb. Réun. Cons. Int. Expl. Mer*, **173**, 117–27.

Smith, R.E.H. and Kalff, J. (1981) The effect of phosphorus limitations on algal growth rates: evidence from alkaline phosphatase, *Can. J. Fish. Aq. Sci.*, **38**, 1421–7.

Smith, R.E.H. and Kalff, J. (1982) Size dependent phosphorus uptake kinetics and cell quota in phytoplankton, *J. Phycol.*, **18**, 275–84.

Smith, R.L. (1978) Poleward propagating perturbations in currents and sea levels along the Peru coast, *J. Geophys. Res.*, **83**, 6083–92.

Smith, S.D. (1970) Thrust-anemometer measurements of wind turbulence, Reynolds stress, and drag coefficient over the sea, *J. Geophys. Res.*, **75**, 6758–70.

Smith, V.H. (1983) Low nitrogen to phosphorus ratios favour dominance by blue-green algae in lake phytoplankton, *Science*, **221**, 669–71.

Smyly, W.J.P. (1952) Observations on the food of the fry of Perch (*Perca fluviatilis* Linn.) in Windermere, *Proc. Zool. Soc. Lond.*, **122**, 407–16.

Smyly, W.J.P. (1976) Some effects of enclosure on the zooplankton in a small lake, *Freshwat. Biol.*, **6**, 241–51.

Smyly, W.J.P. (1978) Further observations on limnetic zooplankton in a small lake and two enclosures containing fish, *Freshwat. Biol.*, **8**, 491–5.

Sommer, U. (1981) The role of *r* and *K* selection in the succession of phytoplankton in Lake Constance, *Acta Oecol. Oecol. Gener.*, **2**, 327–42.

Sommer, U. (1982) Vertical niche separation between two closely related planktonic flagellate species (*Rhodomonas lens* and *R. munita v. nanoplanktonica*) *J. Plankton Res.*, **4**, 137–42.

Sournia, A. (1982) Form and function in marine phytoplankton, *Biol. Rev.*, **57**, 347–94.

Southward, A.J. (1974) Changes in the plankton community of the Western English Channel, *Nature*, **249**, 180–1.

Southward, A.J. (1980) The Western English Channel – an inconstant system? *Nature*, **285**, 361–6.

Southward, A.J., Butler, E.I. and Pennycuick, L. (1975) Recent cyclic changes in climate and in abundance of marine life, *Nature*, **253**, 714–7.

Southwood, T.R.E. (1977) Habitat, the template for ecological strategies? *J. Anim. Ecol.*, **46**, 337–65.

Spigel, R.H. (1980) Coupling of internal wave motion with entrainment at the density interface of a two layer lake, *J. Phys. Oceanogr.*, **10**, 144–55.

Spigel, R.H. and Imberger, J. (1980) The classification of mixed layer dynamics in lakes of small to medium size, *J. Phys. Oceanogr.*, **10**, 1104–21.

Spodniewska, I. (1969) Day to day variations in primary production of phytoplankton in Mikolajskie Lake, *Ekol. Pol. Ser. A.*, **17**, 503–14.

Sprules, W.G. (1977) Crustacean zooplankton communities as indicators of limnological conditions: an approach using principal component analysis, *J. Fish. Res. Bd Can.*, **34**, 962–75.

Sprules, W.G., Casselman, J.M. and Shuter, B.J. (1983) Size distribution of pelagic particles in lakes, *Can. J. Fish. Aq. Sci.*, **40**, 1761–9.

Stanley, S.M. (1979) *Macroevolution: Pattern and Process*, Freeman, San Francisco.

Stanley, S.M. (1981) *The New Evolutionary Timetable*, Basic Books, New York.

Stearns, S.C. (1976) Life history tactics: a review of the ideas, *Quart. Rev. Biol.*, **51**, 3–47.

Stearns, S.C. (1977) The evolution of life history traits: a critique of the theory and a review of the data, *Ann. Rev. Ecol. Syst.*, **8**, 145–71.

Steele, J.H. (1962) Environmental control of photosythesis in the sea, *Limnol. Oceanogr.*, **7**, 137–50.

Steele, J.H. and Frost, B.W. (1977) The structure of plankton communities, *Phil. Trans. Roy. Soc. Lond. B.*, **280**, 485–485–534.

Steemann–Nielsen, E. (1952) The use of radio-active carbon(^{14}C) for measuring organic production in the sea, *J. Cons. perm. int. Explor. Mer.*, **18**, 117–40.

Steemann-Nielsen, E. (1954) On organic production in the ocean, *J. Cons. perm. int. Explor. Mer.*, **19**, 309–28.

Steemann-Nielsen, E. (1963) Productivity, definition and measurement, in *The Sea* (ed. M. N. Hill), Vol. 2, Wiley Interscience, New York, pp. 129–59.

Stein J.A. (ed.) (1973) *Handbook of Phycological Methods. Culture Methods and Growth Measurements*, Cambridge University Press, Cambridge.

Stewart, F.M. and Levin, B.R. (1973) Partitioning of resources and the outcome of interspecific competition, a model and some general conclusions, *Am. Nat.*, **107**, 171–98.

Stewart, W.D.P. (1974) *Algal Physiology and Biochemistry*, (Botanical Monog. 10), Blackwell, Oxford.

Stewart, W.D.P., Rowell, P., Codd, G.A. and Apte, S.K. (1978) Nitrogen fixation and photosynthesis in photosynthetic prokaryotes, in *Proc. 4th Int. Cong. Photosynthesis* (eds D. Hall, J. Coombs and A. D. Greenwood), Biochemical Society, London, pp. 133–146.

Strickland, J.D.H. (1960) Measuring the production of marine phytoplankton, *Bull. Fish. Res. Bd. Can.*, **122**, 1–172.

Stoermer, E.F. (1978) Phytoplankton assemblages as indicators of water quality in the Laurentian Great Lakes, *Trans. Am. Micros. Soc.*, **97**, 2–16.

Strong, A.E. and Eadie, B.J. (1978) Satellite observations of calcium carbonate precipitations in the Great Lakes, *Limnol. Oceanogr.*, **23**, 877–87.

Strong, D.R. and Simberloff, D.S. (1981) Straining at gnats and swallowing ratios: character displacement, *Evolution*, **35**, 810–12.

Strong, D.R., Szyska, L.A. and Simberloff, D.S. (1979) Tests of community-wide character displacement against null hypothesis, *Evolution*, **33**, 897–913.

Stross, R.G. and Pemrick, S.M. (1974) Nutrient uptake kinetics in phytoplankton: a basis for niche separation, *J. Phycol.*, **10**, 164–9.

Stull, E.A., de Amezaga, E. and Goldman, C.R. (1973) The contribution of individual species of algae to primary productivity of Castle Lake, California, *Verh. Int. Verein. Theor. Angew. Limnol.*, **18**, 1776–83.

Stumm, W. (ed.) (1977) *Global Chemical Cycles and their Alterations by Man*, Dahlem Konferenzen, Berlin Bernhardt, Berlin.

Stumm, W. and Morgan, J.J. (1981) *Aquatic Chemistry*, 2nd edn, Wiley-Interscience, New York.

Suess, E. and Ungerer, C.A. (1981) Element and phase composition of particulate matter from the circumpolar between New Zealand and Antarctica, *Oceanol. Acta.*, **4**, 151–160.

Sugihara, G. (1980) Minimal community structure: an explanation of species abundance patterns, *Am. Nat.*, **116**, 770–87.

Sutcliffe, D.W., Carrick, T.R., Heron, J. *et al.* (1982) Long term and seasonal changes in the chemical composition of precipitation and surface waters of lakes and tarns in the English Lake District, *Freshwat. Biol.*, **12**, 451–506.

Sutcliffe Jr, W.H., Sheldon, R.W. and Prakash, A. (1970) Certain aspects of production and standing stock of particulate matter in the surface waters of the N.W. Atlantic Ocean, *J. Fish. Res. Bd Canada.*, **27**, 1917–26.

Sverdrup, H.U. (1953) On conditions for the vernal blooming of phytoplankton, *J. Cons. perm. int. Expl. Mer*, **18**(3), 287–95.

Swift, E., Stuart, M. and Meunier, V. (1976) The *in situ* growth rates of some deep living oceanic dinoflagellates: *Pyrocystis fusiformis* and *Pyrocystis noctiluca*, *Limnol. Oceanogr.*, **21**, 418–26.

Syrett, P.J. (1981) Nitrogen metabolism of microalgae, in *Physiological Bases Phytoplankton Ecology* (ed. T. Platt), *Can. Bull. Fish. Aq. Sci.*, **210**, 182–210.

Talling, J.F. (1951) The element of chance in pond populations, *The Naturalist* (Oct–Dec), 157–170.

Talling, J.F. (1955a) The relative growth rates of three plankton diatoms in relation to U.W. radiation and temperature, *Ann. Bot. N.S.*, **19**, 329–41.

Talling, J.F. (1955b) The light relations of phytoplankton populations, *Verh. Int. Vereim. Limnol.*, **12**, 141–2.

Talling, J.F. (1957a) Photosynthetic characteristics of some freshwater plankton diatoms in relation to underwater radiation, *New Phytol.*, **56**, 29–50.

Talling, J.F. (1957b) The phytoplankton population as a compound photosynthetic system, *New Phytol.*, **56**, 133–49.

Talling, J.F. (1969) The incidence of vertical mixing, and some biological and chemical consequences, in tropical African Lakes, *Verh. Int. Verein. Limnol.*, **17**, 998–1012.

Talling, J.F. (1970) Generalized and specialized features of phytoplankton as a form of photosynthetic cover, in *Prediction and Measurement of Photosynthetic Productivity, Proc. 113P/PP Technical Meeting, Trebon*, pp. 431–45.

Talling, J.F. (1971) The underwater light climate as a controlling factor in the production ecology of freshwater phytoplankton, *Mitt. Int. Verin. Limnol.*, **19**, 214–43.

Talling, J.F. (1973) The application of some electrochemical methods to the measurement of photosynthesis and respiration in fresh waters, *Freshwat. Biol.*, **3**, 335–62.

Talling, J.F. (1976) The depletion of carbon dioxide from lake water by phytoplankton, *J. Ecol.*, **64**, 79–121.

Talling, J.F. (1979) Factor interactions and implications for the prediction of lake metabolism, *Arch. Hydrobiol. Beih. Ergeb. Limnol.*, **13**, 96–109.

Talling, J.F. and Talling, I.B . (1965) The chemical composition of African lake waters, *Int. Rev. Ges. Hydrobiol.*, **50**, 421–63.

Taniguchi, A. (1973) Phytoplankton-zooplankton relationships in the western Pacific Ocean and adjacent seas, *Mar. Biol.*, **21**, 115–21.

Tarapchak, S.J., Bigelow, S.M. and Rubitschun, C. (1982) Overestimation of orthophosphate concentration in surface waters of southern L. Michigan: effects of acid and ammonium molybdate, *Can. J. Fish. Aq. Sci.*, **39**, 296–304.

Taylor, F.J.R. (1980) Basic biological features of phytoplankton cells, in *The Physiological Ecology of Phytoplankton* (ed. I. Morris), Blackwell, Oxford, pp. 3–56.

Taylor, P.A. and Williams, P.J.LeB. (1975) Theoretical studies on the coexistence of competing species under continuous flow conditions, *Can. J. Microbiol.*, **21**, 90–8.

Thom, R. (1972) *Stabilité Structurelle et Morphogénèse*, Benjamin, New York.

Thompson, J.M., Ferguson, A.J.D. and Reynolds, C.S. (1982) Natural filtration rates of zooplankton in a closed system: the derivation of a community grazing index, *J. of Plankton Res.*, **4**, 545–60.

Thorpe, J.E. (1977) Morphology, physiology, behaviour and ecology of *Perca fluviatilis* L. and *P. flavescens* Mitchill, *J. Fish. Res. Bd Can.*, **34**, 1504–14.

Thorpe, J. (1977b) Synopsis of biological data on the Perch *Perca fluviatilis* (Linn) 1758 and *Perca flavescens* (Mitchill) 1814, *FAO Fisheries Synopsis*, No. 113.

Thorpe, S.A. (1977) Turbulence and mixing in a Scottish loch, *Phil. Trans. R. Soc. Lond. A.*, **386**, 125–81.

Tilman, D. (1977) Resource competition between planktonic algae: an experimental and theoretical approach, *Ecology*, **58**, 338–48.

Tilman, D. (1980) Resources: a graphical-mechanistic approach to competition and predation, *Am. Nat.*, **116**, 362–93.

Tilman, D. (1982) *Resource Competition and Community Structure*, Princeton University Press, Princeton, N.J.

Tilman, D. and Kilham, S.S. (1976) Phosphate and silicate growth and uptake kinetics of the diatoms *Asterionella formosa* and *Cyclotella meneghiniana* in batch and semi-continuous culture, *J. Phycol.*, **12**, 375–83.

Tilman, D., Kilham, S.S. and Kilham, P. (1982) Phytoplankton community ecology: the role of limiting nutrients, *Ann. Rev. Ecol. Syst.*, **13**, 349–72.

Tilman, D., Mattson, M. and Langer, S. (1981) Competition and nutrient kinetics along a temperature gradient: An experimental test of a mechanistic approach to niche theory, *Limnol. Oceanogr.*, **26**, 1020–33.

Tilzer, M.M. (1984) Estimation of phytoplankton loss rates from daily photosynthetic rates and observed biomass changes in Lake Constance, *J. Plankton Res.*, **6**, 309–324.

Titman, D. (1976) Ecological competition between algae: experimental confirmation of resources-based competition theory, *Science*, **192**, 463–5.

Trimbee, A.M. (1983) The phytoplankton population dynamics of a small reservoir, PhD Thesis, McMaster University, Hamilton, Ontario.

Trimbee, A.M. and Harris, G.P. (1983) Use of time-series analysis to demonstrate advection rates of different variables in a small lake, *J. Plankton Res.*, 5 819–34.

Trimbee, A.M. and Harris, G.P. (1984) Phytoplankton population dynamics in a small reservoir: use of sedimentation traps to study the loss of diatoms and the recruitment of summer bloom-farming blue-greens, *J. Plankton Res.*, **6**, 897–918.

Trudinger, P.A. and Swaine, D.J. (eds) (1979) *Biogeochemical Cycling of Mineral Forming Elements*, Elsevier, Amsterdam.

Trudinger, P.A., Swaine, D.J. and Skyring, G.W. (1979) Biogeochemical cycling of elements – general considerations, in *Biogeochemical Cycling of Mineral Forming Elements* (eds P. A. Trudinger and D. J. Swaine), Elsevier, Amsterdam, pp. 1–28.

Tunnicliffe, V. (1981) High species diversity and abundance of epibenthic community in an oxygen deficient basin, *Nature*, **294**, 354–6.

Turner, J.S. and Kraus, E.B. (1967) A dimensional model of the seasonal thermocline. 1. A laboratory experiment and its interpretation, *Tellus*, **19**, 88–97.

Turpin, D.H. and Harrison, P.J. (1979) Limiting nutrient patchiness and its role in phytoplankton ecology, *J. Exp. Mar. Biol. Ecol.*, **39**, 151–66.

Turpin, D.H. and Harrison, P.J. (1980) Cell size manipulation in natural marine, planktonic, diatom communities, *Can. J. Fish. Aq. Sci.*, **37**, 1193–5.

Turpin, D.H., Parslow, J.S. and Harrison, P.J. (1981) On limiting nutrient patchiness and phytoplankton growth: a conceptual approach, *J. Plankton Res.*, **3**, 421–31.

US Environmental Protection Agency (1975) A compendium of lake and reservoir data collected by the national eutrophication survey in North-Eastern and North-Central United States, Working paper 474, Corvallis Environmental Research Laboratory, Corvallis, Oregon.

US Environmental Protection Agency (1978a) A compendium of lake and reservoir data collected by the national eutrophication survey in Eastern, North-Central and South-Eastern United States, Working paper 475, Corvallis Environmental Research Laboratory, Corvallis, Oregon.

US Environmental Protection Agency (1978b) A compendium of lake and reservoir data collected by the national eutrophication survey in Central United States, Working paper 476, Corvallis Environmental Research Laboratory, Corvallis, Oregon.

US Environmental Protection Agency (1978c) A compendium of lake and reservoir data collected by the national eutrophication survey in Western United States, Working paper 477, Corvallis Environmental Research Laboratory, Corvallis, Oregon.

Vaccaro, R.F. (1965) Inorganic nitrogen in sea water, in *Chemical Oceanography I*, (eds J. P. Riley and G. Skirrow), Academic Press, London, pp. 365–408.

Vandermeer, J.H. (1972) Niche theory, *Ann. Rev. Ecol. Syst.*, **3**, 107–32.

Vaulot, D. and Chisholm, S.W. (1982) Population dynamics in phytoplankton cultures, the effect of generation time variability, *EOS; Trans. Am. Geophys. Union*, **63**, 97.

Venrick, E.L. (1982) Phytoplankton in an oligotrophic ocean: observations and questions, *Ecol. Monogr.*, **52**, 129–54.

Venrick, E.L., Beers, J.R. and Heinbokel, J.F. (1977) Possible consequences of containing microplankton for physiological rate measurements, *J. Exp. Mar. Biol. Ecol.*, **26**, 55–76.

Verduin, J. (1975) Fate of carbon dioxide transport across air-water boundaries in lakes, *Limnol. Oceanogr.*, **20**, 1052–3.

Vinogradov, A.P. (1953) *The Elemental Composition of Marine Organisms*, Memoir No. 2, Sears Foundation for Marine Research, Yale University, New Haven, Conneticut, pp. 1–647.

Vollenweider, R.A. (1948) Zum Gesellschaftsproblem in der Limnobiocoenologie, *Schweiz. Z. Hydrol.*, **10**(4) 1–12.

Vollenweider, R.A. (1950) Oklologische Untersuchungen von planktischen Algen auf experimenteller Grundlage, *Schweiz. Z. Hydrol.*, **12**(2), 193–262.

Vollenweider, R.A. (1953) Einige Bemerkungen zur ökologischen Valenzanalyse, *Schweiz. Z. Hydrol..*, **15**(1), 190–7.

Vollenweider, R.A. (1960) Beiträge zur Kenntnis optischer Eigenschaften der Gewässer und Primärproduktion, *Mem. Ist Ital. Idrobiol.*, 12, 201–44.

Vollenweider, R.A. (1965) Calculation models of photosynthesis – depth curves and some implications regarding day rate estimates in primary production measurements, *Mem. Ist Ital. Idrobiol.*, 18, 425–57.

Vollenweider, R.A. (1968) Scientific fundamentals of the eutrophication of lakes and flowing waters, with particular reference to nitrogen and phosphorus as factors in eutrophication, *OECD Tech. Rep.*, DAS/CSI/68.27, Paris.

Vollenweider, R.A. (1969) Moglichkeiten und grenzen elementarer Modelle der Stoffbilanz von Seen, *Arch. Hydrobiol.*, 66, 1–36.

Vollenweider, R.A. (1970) Models for calculating integral photosynthesis and some implications regarding structural properties of the community metabolism of aquatic systems, in *Prediction and Measurement of Photosynthetic Productivity*, *Proc. 1BP/PP Meeting Trebon*, pp. 455–72.

Vollenweider, R.A. (1975) Input–output models with special reference to the phosphorus loading concept in limnology, *Schweiz. Z. Hydrol.*, 37, 53–84.

Vollenweider, R.A. (1976) Advances in defining critical loading levels for phosphorus in lake eutrophication, *Mem. Ist Ital. Idrobiol.*, 33, 53–83.

Vollenweider, R.A. and Dillon, P.J. (1974) The application of the phosphorus loading concept to eutrophication research, *NRCCC 13690*, National Research Council, Ottawa, Canada.

Vollenweider, R.A. and Harris, G.P. (1985) Elemental ratios in marine and freshwater plankton, (manuscript in preparation).

Vollenweider, R.A. and Kerekes, J. (1980) The loading concept as a basis for controlling eutrophication: philosophy and preliminary results of the OECD programme on eutrophication, *Prog. Wat. Technol.*, 12, 5–38.

Vollenweider, R.A. and Kerekes, J. (1981) OECD cooperative programme on monitoring of inland waters (eutrophication control). Condensed synthesis report, OECD, Paris.

Vollenweider, R.A. and Nauwerck, A (1961) Some observations on the ^{14}C method for measuring primary production, *Verh. Int. Verein. Limnol.*, 14, 134–9.

Vollenweider, R.A., Rast, W. and Kerekes, J. (1980) The phosphorus loading concept and Great Lakes eutrophication, in *Proc. 11th Conf. Cornell University, Phosphorus management strategies for the Great Lakes*, (eds R. C. Locke, C. S. Martin and W. Rast) Ann Arbor Science, Rochester, NY, pp. 207–34.

Volterra, V. (1926) Fluctuations in the abundance of a species considered mathematically, *Nature*, 188, 558–60.

Von Bertalanffy, L. (1952) *Problems of Life*, Watts and Co., London.

Wall, D. and Briand, F. (1980) Spatial and temporal overlap in lake phytoplankton communities, *Arch. Hydrobiol.*, 88, 45–57.

Walsby, A.F. and Reynolds, C.S. (1980) Sinking and floating, in *The Physiological Ecology of Phytoplankton* (ed. I. Morris), Blackwell, Oxford, pp. 371–412.

Waterbury, J.B., Watson, S.W., Guillard, R.R.L. and Brand, L.E. (1979) Widespread occurrence of a unicellular, marine, planktonic, cyanobacterium, *Nature*, 277, 293–4.

Watson, S. and Kalff, J. (1981) Relationships between nanoplankton and lake trophic status, *Can. J. Fish. Aq. Sci.*, 38, 960–7.

Watt, W.D. (1971) Measuring the primary production rates of individual phyto-plankton species in natural mixed populations, *Deep Sea Res.,* **18**, 329–39.

Weatherley, A.H. (1972) *Growth and Ecology of Fish Populations,* Academic Press, London.

Wells, L. (1970) Effects of Alewife predation on zooplankton populations in Lake Michigan, *Limnol. Oceanogr.,* **15**, 556–65.

West, G.S. and Fritsch, F.E. (1927) *A Treatise on the British Freshwater Algae,* Cambridge University Press, Cambridge.

Wetzel, R.G. (1975) *Limnology,* Saunders, Philadelphia.

Wiens, J.A. (1976) Population responses to patchy environments, *Ann. Rev. Ecol. Syst.,* **7**, 81–120.

Wiens, J.A. (1977) On competition and variable environments, *Am. Sci.,* **65**, 590–7.

Whillans, T.H. (1977) Fish community transformations in three bays within the lower Great Lakes, PhD thesis, University of Toronto, Toronto, Ontario, Canada, 328 pp.

White, E., Payne, G., Pickmere, S. and Pick, F.R. (1981) Orthophosphate and its flux in lake waters, *Can. J. Fish. Aq. Sci.,* **38**, 1215–9.

White, E., Payne, G., Pickmere, S. and Pick, F.R. (1982) Factors influencing ortho-phosphate turnover times: a comparison of Canadian and New Zealand lakes, *Can. J. Fish. Aq. Sci.,* **39**, 469–74.

Whittaker, R.H. (1965) Dominance and diversity in land plant communities, *Science,* **147**, 250–60.

Whittaker, R.H. and Levin, S.A. (1975) *Niche, Theory and Application (Benchmark Papers in Ecology,* Vol. 3), Dowdem, Hutchinson and Ross, Stroudsberg, Pa.

Williamson, M.H. (1981a) *Island Populations,* Oxford University Press, Oxford.

Williamson, P.G. (1981b) Palaeontological documentation of speciation in Cenozoic molluscs from Turkana Basin, *Nature,* **293**, 437–43.

Wofsy, S.C. (1983) A simple model to predict extinction coefficients and phyto-plankton biomass in eutrophic waters, *Limnol. Oceanogr.,* **28**, 114–55.

Woods, J.D. (1980) Do waves limit turbulent diffusion in the oceans? *Nature,* **288**, 219–24.

Woodwell, G.M. and Pecan, E.V. (eds) (1973) Carbon and the biosphere, *AEC Symp. Ser. 30* NTIS, Conf. 720/510, Springield, VA.

Wyrtki, K. (1975) El Nino – the dynamic response of the equatorial Pacific Ocean to atmospheric forcing, *J. Phys. Oceanogr.,* **5**, 572–84.

Wyrtki, K., Stroup, E., Patzert, W. *et al,* (1976) Predicting and observing El Nino, *Science,* **191**, 343–346.

Yentsch, C.S. (1980) Phytoplankton growth in the sea. A coalescence of disciplines, in *Primary Productivity in the Sea* (ed. P. Falkowski), Plenum, New York, pp. 17–32.

Yentsch, C.S. and Ryther, J.H. (1959) Relative significance of the net phytoplankton and nanoplankton in the waters of Vineyard Sound, *J. Cons. perm, int. Explor Mer.,* **24**, 231–8.

Yentsch, C.M., Yentsch, C.S. and Strube, L.R. (1977) Variations in ammonium enhancement, an indication of nitrogen deficiency in New England coastal phy-toplankton populations, *J. Mar. Res.,* **35**, 537–55.

Youngs, W.D. and Heimbuch, D.G. (1982) Another consideration of the morphoe-daphic index, *Trans. Am. Fish. Soc.,* **111**, 151–3.

Index